Scientific Perspectives on Divine Action

Twenty Years of Challenge and Progress

Inaugural Volume – 1988
Physics, Philosophy, and Theology:
A Common Quest for Understanding
Edited by Robert John Russell, William R. Stoeger, S.J.,
and George V. Coyne, S.J.

Published by the Vatican Observatory in commemoration
of the 300th anniversary of the publication of Isaac Newton's
Philosophia Naturalis Principia Mathematica

A Series on "Scientific Perspectives on Divine Action"
Robert John Russell, General Editor

First Volume – 1993 (Revised Edition 1996)
Quantum Cosmology and the Laws of Nature:
Scientific Perspectives on Divine Action
Edited by Robert John Russell, Nancey Murphy,
and C.J. Isham

Second Volume – 1995
Chaos and Complexity:
Scientific Perspectives on Divine Action
Edited by Robert John Russell, Nancey Murphy,
and Arthur Peacocke

Third Volume – 1998
Evolutionary and Molecular Biology:
Scientific Perspectives on Divine Action
Edited by Robert John Russell, William R. Stoeger, S.J.,
and Francisco Ayala

Fourth Volume – 1999
Neuroscience and the Person:
Scientific Perspectives on Divine Action
Edited by Robert John Russell, Nancey Murphy,
Theo C. Meyering, and Michael A. Arbib

Fifth Volume – 2001
Quantum Mechanics:
Scientific Perspectives on Divine Action
Edited by Robert John Russell, Philip Clayton,
Kirk Wegter-McNelly, and John Polkinghorne

Sixth Volume – 2007
Scientific Perspectives on Divine Action:
Twenty Years of Challenge and Progress

Jointly published by the Vatican Observatory and
the Center for Theology and the Natural Sciences

Supported in part by a grant from the
Wayne and Gladys Valley Foundation

Scientific Perspectives on Divine Action

Twenty Years of Challenge and Progress

Robert John Russell
Nancey Murphy
William R. Stoeger, S.J.

Editors

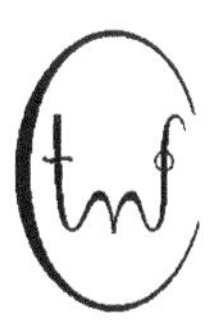

Vatican Observatory
Publications
Vatican City State

The Center for Theology
and the Natural Sciences
Berkeley, California

2008

Robert John Russell (General Editor) is The Ian G. Barbour Professor of Theology and Science in Residence at the Graduate Theological Union, and Founder and Director of the Center for Theology and the Natural Sciences, in Berkeley, California, USA

Nancey Murphy is Professor of Christian Philosophy, Fuller Theological Seminary, Pasadena, California

William R. Stoeger, S.J. is Staff Astrophysicist at the Vatican Observatory, Vatican Observatory Research Group, Steward Observatory, The University of Arizona, Tucson, Arizona. He is also Adjunct Associate Professor of Astronomy at the University of Arizona.

Jointly published by the Vatican Observatory and
the Center for Theology and the Natural Sciences

Distributed (except in Italy and the Vatican City State) by
The University of Notre Dame Press
Notre Dame, Indiana 46556
U.S.A.

Distributed in Italy and the Vatican City State by
Libreria Editrice Vaticana
V-00120 Citta del Vaticano
Vatican City State

ISBN 978-88-209-7961-4 (pbk.)

***Cover* – Design by Bonnie Johnston**

ACKNOWLEDGMENTS

The editors would like to express their gratitude to the Vatican Observatory and the Center for Theology and the Natural Sciences for co-sponsoring this research. We would also like to thank George Coyne in particular for his leadership at the Vatican Observatory which made this research and volume possible. Thanks also to the staff at the Observatory and at CTNS for their work in organizing the conference and the numerous details related to the publication of this volume.

The editors thank the conference participants for their efforts in bringing this project to its culmination.

We would also like to express thanks to Joshua M. Moritz at CTNS who managed the final editing of this volume and who, along with Melissa Moritz at CTNS, copyedited, formatted, proofread and typeset the text, and created the indices and back cover for the volume. We appreciate your diligence, patience, and expertise in these tasks. We are also grateful to James Haag for his work on clarifying the footnotes, and to Bonnie Johnston for her work on the cover design.

DEDICATION

We dedicate this "Capstone" volume to Dr. George V. Coyne, S. J., Director of the Vatican Observatory (the Specola), 1978-2006. Without George's pioneering leadership and unending service as a scientist, a Roman Catholic priest, and Director of the Specola, the five volumes of research in theology and science assessed here would never have been possible. Dr. Coyne's efforts ensured that the mandate of His Holiness, Pope John Paul II—that science and religion come together in service to the wider human community—was carried out and embodied in two decades of research conferences among over sixty distinguished scientists, philosophers, historians and theologians. To each of us, George offered the rich hospitality of the Specola perched above beautiful Lake Albano and an endless bounty of joy, humor, critical analysis and abiding and profound Christian faith, gifts which truly blessed our decades together and bequeathed us a lasting sense of community.

George, our dear friend, colleague, and brother in the Christian faith, we deeply thank you.

Bill, Nancey, and Bob.

FOREWORD

When the adventure of exploration which is revisited in this book began I was privileged to be a party to it and, in fact, wrote the Preface to the book, *Physics, Philosophy and Theology: A Common Quest for Understanding* (PPT), whose publication became the inspiration and the guiding beacon for the series of conferences which are revisited in this current Capstone volume (called Capstone because it seeks to evaluate in a thematic fashion the entire series of five conferences). Therefore, in the spirit of revisiting, I would like to reflect now by hindsight upon some thoughts that I presented in that Preface to the PPT volume in light of the adventure that ensued over a period of twenty years with a series of five conferences, dedicated to the overarching theme of *Scientific Perspectives on Divine Action*. From each of those conferences there resulted a volume, the whole series of which constitutes the substance which is evaluated in this current Capstone publication.

In that PPT Preface I insisted upon the fact that we were about to engage in a "small beginning." That small beginning was a conference held at the Vatican Observatory from 21 to 26 September 1987 at the request of Pope John Paul II to commemorate the 300th anniversary of the publication of Isaac Newton's *Principia*. This "is a small beginning" said I, "with more questions . . . than answers. It, therefore, leans into the future. It is a promise, a pledge and a challenge to continue the 'Common Quest.'" Furthermore, that first step into the future was inspired by the publication in that PPT volume of an outstanding message of John Paul II on the interaction between the culture of religious faith and that of science. In fact, the Papal message itself participates in our adventure of exploration and leans into the future. The Pope says, for instance, directing himself to the scientists, philosophers and theologians whose research is contained in the PPT volume: "You are called to learn from one another, to renew the context in which science is done and to nourish the inculturation which vital theology demands. Each of you has everything to gain from such an interaction, and the human community which we both serve has a right to demand it from us."

And so from that small beginning where have we arrived? From the scholarly evaluation given in this Capstone volume of the series of five volumes which followed upon and were inspired by PPT I would propose that, having accomplished a great deal in defining the principal issues in coming to understand God's action in the universe in light of our scientific knowledge in areas ranging from cosmology to the neurosciences and having begun to refine a research methodology to address those issues, we are at a new beginning. A sampling of the issues which came to be more acutely defined as the series progressed and are identified in the current volume would include: diversity in the metaphysics employed in understanding the nature of God, the interplay between general divine action and special divine action as well as between God's activity in nature and in history, the status of natural theology, interventionist versus non-interventionist approaches to understanding God's action, the limits of science, the nature of the laws of nature, the understanding of quantum indeterminacy, physical evil, reductionist versus an emergent philosophy of nature, the role of information theory, and so on. They are very fundamental issues but as you study this Capstone volume you will see that all of them have been very wisely defined and that ways to pursue them have been proposed. The adventure continues and the excitement of discovery is there. This volume presents a rather well defined road for the discoveries that lie ahead.

It must be said that this volume is an extremely valuable tool for those who wish to research further in any or all of the five topics which the series of conferences has covered. In this respect the two tables prepared by Robert Russell and published by him in his chapter in the overview section of this volume will be especially useful. They list the primary areas in science, philosophy and theology found in the series and indicate the authors and papers that focus on each area. This volume, I suspect, will also be an indispensable tool to students who wish to pursue research in the ever more intense interplay between the culture of science and that of religious faith. I suggest that the usefulness of this volume to both scholars and students will be especially, but not exclusively, applicable in the current situation of a not very healthy interaction among the science of evolution at all levels, religious beliefs and the political and educational scene in the United States. By this suggestion I do not intend to say that a careful study of this volume and the series that it evaluates will readily solve all of the issues related to the situation I have just described; but it will surely help, especially with reference to the third volume on evolutionary and molecular biology, to raise the discussions to a more rational level than that which not infrequently characterizes them. I refer to a not insignificant part of American society that sees an incompatibility between scientific evolution and religious beliefs or even substitutes religious creeds for well-established scientific explanations of the origins of the universe and of ourselves. The spirit of the research reviewed in this Capstone volume is diametrically opposite to those positions.

In 1988, with the success of the conference to commemorate Isaac Newton and the subsequent PPT publication, I asked Nancey Murphy (Professor of Christian Philosophy at Fuller Theological Seminary in Pasadena, California), Robert J. Russell (Founder and Director of the Center for Theology and the Natural Sciences at Berkeley, California) and William R. Stoeger, S.J. (Astronomer and Cosmologist at the Vatican Observatory) to consider serving as a core organizing group for a series of conferences to build on the adventure begun with PPT. This volume presents a Capstone of what they, with their colleagues, have accomplished. We owe a very deep expression of gratitude to them and to all of their colleagues who contributed to the series of conferences represented in this volume for having set us off on an adventure which now has a new beginning with no end in sight.

George V. Coyne, S. J
Vatican Observatory, Castel Gandolfo (Rome)

CONTENTS

INTRODUCTION

Robert John Russell

1 Background to the Volume

In early September, 2003, fifteen scholars with cross-disciplinary expertise in physics, astronomy, cosmology, evolutionary biology, mathematics, philosophy of religion, philosophy of science, and philosophical and systematic theology, met for a week-long conference at the Vatican Observatory (VO) located in Castel Gandolfo, Italy. The primary purpose of the conference was to review and assess the series of five scholarly publications resulting from five research conferences which the Observatory and the Center for Theology and the Natural Sciences (CTNS) had co-sponsored during the 1990s. The goal of this series was to engage theology, philosophy, and natural science in a process of constructive and creative mutual interaction, highlighting both the implications of science for philosophical and systematic theology and the philosophical and theological elements implicitly found within science. The overarching theme of the series was a topic drawn from philosophical theology—divine action—and placed within the context of theology and science. The result, "scientific perspectives on divine action," surfaced in a variety of ways in the theologies represented in the series. An additional purpose of the conference was to suggest an overarching theme for future conferences that would build upon and expand beyond the achievements of the series.[1] Eleven of the papers discussed at the conference and revised in light of these discussions are published here, together with an overview of the series.

2 Background to the Five Volume Series on "Scientific Perspectives on Divine Action"

Of the many remarkable events and publications that marked the decade of the 1990s as a watershed in the interdisciplinary field of "theology and science," one of the most significant was the series of five international and ecumenical research conferences co-sponsored by the Vatican Observatory and the Center for Theology and the Natural Sciences. Some fifty scholars participated in the series, many with cross-disciplinary expertise in physics, astronomy, cosmology, mathematics, evolutionary and molecular biology, the neurosciences and cognitive sciences, philosophy of science, history of science, philosophy of religion, history of religion, Old and New Testament, philosophical and systematic theology, and theological ethics. Ninety-one essays were published in the five volumes, along with detailed analytic introductions to each volume. The overarching goal was to engage theology, philosophy, and natural science in a process of constructive dialogue and creative mutual interaction.

[1] Because this conference stood separate from, and functioned as an evaluation of, the entire series, considering the publications not in a historical sequence but as a unified whole, it came to be referred to as the "Capstone conference."

The immediate background and context of the series was the international conference on theology and science organized by George Coyne, S. J., the Director of the Vatican Observatory, at the request of Pope John Paul II to celebrate the 300th anniversary of the publication of Isaac Newton's *Principia.* The resulting publication, *Physics Philosophy and Theology* (*PPT*), included a remarkable statement by the Pope on science and religion. The date of the *PPT* conference (1987) marks the twenty-year span under review in the present publication.

Based on the accomplishments of *PPT* and the vision offered by the Papal Message, George Coyne took steps to initiate a new series of conferences on theology and science. He asked Nancey Murphy (Professor of Christian Philosophy, Fuller Theological Seminary), to join Bill Stoeger, S. J. (the senior cosmologist at the Observatory) and me in forming the long-term steering committee for the series. Our task was to build on the accomplishments of *PPT* by moving further into areas in the physical and biological sciences already touched on in *PPT* as well as to expand the basis of research in science into new areas such as the neurosciences and cognitive sciences. Together we initiated a new method for research in theology and science designed to help move the participants beyond terminological confusions, clarify the role of philosophy in mediating the discussions, and promote a two-way interaction between scientists and theologians. The method included identifying a guiding theme for the entire series of conferences, one that would underlie the diversity of theological perspectives, draw out the implications of the sciences, and attract the attention of the scientists to further reflect on their own work. In June 1990, the committee chose the topic of divine action as the most promising candidate. The method called for at least some participants to have rigorous cross-disciplinary expertise. All participants were asked to read and respond critically to preconference papers. The conferences were devoted to discussing these papers whose revised versions were, in most cases, published in the series.

The first conference focused on quantum cosmology and the origin of the laws of nature, building on the initial exploration of cosmology in *PPT.* Next came conferences exploring the sciences of chaos and complexity, evolutionary and molecular biology, and the neurosciences, all of which expanded the scope of research presented in *PPT.* The fifth conference returned to one of the central themes of *PPT*: quantum mechanics. In summary, the five volumes are titled:

1. *Quantum Cosmology and the Laws of Nature* (*QC*)
2. *Chaos and Complexity* (*CC*)
3. *Evolutionary and Molecular Biology* (*EMB*)
4. *Neuroscience and the Person* (*NP*)
5. *Quantum Mechanics* (*QM*)

Each was subtitled *Scientific Perspectives on Divine Action.*

3 Overview of the Volume

The following is a brief overview of the present volume, followed by a summary in bullet form of some of the comments and recommendations contained in it.

3.1 Part I: Critical Appraisals of the Series as a Whole

The volume opens with two detailed appraisals of the entire VO/CTNS series of publications.

The first, by **Robert John Russell**, begins with a brief account of the history that preceded the VO/CTNS series of conferences, including the *PPT* conference in 1987 and the Papal Message. Next he recounts the development of a new methodology for research in theology and science that was deployed in the series. He then turns to the ninety-one chapters published in the series, using a table to list the primary areas in science, philosophy, and theology found in the series and indicating those authors and papers that focus on each area. (This table, together with summaries of all ninety-one chapters, is also available online at www.ctns.org/books.html.) He provides a similar table for the 25 chapters, drawn from the previous table, which most clearly focus on "scientific perspectives on divine action." These tables are meant to help readers use the series as a resource for their own research and to explore a variety of questions about the series such as the relative attention given to specific areas in science, philosophy, and theology, the proportion of papers that explicitly engage the sciences through their philosophical implications, etc.

He then analyzes areas of progress in the series as a whole as well as in areas focused on divine action. The former includes the effectiveness of the new methodology; the inclusion of landmark publications in the series; the variety of introductory resources in science and philosophy; jointly-authored essays and coordinated, separately-authored essays on interdisciplinary research topics in theology, philosophy, and science; novel directions in research in theology and science; and an empirical assessment of the impact of the series on scholars and the general public. The latter include clarification on terminology on divine action; distinguishing among six approaches to divine action; and the development of the theme of non-interventionist objective divine action (NIODA) in light of science. He closes by listing challenges, problems, and insights that have emerged in the discussion and whose sustained analysis is pivotal in making further progress. These include differences in the doctrine of God; the relative merits of various metaphysical commitments; compatibilist vs. incompatibilist views of divine action; the ontological status of the laws of nature; criteria for a successful theory of divine action; natural theodicy or suffering in nature; and eschatology. For each of these he offers a recommendation for directions in future collaborative research by CTNS and the VO. Appendices include the Table of Contents of *PPT*, a sample of summaries of chapters in the series, and an annotated typology of approaches to divine action.

The appraisal by **George Ellis** focuses specifically on the scientific dimensions of the VO/CTNS series. According to Ellis, the series made a very significant contribution to the field of theology and science by the ways the diverse sciences represented in the series were given an integrated treatment and placed in relation to each other. Ellis pointed specifically to three major areas in physics—the physics of the large (the cosmos), the physics of the small (quantum physics), and the physics underlying the complex (life and the brain)—and their multiple interactions. The series also identified a number of important philosophical issues that cut across these scientific areas: issues of ontology and epistemology, of reductionism and emergence, and of information. Finally, Ellis described how the approach taken in the series emphasized the complex and interacting nature of the causal nexus linking these areas. This includes strong causal links in action at any time ('laws of nature'), such as fundamental and effective laws of physics, constitutional equations, initial and boundary conditions, and bottom-up and top-down interactions between causal levels in the hierarchy of structure. The implication, for Ellis, is that "any attempt to claim that only one factor is the cause of some physical or biological effect will always be highly misleading and indeed a profoundly inadequate basis on which to make philosophical or metaphysical claims (as some attempt)." This in turn means that "the primary usefulness of the series was the refusal to be boxed in by the boundaries of any specific scientific discipline." As evidence of this usefulness, Ellis cites further meetings that have taken up the issue of emergence and his own cross-disciplinary look at the natural and life sciences.

Ellis concludes with a powerful argument for the limits of science. "There are aspects of existence that science per se cannot handle (although it may help us understand their context and implications)—they lie outside the domain of science." This includes ethics and aesthetics, where results or quantities cannot be defined scientifically even though these areas deal with causally effective processes. The limits of science also show up in discussing the question of the ontology of the laws of nature, leading one beyond the domain of empirical science to metaphysics. He recommends that we consider deploying future VO/CTNS research into such areas as the social and human sciences. "However, that will be a difficult task, because those subjects are so diverse and so disunited in themselves."

3.2 Part II: Philosophical Analysis of Specific Issues in the Series

The volume continues with three essays dealing with the philosophical analysis of specific issues in the series.

At the core of his essay, **Philip Clayton** poses a crucial test for any theory of divine action in terms of what he calls "traction." According to Clayton, traction requires that theology "can be impacted, either positively or adversely, by the results of philosophical critique, historical-critical research, or scientific knowledge." Traction comes in three forms: the first, when a theory of divine action is derived from science, history, or philosophy; the second when it can be falsified by them; and the third when it can be shown to be consistent with them. He claims that traditional Christian theism has failed to achieve the first kind of traction (i.e., that the project of natural theology has failed), while the third kind is not helpful in evaluating theories of divine action because most forms of Christian theism, and many other religious worldviews, pass the test it poses. The second kind of traction, however, poses a fruitful and valuable test for various forms of Christian theism. Clayton lists several theories of divine action which might have "traction 2" including non-interventionist divine action based on quantum mechanics as discussed in the VO/CTNS series.

Clayton then explores what he calls "the core problem: the relationship between science, metaphysics and theology." Rather than isolating or equating them, Clayton describes a series of transitions that lead from science to metaphysics. He identifies these transitions in many of the writings in the VO/CTNS series, calling this approach "scientific metaphysics." He lists eight distinct theories of divine action represented in the series and introduces the "counterfactual principle" by which they can best be evaluated: "the event in question must be in some way different than it would have been had divine action not been involved." VO/CTNS authors would reject as "interventionist" any counterfactual theory which requires God to violate the conservation of energy. But what if God brought such events about through the input of information? Clayton sees this approach as problematic, too. In closing Clayton briefly presents his work on emergence as a model for divine action. This model suggests that "divine action is to be located not within specific scientific disciplines, but in the interrelationships between them." Divine action becomes a "meta-scientific" position made possible by the combination of various scientific disciplines into an overarching worldview.

Nancey Murphy begins by listing over a dozen philosophical issues that have arisen in the VO/CTNS series, including linguistic, methodological, epistemological and metaphysical issues. She then focuses on reductionism and its counter-theses. She cites the works of Ian Barbour and Arthur Peacocke as introducing her to a non-reductive concept of the hierarchy of the sciences. She then describes Peacocke's nonreductionist approach in terms of downward causation and Philip Clayton's approach via emergence. Her assessment of their work turns on its ability to counter the most severe form of reductionism: causal reductionism. While affirming Peacocke's focus on top-down causation she does not see how his approach avoids causal over-determination. By generalizing previous accounts of whole-part constraint, Peacocke's proposal "loses just the specificity needed to understand how to reconcile" top-down and bottom-up causation. She offers similar criticisms of Clayton's position.

She then moves to her own work on supervenience and incorporates it into an account of downward causation in order to meet the challenge of causal reductionism. Here she finds Robert Van Gulick's emphasis on selection among lower-level causal processes particularly helpful in responding to the problem of generalization in Peacocke's work, especially when she adds to it her previous work on the context-dependency of supervenience. She also values Terrence Deacon's account of three orders of emergence based on his concept of amplification processes. But her use of Van Gulick's focus on selection, together with the theological view of the ontological distinction between God and creation, make her once again critical of Peacocke's approach: how does God as a "higher-level agent" causally affect the events of the world as "lower-level entities?" Even if the concept of information is deployed, the problem still remains: how does God cause one pattern rather than another?

The basic aim of **Wesley Wildman's** chapter is to provide a detailed overview and a robust assessment of the results of the "Divine Action Project (DAP)" pursued by a subgroup of scholars in the VO/CTNS series. Wildman's chapter includes materials drawn from the present publication as they were available in the capstone meeting and from the published output of the project as a whole. He provides an introduction to the divine action project, including its purpose and history; a discussion of the terminology used in it; a classification of views of special divine action; and an extensive analysis of the proposals based on chaos theory and quantum mechanics. His chapter includes a complex and illuminating diagram presenting a "decision tree" for theories of special divine action.

Wildman then evaluates criticisms of the DAP, especially those of Nicholas Saunders. According to Wildman, Saunders does not provide a compelling argument for his pessimistic conclusions about the project. Saunders's criticisms of the DAP theories by Ellis, Murphy, Russell, and Thomas Tracy depend on an interpretation of the laws of nature that these scholars reject. By referring to Immanuel Kant's claim that it is impossible to demonstrate freedom in terms of categories of causation, Wildman offers his own diagnosis of the peculiar challenges facing the DAP, an analysis of how participants handled those challenges, and how Saunders's criticism fits into this wider context. In particular, Wildman believes that intentional divine action, if it were valid, would exacerbate the problem of theodicy to such a degree that we are justified in rejecting it for moral and theological reasons. Thus, although he is persuaded that the DAP has succeeded in demonstrating the coherence and technical feasibility of several theories of intentional divine action, Wildman strongly rejects the view of God which he claims underlies these proposals.

3.3 Part III: Theological Analysis of Specific Issues in the Series

The volume concludes with seven essays dealing with the theological analysis of specific issues in the series.

In his chapter, **Niels Henrik Gregersen** points to the theological context of the project and the epistemological program that constitutes its core vision: "how can a scientifically informed Christian believer conceive of God's objective interaction with the world of nature within the constraints and opportunities offered by the natural sciences?" This in turn requires ways of relating divine action to our scientific understanding of nature that offer "traction" between theology and science and that focus on non-interventionist divine action as objective, special, and mediated. The project has crystallized around three proposals: divine action in relation to quantum physics, to complexity, and to chaos theory.

Gregersen then focuses on the core vision from a theological perspective: Can general divine action (GDA) and special divine action (SDA) be separated theologically, as is often assumed in the VO/CTNS series? In response he argues that a strict distinction between GDA and SDA may be part of the problem wrestled with in the series, instead of being part of the solution. From Gregersen's perspective, although we should overcome uniformitarian and deistic views of divine action that collapse the distinction between SDA and GDA, we should not make this distinction into a stark contrast. Rather, all divine action must be treated as both special, such as in response to prayer, and general, thus falling within the over-all pattern of divine self-consistency. The keys to this are the philosophical assumptions underlying the concept of the "laws of nature" which Gregersen finds embedded in the very concept of non-interventionist SDA in the VO/CTNS series.

Gregersen proposes that we take SDA as fundamental rather than as something additional to GDA. In so doing, SDA opens us to a view of the living

God who is always self-giving, both as the ground of order (Logos) and as responsive presence (Spirit). Gregersen then challenges us to address the idea of divine causation within a richer concept of natural law as made possible through recent developments in the philosophy of science, including the work of Nancy Cartwright, Peter Lipton, and Ronald Giere.

Arthur Peacocke begins with God's action as creator ex nihilo, sustaining in existence processes endowed with the capacity for the evolution of biological complexity. But does God act in particular events beyond this sustaining activity—is there special providence along with general providence? In a deterministic understanding of nature this would require divine intervention, which Peacocke rejects. The VO/CTNS series, however, has focused on new areas in science which might allow for non-interventionist special divine action: chaotic systems, quantum mechanics, and whole-part / top-down causation operating on the world-as-a-whole, and it is to these he turns.

Chaos: Although the future behavior of chaotic systems is inherently unpredictable, these systems fall within Newtonian determinism, and special divine action would still be interventionist. John Polkinghorne has made a metaphysical proposal that views such systems as "downward emergent" from higher-order systems which themselves might be open to non-interventionist special divine action. While Peacocke finds this idea attractive, the appeal to unpredictability in itself as a basis for such action is a cul-de-sac.

Quantum events: Perhaps the indeterminism characterizing quantum events offers a route to non-interventionist divine action. Peacocke rejects one option—that God acts in all quantum events—as being a version of occasionalism. The alternative, that God acts in only some quantum events, raises a variety of problems depending on whether the laws of nature have an ontological status, whether God can know in advance the outcome of statistical processes, whether stochastic laws apply only to ensembles or to individual events in the ensemble, etc. All this leads Peacocke to view the appeal to quantum events as "not the most fruitful path to follow."

Whole-part influence: Peacocke now focuses on his own preferred approach which draws on the fact that complex systems allow for both bottom-up and whole-part forms of causality. In complex systems, the units behave as they do because of the behavior of the overall system. If the "world-as-a-whole" is regarded as an overall "System-of-systems" then it will influence its constituent parts. Since Peacocke understands God panentheistically to be "the circumambient Reality of all-that-is," God can be thought of as interacting with the complete world-system, bringing about events of special providence in the world without intervening in it. Peacocke adds to this view the mind-brain-body complex as yet another exemplification of the whole-part model of influence. This resource for modeling God's interaction with the world allows him to recapture personal and Biblical categories to describe non-interventionist special divine agency.

William R. Stoeger, S. J., writes that significant progress has been made through the VO/CTNS series in two key areas: in framing the question of divine action in light of the natural sciences and in modeling both God's creative action in nature and God's special action in history in ways coherent with science and Christian theology. He then lays out the main features of the conception of divine action which has emerged in the series and refines it further in light of several other issues which he claims were not adequately emphasized, addressed, or resolved in the series.

The key distinction regarding divine action which was employed extensively in the series is that between God's universal creative action in nature and God's special action in history. The former has been described predominantly in terms of *creatio ex nihilo* and *creatio continua* as models of the world's ontological dependence on God and God's continuing creation in and through natural processes. According to Stoeger, the VO/CTNS participants were in agreement that the laws of nature investigated by the sciences "require an ontological grounding which they themselves cannot provide, and that they are the channel of God's continuing creative action in the universe." Stoeger then carefully examines the transcendental character of divine action in the conceptual framework of creation *ex nihilo,* its fundamental resistance to any conceptualization, and the protocols which must be enforced in describing it by analogy as "cause," "act," or "ground." His reasons are twofold: *creatio ex nihilo* provides the fundamental basis for properly understanding both God's universal, creative action, and God's special action, and the implications of science reinforce and nuance this articulation of divine action. Stoeger believes that the issues surrounding creatio ex nihilo were not adequately engaged in the VO/CTNS series but that they should be addressed in future research.

Stoeger next focuses on how "the laws of nature" are most helpfully conceived and how God's universal creative action is manifested in and through them. He then lists six approaches for conceiving special divine action in relation to God's universal creative action in the VO/CTNS series: (1) conceiving of God's special acts as flowing "bottom-up" via quantum mechanics; (2) postulating that "the laws of nature" are in reality much richer and deeper than what we have modeled so far; (3) seeing God's special acts as flowing from the "top-down" or "whole-part"; (4) opening nature at higher levels of complexity to God's intentional active information input (Polkinghorne); (5) employing the process idea of God presenting new possibilities to nature and inviting it to respond (Barbour, Birch and Haught); and (6) showing simply that there is no determinism which rules out God's special acts.

These approaches raise crucial issues whose resolution strongly influences the conclusions one can draw here. Stoeger also discusses the meaning and limits of determinism in nature, whether or not it holds, or can be shown to hold, and whether special divine action requires windows of indeterminacy in order to operate (what he calls "the compatibilism/incompatibilism divide"). In closing, Stoeger suggests portraying special divine action as a manifestation or mode of God's universal creative action, properly conceived.

Thomas F. Tracy distinguishes between three senses in which we may speak of events as special divine actions: subjectively special, materially special, and objectively special. The third type has been particularly problematic for theology both for internal reasons and because it seems to place theology in conflict with science. Tracy argues that the conflict actually arises from a deterministic interpretation of science, and the solution is to offer an indeterministic interpretation in which nature includes "events that have necessary but not sufficient natural causes." Objectively special divine action in relation to these events would affect the course of nature without intervention. The question, then, is which, if any, areas in contemporary science provides such a view of nature.

Before pursuing this question, Tracy first responds to several general objections. One is that we should not take a "zero-sum" view of divine and created agency. Tracy agrees with this objection as it applies to God's fundamental activity as creator, but holds that there are modes of divine action that require such a trade-off, particularly in two cases: objectively special divine action and incompatibilist human freedom. Another objection, one voiced by Peacocke, is that a non-interventionist account of God's objectively special action can be constructed without reliance on indeterminism in nature. The central idea is to model God's relation to the world by a whole-part analogy drawn from hierarchically-nested complex systems. In such systems, the whole, Peacocke argues, has causal efficacy because its structure constrains the dynamics of the parts. Peacocke then places this model within a panentheistic view of God's relation to the world as a whole, depicting God as the highest-level whole. While Tracy finds this "theologically appealing and conceptually artful," he claims, against Peacocke, that it still requires ontological openness at some level in nature. A third objection is that we need a full metaphysical system in order to relate scientific theories to a theology of divine action, and one prominent candidate is Alfred North Whitehead's metaphysics. Again, while acknowledging its apparent advantages, Tracy parts company with it in light of what he sees as its vulnerabilities both in interpreting the natural sciences and in guiding theological construction.

Tracy then turns to three specific proposals for non-interventionist objective divine action, beginning with those based on chaos theory, which he argues do not support an indeterministic interpretation. He turns next to proposals that make use of quantum mechanics, assessing their strengths and responding to criticisms involving puzzles of multiple interpretation, the measurement problem, and the amplification of the effects of quantum events. He poses a sharp reply to criticisms by Nicholas Saunders and Wesley Wildman. Finally, he examines the prospects for appealing to mental life as the locus of special divine action. After a detailed analysis of the accounts of mental causation by Theo Meyering and Murphy, he discusses the relation between Murphy's "nonreductive physicalism" and Clayton's "emergent monism." Tracy concludes by listing scientific, metaphysical and theological issues requiring continued research.

Keith Ward begins with the challenge raised by a deterministic account of physical laws to the reality of free action, human or divine. One possibility is to argue for a non-deterministic cosmos, not by focusing specifically on quantum indeterminism, but on the claim that "initial physical states and general physical laws do not always sufficiently determine subsequent physical states." Ward then argues that science cannot rule out indeterminism for several reasons.

First, God, as the creator of the universe, is an immaterial consciousness. This implies that there is at least one case of immaterial and indeterministic action that results in physical states—divine creation per se. But does this imply that other, particular states or laws are indeterministic? Ward marshals arguments from Stoeger, Polkinghorne, Bernard d'Espagnat, Wildman and Russell that challenge the assumptions that science supports determinism. For example, while the striking predictability of closed causal systems implies that they are deterministic, open systems interacting with their environment may not be completely governed by the known set of physical laws. Here, then, the intentional acts of God might have a particular causal influence on the world.

Is this just a "God of the gaps" argument? Ward rejects this charge, urging instead that we view the universe as emergent and open to divine action. A more serious concern about divine action is the wastefulness and suffering entailed in the evolution of life. Ward's response is to see this as an inevitable outcome of the probabilistic nature of physical laws, which are themselves required for the emergence of freedom and moral action in a self-organizing world, as Peacocke and John Haught suggest. Thus God may intend there to be general laws but God does not intend every specific event that happens. God's particular actions, when they occur, will be set within the long-term context of divine intentions and continuing influence on the world. Science can ignore such divine action, assigning the resulting events to chance, but theists can understand God's actions as progressively realizing God's goals throughout the history of the universe. This leads Ward to view God as "the spiritual basis or macro-environment of the cosmos." Finally, like the resurrection of Jesus, some divine actions will indeed be miraculous, not in David Hume's sense of them violating the laws of nature but as law-transcending events that "show the power of Spirit to relate matter to Spirit." Through these events God perfects physical processes and reveals God's ultimate purpose for them.

Kirk Wegter-McNelly first offers some general observations on the series' strengths and weaknesses. He highlights the participation by many leading scientists and the increasing willingness of theologians to take particular scientific theories seriously, but he also cites the lack of collaborative essays by scientists and theologians. He views *EMB* and *QC* as showing the most "convergence" regarding the importance of science for theology. Wegter-McNelly then turns to the core of his essay, focusing attention on the various strategies for constructing a NIODA account and raising a crucial question: does NIODA require nature to be indeterministic? According to Wegter-McNelly, some of the VO/CTNS contributors do make this assumption. Objectively special divine acts and physical processes stand in a "zero-sum fashion" in which the operation of one necessarily precludes the operation of the other. Calling this "theo-physical incompatibilism," Wegter-McNelly draws on the work of William Placher to describe this as a distinctively modern assumption which "domesticates" the concept of divine transcendence. Other contributors adopt what Wegter-McNelly calls theo-physical compatibilism, following such pre-modern thinkers such as Thomas Aquinas, for whom objectively special divine action is compatible with physical determinism and yet not interventionist. To clarify this issue Wegter-McNelly develops two related distinctions, "anthropo-physical in/compatibilism" and "anthropo-theological in/compatibilism." He supports anthropo-physical incompatibilism but not anthropo-theological and theo-physical compatibilism, claiming that special divine action is not analogous to human agency and it is not undermined by determinism.

Wegter-McNelly then expands the typology that was published in *CC* and *QM* (see Appendix D of this volume) by adding a column under non-interventionist approaches to objectively special divine action that designates theo-physical compatibilists and incompatibilists. According to his analysis, theo-physical compatibilist strategies include those of Peacocke (via top-down or whole-part causality), Stephen Happel and Stoeger (via the distinction between primary and secondary causes), and Tracy (via what Wegter-McNelly terms Tracy's "functionalism" in *QM*). Incompatibilist strategies include those of Russell, Ellis, Tracy, and Murphy (via bottom-up causality), Polkinghorne (via lateral amplification), and Barbour, Charles Birch and Haught (via process philosophy). Wegter-McNelly believes that this distinction between compatibilism and incompatibilism reveals a significant but previously unnoticed rift that has influenced and even obscured the debate over the relative merits of different NIODA proposals. He concludes by arguing that future work on the problem of special divine action should adopt a theo-physical compatibilist framework and seek to show the meaning and significance of physical indeterminism from this perspective.

According to **Mark Worthing**, affirming special divine action might seem to undermine science, while denying it might "erode the very core of Christian faith." In response, Worthing cites the striking consensus among many VO/CTNS participants that God acts in the world and that this action is noninterventionist. Nevertheless, he underscores the complexities that this claim entails as seen in essays by Tracy, Michael Heller, Willem Drees, and Russell. Worthing focuses attention on the variety of meanings of "non-intervention" both within the VO/CTNS series and beyond it. Some theologians, for example, may still use the term "intervention" for any objectively special divine act whether or not it entails a violation of a law of nature. "What we seem to have arrived at in the VO/CTNS series is not so much a theologically noninterventionist model...but a scientifically non-interventionist model." He discusses the subtleties in the concept of the laws of nature, the extent to which the laws we formulate relate to the actual laws, whether the laws are prescriptive or descriptive, and the implications of compatibilist versus incompatibilist views about divine action in relation to the laws of nature, noting positions taken by Stoeger, Peacocke, Russell, and Murphy. He then turns to the question of openness/indeterminism in nature, drawing on VO/CTNS essays by James Cushing, Polkinghorne, Heller, and Wildman, as well as the publications of Roger Penrose. Here he focuses theological attention on God's knowledge of the future and on predestination in relation to human freedom. "While current discussions among physicists cannot possibly provide answers to these theological questions, they may very well provide useful paradigms" for discussing them.

Worthing concludes with a general assessment of the VO/CTNS series. He is appreciative of the fact that such an in-depth and sustained series was held together by an explicitly theological theme, divine action. Future research could profit from the inclusion of "reputable theologians (who) hold moderate and nuanced interventionist views." He suggests more representation for the Roman Catholic tradition, with its focus on the distinction between primary and secondary causes, as seen here in chapters by Stoeger and Denis Edwards. He recommends exploring other areas in the sciences beyond those that typify the dialogue. He also urges the inclusion of more theologians who have not yet been involved in the dialogue. They should be encouraged to bring their own terminology and categories to the issues and to reflect on the ways theological differences shape how nature is viewed even aside from the conversation with science.

4 Closing Comments: Assessing and Celebrating the Series

This volume offers a rich and diverse tapestry of analysis and assessment of the VO/CTNS series of publications. It presents topics for continued discussion, issues whose importance has been brought out by our work so far, and areas on the growing edge of research. From these chapters one can draw out a very helpful set of recommendations for future research in the relations among philosophy, theology, and science. These recommendations include the following:

4.1 Philosophy

- more extensive discussion of the limits of science (Ellis)
- detailed differentiation among a variety of arguments against causal reductionism: those involving the hierarchy of the sciences, downward (top-down/whole-part) divine causation, emergence, supervenience and selection (Peacocke, Clayton, Murphy, Tracy)
- further examination of the concept of "laws of nature" and their ontological status (Ellis, Gregersen, Russell, Stoeger, Worthing)
- a detailed assessment of the relative merits of proposals on the table: top-down, whole-part, lateral, bottom-up, primary/secondary, and process metaphysics (Gregersen, Peacocke, Wildman)
- further discussion of the relative merits of differing metaphysical systems employed in theology/science research (Russell)

4.2 Theology

- further examination of the relation between special and general providence (Gregersen)
- further discussion of the Trinitarian doctrine of God (Russell) and of the doctrine of creation *ex nihilo* (Stoeger)
- further discussion of the divine attributes and God's relation to creatures and to time (Tracy)

4.3 Divine Action and Science

- assessing the meaning of and need for traction between theology and science (Clayton, Gregersen)
- clarification on a counterfactual understanding of divine action in the context of an indeterministic view of natural processes (Clayton)
- the wider meaning of divine intervention as God's action whether or not it involves a violation of the laws of nature (Worthing);
- the relation between divine action and theodicy (Russell, Tracy, Wildman)
- a careful examination: of compatibilism vs. incompatibilism (Tracy), of determinism vs. indeterminism (Ward), of the relation between these two sets of options in the context of divine action

(Stoeger, Worthing), and of a compatibilist interpretation of divine action in the context of an indeterministic interpretation of nature (Wegter-McNelly)

- the liability of a "zero-sum" approach to the relation between divine and created agencies and their alternatives (Tracy, Wegter-McNelly)
- the need to expand the concept of agency in the God-world context to include agency in the God-human context and in the human-world context (Wegter-McNelly)
- further clarification regarding the relation among the divine attributes, God's relation to creatures and to time, and questions of divine action in nature (Tracy)
- the relation between non-interventionist objective divine action and eschatology (Russell)

5 Celebrating the Conclusion of the Series

With the publication of this volume, the VO/CTNS series on "scientific perspectives on divine action" and its scholarly assessment comes to a conclusion—or at least a resting point before the adventure continues. It is safe to say that, thanks to the hard work and good will of all involved—including the staff of CTNS and the VO, which facilitated the conferences and brought the publications to light—this series represents a milestone in the burgeoning scholarly interaction among science, philosophy, and theology. In my view, the series has clarified and in many cases settled issues previously dominating, and often cluttering, the interdisciplinary landscape. It has marshaled diverse resources to move the frontiers forward in creative new ways. It has provided a lasting resource for future scholars, especially those entering the field. And it has discovered new, or enhanced existing, problems which may now be more effectively addressed—which I take to be a true mark of intellectual progress.

I firmly believe that this series of publications has taken all of us in, through, and beyond many of the peaks and valleys that twenty years ago lay only in the dim distance, barely if at all perceived, let alone understood. My fervent hope is that the next twenty years are equally fruitful. If they prove to be, it will be so because of the extraordinary gifts of time, talent, vision, erudition, honesty, self-criticism, dedication, and joy which the participants in this series have brought to the process, and underlying it all, because of the surpassing grace of God whose special providence leads us all with sublime hope into our true future.

I. CRITICAL APPRAISALS OF THE SERIES AS A WHOLE

Robert John Russell
George Ellis

CHALLENGE AND PROGRESS IN "THEOLOGY AND SCIENCE": AN OVERVIEW OF THE VO/CTNS SERIES

Robert John Russell

> Science can purify religion from error and superstition; religion can purify science from idolatry and false absolutes. Each can draw the other into a wider world, a world in which both can flourish.[1]

1 Introduction and Historical Background

Of the many remarkable events and publications that marked the decade of the 1990s as a watershed in the burgeoning interdisciplinary field of "theology and science," one of the most significant was the series of five international and ecumenical research conferences co-sponsored by the Vatican Observatory (VO) and the Center for Theology and the Natural Sciences (CTNS). Some fifty scholars participated in the series, many with cross-disciplinary expertise in physics, astronomy, cosmology, mathematics, evolutionary and molecular biology, the neurosciences and cognitive sciences, philosophy of science, history of science, philosophy of religion, history of religion, Old and New Testament, philosophical and systematic theology, and theological ethics. Ninety-one essays were published in the five volumes, along with detailed analytic introductions to each volume. The overarching goal was to engage theology, philosophy, and natural science in a process of constructive dialogue and creative mutual interaction. The purpose of this chapter is to provide an overview of the topics addressed, to offer a brief assessment of the "divine action project" represented more specifically by two dozen chapters in the series, and to conclude with a survey of the problems and progress achieved.

First, though, we will take a brief look at the historical background of the series. The Vatican Observatory, or "Specola Vaticana," is housed in the Papal Palace in the picturesque town of Castel Gandolfo overlooking Lake Albano thirty miles southeast of Rome. Since 1935 it has been the site of basic research in both observational and theoretical astronomy. It is also here that Pope John Paul II often resided during the summer. In earlier years the Pope, then Cardinal Archbishop of Krakow, had regularly entered into conversations on cosmology and philosophy with Polish friends and colleagues. On becoming Pope in 1978, he continued his interest in this dialogue and sought to improve the relationships between the Church and the scientific community. In 1979, in an address to commemorate the 100th anniversary of the birth of Albert Einstein, John Paul II said:

> I hope that theologians, scholars, and historians, animated by a spirit of sincere collaboration, will study the Galileo case more deeply and, in loyal recognition of the wrongs from whatever side they come, will dispel the mistrust that still opposes, in many minds, the fruitful concord between science and faith, between the Church and the world. I give my support to this task which will be able to honor the truth of faith and of science and open the door to future collaboration.[2]

---

[1] John Paul II, "Message to George Coyne," in *PPT*, M 13.

[2] *Discourses of the Popes from Pius XI to John Paul II to the Pontifical Academy of*

In response George Coyne, S. J., the Director of the Vatican Observatory, together with Michael Heller, a member of the Philosophy Faculty at the Pontifical Academy of Theology and the Center for Interdisciplinary Studies in Krakow, organized a conference in Poland in 1984 which resulted in a major publication on the Church and the Galileo case.[3] Next, George Coyne, together with Bill Stoeger, S. J., the senior cosmologist at the Specola, and Michael Heller, invited me to help plan a conference in the late spring, 1987, also held in Poland, on the theological implications of the rise of modern science.[4]

Following the success of these conferences and publications, the Pope asked the Vatican Observatory to organize a major international conference to further the science-faith dialogue on the occasion of the 300th anniversary of the publication of Isaac Newton's *Principia.* It was held at the Specola in September 1987. It concluded with a remarkable Papal Audience at the Vatican. The resulting publication, *PPT,* has been frequently used in courses and conferences on theology and science. (See section 4.1.1.6 below for the data on its usage and Appendix A for the Table of Contents of *PPT.*) It includes a "message" by the Pope given during the audience on the relations between the church and the scientific communities. As the first major Pontifical statement on science and religion in three decades, the Message has been widely discussed and quoted. In 1990, the Papal Message was the centerpiece of *John Paul II on Science and Religion: Reflections on the New View from Rome*[5] which included nineteen responses by scientists and theologians. (See Appendix B for the Table of Contents).

Based on the accomplishments of *PPT* and the vision offered by the Papal Message, George Coyne took steps to initiate a major new series of conferences on theology and science. He convened a week-long meeting at the Specola in June 1990, to plan the overall direction of research. During this meeting he asked Nancey Murphy, from Fuller Theological Seminary, to join Bill and me in forming the long-term steering committee for the series. Our task was to build on the accomplishments of *PPT* by moving further into areas in the physical and biological sciences already touched on in *PPT* as well as to expand the basis of research in science into new areas such as the neurosciences and cognitive sciences (see Table 1 below). Coyne invited CTNS to co-sponsor the series and co-publish the resulting volumes and asked me to serve as General Editor.[6]

Sciences (Vatican City State: Pontificia Accademia Scientiarum, 1986), Scripta Varia 66, 73-84.

[3] G. V. Coyne, M. Heller and J. Zycinski, eds., *The Galileo Affair: A Meeting of Faith and Science* (Libreria Editrice Vaticana: Vatican City State, 1985).

[4] G. V. Coyne, M. Heller, and J. Zycinski, eds., *Newton and the New Direction in Science* (Libreria Editrice Vaticana: Vatican City State, 1988).

[5] Robert John Russell, William R. Stoeger, and George V. Coyne, eds. (Vatican Observatory Publications, 1990).

[6] CTNS was able to accept the invitation thanks to a generous grant from a local Bay Area family foundation which supported our participation for the entire series of conferences.

2 An Overview of the Series

2.1 The Invention and Deployment of a New Method for "Theology and Science"

Looking back from the vantage point of 2008, it is evident how far the theology and science dialogue has come since 1990. In those days several major issues loomed over the entire discussion and impeded progress and they all focused on methodology. Granted many scholars had already moved beyond the sterile conflict or independence models of the relation between theology and science. Still the first issue that impeded progress regarded the role science should play in the conversations. Too often scientists were asked to make the first presentations at a conference with the unspoken assumption that the results they described were to be taken verbatim and that the theologians really had nothing to say of interest to the scientists. In practice this usually meant that after the science presentations were finished, the philosophers, theologians and religious scholars were left to try to decide what had been said and why it was significant to them. Often the conversations got bogged down over terminology (when a physicist speaks about causes is it the same thing as when a theologian does?). If that obstacle was surmounted the real challenge arose: can scientific results, like the details of Big Bang cosmology or the role of DNA in molecular biology, be taken directly into theology or should they be mediated by a philosophical discussion of their meaning and significance? If philosophy is needed, does this require the adoption of an entire metaphysical system, such as process philosophy or contemporary Catholic philosophy, within which both science and theology can be situated or is a topic by topic philosophical analysis sufficient for the purpose of theological appropriation?

From the beginning it was the clear intention of the steering committee that our research methodology should take us beyond these obstacles and insure a two-way interaction between scientists and theologians. In order to achieve this goal we created whole-cloth a new, four-fold strategy.

2.1.1 Guiding Theme of the Series of Conferences: "Scientific Perspectives on Divine Action"

First, we searched for a topic in philosophical theology to thematize the entire series of conferences and to inspire interdisciplinary discussion by both scientists and theologians. The topic would therefore have to satisfy two broad criteria: a) It should function at the presuppositional level underlying the spectrum of particular doctrines in and approaches to systematic theology. This would allow scholars from a variety of perspectives and denominations to pursue their individual theological interests and at the same time gain from their interactions with each other as they engaged with specific scientific topics through the lens of the trans-conference topic in philosophical theology. It should also serve to draw out the philosophical and theological implications of the variety of sciences to be explored and it should do so in a way that would allow for a diversity of theological approaches to the way science is appropriated (e.g., natural theology, theology of nature, etc.), replacing the usual debates over which one is preferable.

b) The topic in philosophical theology should be of interest to scientists, thereby making the conversations genuinely "two-way." As scientists at the conference saw the impact of their discoveries on the theologians' work, they might, in turn, be led to examine their own presuppositions about and conceptions of nature, a process which potentially might inspire them to ask new questions and develop promising new lines of scientific research. During that fateful meeting in June 1990, the topic of divine action—God's action in and interaction with the world—was eventually singled out as a promising candidate for the thematization of the series of conferences since it met both of these criteria nicely.

2.1.2 Cross-disciplinary Expertise of the Participants

To overcome some of the terminological issues and to increase the genuine interaction between science and theology we gave strong preference to participants who already had achieved solid expertise in both fields. This meant inviting cutting-edge scientists who were versed in philosophy and theology and leading theologians who were passionate about the issues raised by science and willing to learn more of the underlying technical material. In some cases we were blessed with scholars who were already steeped in all three fields.

2.1.3 Pre-conferences and Papers Read in Advance

We agreed to hold regional pre-conferences to provide an introduction for participants to relevant technical issues in science, philosophy and theology and to foster joint research and collaboration among participants prior to the conference. Participants would circulate pre-conference drafts for written responses and these drafts, in turn, would be revised and recirculated before the conference. During the conference papers would not be read; instead each paper was critically discussed during a designated session. To be published in the conference volume, post-conference revisions had to reflect these discussions.

2.1.4 The Results

With these strategies in place the organizing committee then planned a series of five conferences to span the decade of the 1990s. Each would involve a two year cycle: the first year for pre-conferences and critical reading of papers, the second year for post-conference revisions, final selection of papers for publication, and the drafting of the analytic introductions, etc. The cycles overlapped, with the post-conference activities of one conference being simultaneous with the pre-conference activities of the following conference, making for a demanding schedule but a very productive result.

The first conference focused on quantum cosmology and the origin of the laws of nature. It built on the initial exploration of cosmology in *PPT* and included such issues as $t = 0$ and the Anthropic Principle.[7] Next came an examination of the sciences of chaos and complexity, followed by evolutionary and molecular biology, and then by the neurosciences, all of which greatly expanded the scope of

[7] See *QC*.

research presented in *PPT*.[8] The fifth conference returned to one of the central themes of *PPT*:[9] quantum mechanics. The first and third conferences were held at Castel Gandolfo in 1991 and 1996; the second was held in Berkeley in 1993. For the fourth conference we gathered at Pasierbiec, Poland, at the invitation of Michael Heller and the Pontifical Academy of Theology in Krakow. We returned to beloved Castel Gandolfo for the final quantum mechanics conference in 2000. In summary, the five volumes are titled:

1. *Quantum Cosmology and the Laws of Nature* (*QC*)
2. *Chaos and Complexity* (*CC*)
3. *Evolutionary and Molecular Biology* (*EMB*)
4. *Neuroscience and the Person* (*NP*)
5. *Quantum Mechanics* (*QM*)

with each containing the subtitle, *Scientific Perspectives on Divine Action*.

2.2 Who Published on What Topics?

Number of authors in the series: 51
Number of chapters in the series: 91

Table 1 lists the variety of areas in science, philosophy and theology treated in the five-volume series. The Table indicates whether these areas build on those discussed in *PPT* or considerably extend beyond *PPT* into new areas.

Table 1: Areas in science, philosophy and theology in the five-volume series and an indication of whether they either build upon or extend beyond areas found in PPT.

Sciences:	*PPT*:
Chaos, Complexity, Thermodynamics and Self-Organization	New
Relativity (Special, General)	Builds
Quantum Mechanics	Builds
Cosmology (Standard Big Bang, Inflation, Quantum)	Builds
Evolutionary and Molecular Biology	Builds
Sociobiology and Biocultural Evolution	New
Neuroscience	New

[8] See *CC*, *EMB*, and *NP*.

[9] See *QM*.

Philosophy:	
Critical Realism (the Comprehensibility of the World)	Builds
Laws of Nature & Ontological (In)determinism	Builds
Teleology in Nature	New
Time in Relativity and its Philosophical Interpretations	Builds
Quantum Mechanics and its Philosophical Interpretations	Builds
Philosophical Anthropology & the Philosophy of Mind	New
Emergence, Supervenience and Downward Causality	New

Theology:	
Methodology in Theology and Science	
(General Relations, Specific Models, Role of Philosophy)	Builds
God (Revelation, Trinity)	Builds
Creation (*ex nihilo*, continuous, the Anthropic Principle)	Builds
Divine Action & Special Providence	Builds
Theological Anthropology	New
Theodicy (Sin and Moral Evil / Natural Evil)	New
Eschatology (Resurrection)	New

Table 2 offers a rough overview of the results. Across the top the name of the author of each chapter is listed. (See Appendix C). Under each author's name an "x" marks the primary areas in science, philosophy and theology discussed in that chapter. The areas are denoted by an abbreviation of the more complete list shown in Table 1. Needless to say, it was often difficult to decide between primary topics and subordinate topics in specific chapters, and I apologize to those authors who will disagree with my assessment!

One very interesting way to use Table 2 is to explore the role philosophy plays in a given author's way of relating science and theology. Of the 91 chapters, and setting aside the eight introductory chapters to the relevant sciences, the explicit use of philosophy in relation to science and theology breaks down as follows:

Number of chapters focusing on science and theology: 35
Number of chapters focusing on science,
philosophy and theology: 30

This means that, of the 65 chapters dealing with science and theology, the authors were almost evenly split over the role of philosophy in the relation between science and theology. (For completeness I would point out that sixteen chapters focus on science and philosophy while two chapters focus on theology and philosophy.) Note: Summaries of all 91 chapters in the series are available online at: www.ctns.org/books.html. They are arranged by book, by author and by the twenty-two areas listed in Table 1 (and repeated in Tables 2 and 3).

	Alston QC	Arbib NP ("Crusoe")	Arbib NP ("Towards")	Ayala EMB ("Darwin's")	Ayala EMB ("Evolution")	Barbour EMB	Barbour NP	Berry QM	Birch EMB	Brothers NP	Butterfield QM	Cela-Conde EMB ("Hominid")	Cela-Conde EMB ("Beyond")	Chela-Flores EMB	Chiao QM	Clarke QM
NATURAL SCIENCE																
Chaos Theory						X		X	X							
Relativity Theory	X															
Quantum Mechanics (QM)	X					X		X			X				X	X
Cosmology	X															
Evolutionary Biology				X	X	X			X			X	X	X		
Sociobiology/Evolutionary Psychology													X			
Neuroscience		X	X				X			X						
PHILOSOPHY																
Critical Realism						X										
Laws of Nature	X															
Teleology				X												
Time																
QM: Philosophical Interpretations											X					X
Anthropology/Mind	X		X				X									
Emergence						X	X	X	X				X			
Freedom	X															X
THEOLOGY																
Methodology						X	X		X							
God	X					X			X							
Creation						X									X	
Divine Action	X					X			X							X
Theological Anthropology	X		X				X									
Theodicy						X										
Eschatology																

	Clayton NP	Clayton QM	Clifford EMB	Coyne EMB	Crutchfield CC	Cushing QM	Davies QC	Davies EMB	Drees QC	Drees EMB	Drees CC	Edwards EMB	Edwards CC	Ellis NP	Ellis CC	Ellis QC ("Theology")
NATURAL SCIENCE																
Chaos Theory					X						X		X		X	
Relativity Theory																
Quantum Mechanics (QM)		X				X									X	
Cosmology							X		X							X
Evolutionary Biology			X	X				X		X		X				
Sociobiology/ Evolutionary Psychology										X		X				
Neuroscience	X													X		
PHILOSOPHY																
Critical Realism							X		X		X					
Laws of Nature																
Teleology								X								
Time									X							
QM: Philosophical Interpretations		X				X										
Anthropology, Mind										X						
Emergence	X						X	X								
Freedom	X	X														
THEOLOGY																
Methodology		X	X	X				X		X						
God	X								X	X			X	X	X	
Creation			X													X
Divine Action		X									X		X		X	
Theological Anthropology				X								X				
Theodicy												X			X	X
Eschatology																

	Ellis EMB	Ellis QC ("Intro.")	Ellis QM	Gilkey CC	Green NP	Grib QC	Hagoort NP	Happel NP	Happel CC	Happel QC	Haught EMB	Hefner EMB	Heller CC	Heller QM	Heller QC	Isham QC
NATURAL SCIENCE																
Chaos Theory				X					X				X			
Relativity Theory		X														X
Quantum Mechanics (QM)			X			X								X		
Cosmology		X				X				X					X	
Evolutionary Biology	X										X	X				
Sociobiology, Evolutionary Psychology												X				
Neuroscience					X		X	X								
PHILOSOPHY																
Critical Realism													X			X
Laws of Nature																
Teleology																
Time		X														X
QM: Philosophical Interpretations																
Anthropology, Mind							X									
Emergence	X		X													
Freedom																
THEOLOGY																
Methodology	X							X			X			X	X	X
God				X		X			X	X	X			X	X	X
Creation											X	X				
Divine Action			X					X								
Theological Anthropology					X			X				X				
Theodicy																
Eschatology								X			X					

	Jeannerod NP ("Are")	Jeannerod NP ("Cognitive")	Kerr NP	Küppers CC	LeDoux NP ("A View")	LeDoux NP ("How")	Lucas QC	McMullin QC	Meyering NP	Moltmann CC	Murphy CC	Murphy QC	Murphy EMB	Murphy NP	Peacocke EMB	Peacocke CC ("Chance")
NATURAL SCIENCE																
Chaos Theory				X						X	X					X
Relativity Theory																
Quantum Mechanics (QM)								X			X					
Cosmology							X	X				X				
Evolutionary Biology													X	X	X	X
Sociobiology/Evolutionary Psychology																
Neuroscience	X	X	X		X	X			X				X			
PHILOSOPHY																
Critical Realism																
Laws of Nature											X	X				
Teleology																
Time																
QM: Philosophical interpretations								X								
Anthropology, Mind			X		X	X										
Emergence				X									X	X		
Freedom																
THEOLOGY																
Methodology								X	X	X	X	X	X	X		
God							X			X						
Creation												X			X	X
Divine Action										X	X					
Theological Anthropology			X										X	X		
Theodicy												X			X	
Eschatology										X					X	

	Peacocke CC ("God's")	Peacocke NP	Peters EMB	Peters NP	Peters QC	Polkinghorne QM	Polkinghorne QC	Polkinghorne CC	Pope John Paul II EMB	Redhead QM	Russell QM	Russell QC	Russell EMB	Shimony QM	Stoeger EMB	Stoeger QC
NATURAL SCIENCE																
Chaos Theory	✓						✓	✓								
Relativity Theory											✓					
Quantum Mechanics (QM)						✓	✓			✓	✓		✓	✓		✓
Cosmology					✓							✓				✓
Evolutionary Biology			✓						✓				✓		✓	
Sociobiology / Evolutionary Psychology																
Neuroscience		✓		✓												
PHILOSOPHY																
Critical Realism							✓	✓								✓
Laws of Nature							✓									✓
Teleology															✓	
Time																
QM: Philosophical Interpretations										✓	✓		✓	✓		
Anthropology, Mind																
Emergence																
Freedom																
THEOLOGY																
Methodology	✓	✓		✓		✓	✓	✓	✓		✓	✓	✓		✓	
God		✓			✓			✓								
Creation									✓			✓				
Divine Action	✓					✓	✓	✓	✓		✓		✓		✓	
Theological Anthropology			✓						✓							
Theodicy											✓		✓			
Eschatology			✓	✓	✓		✓				✓		✓			

	Stoeger QM	Stoeger NP	Stoeger CC	Tracy QM	Tracy EMB	Tracy CC	Ward QC	Watts NP	Wildman EMB	Wildman NP	Wildman CC
NATURAL SCIENCE											
Chaos Theory			X			X					X
Relativity Theory											
Quantum Mechanics (QM)	X			X		X	X				
Cosmology		X					X				
Evolutionary Biology					X				X		
Sociobiology; Evolutionary Psychology											
Neuroscience		X						X		X	
PHILOSOPHY											
Critical Realism			X				X				
Laws of Nature		X									X
Teleology									X		
Time											
QM: Philosophical Interpretations	X			X							
Anthropology; Mind		X									
Emergence											
Freedom											
THEOLOGY											
Methodology			X	X	X	X	X		X		
God							X				
Creation						X					
Divine Action			X	X	X	X			X		X
Theological Anthropology								X		X	
Theodicy					X						
Eschatology											

3 Special Focus on the "Divine Action Program"

As stated above, the central if often only implicit topic for the chapters in the series is that of divine action. On the one hand, almost every theological topic discussed in the series from God, revelation and creation to theodicy and eschatology, and by authors regardless of their confessional commitments, bears at least indirectly on divine action. On the other hand, several of the chapters in the series concentrated specifically on it. This section of my paper focuses on these chapters.

3.1 Who Published on Divine Action?

Number of authors focusing specifically on divine action:	19
Number of chapters focusing specifically on divine action:	25
Number of chapters in the series:	91

→ 27% of the total number of chapters were on divine action. Of these 25 total chapters, the use of philosophy along with science and theology breaks down as follows:

Number of chapters focusing on science and theology:	10
Number of chapters focusing on science, philosophy and theology:	14

One chapter focused on philosophy and theology.

Table 3 lists those chapters specifically on divine action. See Appendix D for a typology of the approaches to divine action.

	Alston QC	Barbour EMB	Birch EMB	Clarke QM	Clayton QM	Drees CC	Edwards CC	Ellis CC	Ellis QM	Happel NP	Moltmann CC	Murphy CC	Peacocke CC ("God's")
NATURAL SCIENCE													
Chaos Theory		X	X			X		X			X	X	X
Relativity Theory	X												
Quantum Mechanics (QM)	X	X		X	X			X	X			X	
Cosmology	X												
Evolutionary Biology		X	X				X						
Sociobiology/Evolutionary Psychology							X						
Neuroscience										X			
PHILOSOPHY													
Critical Realism		X				X							
Laws of Nature	X											X	
Teleology													
Time													
QM: Philosophical Interpretations				X	X								
Anthropology, Mind	X												
Emergence		X	X						X				
Freedom	X			X	X								
THEOLOGY													
Methodology		X	X		X					X	X	X	X
God	X	X	X					X			X		
Creation		X											
Divine Action	X	X	X	X	X	X		X	X	X	X	X	X
Theological Anthropology	X						X			X			
Theodicy		X					X	X					
Eschatology										X	X		

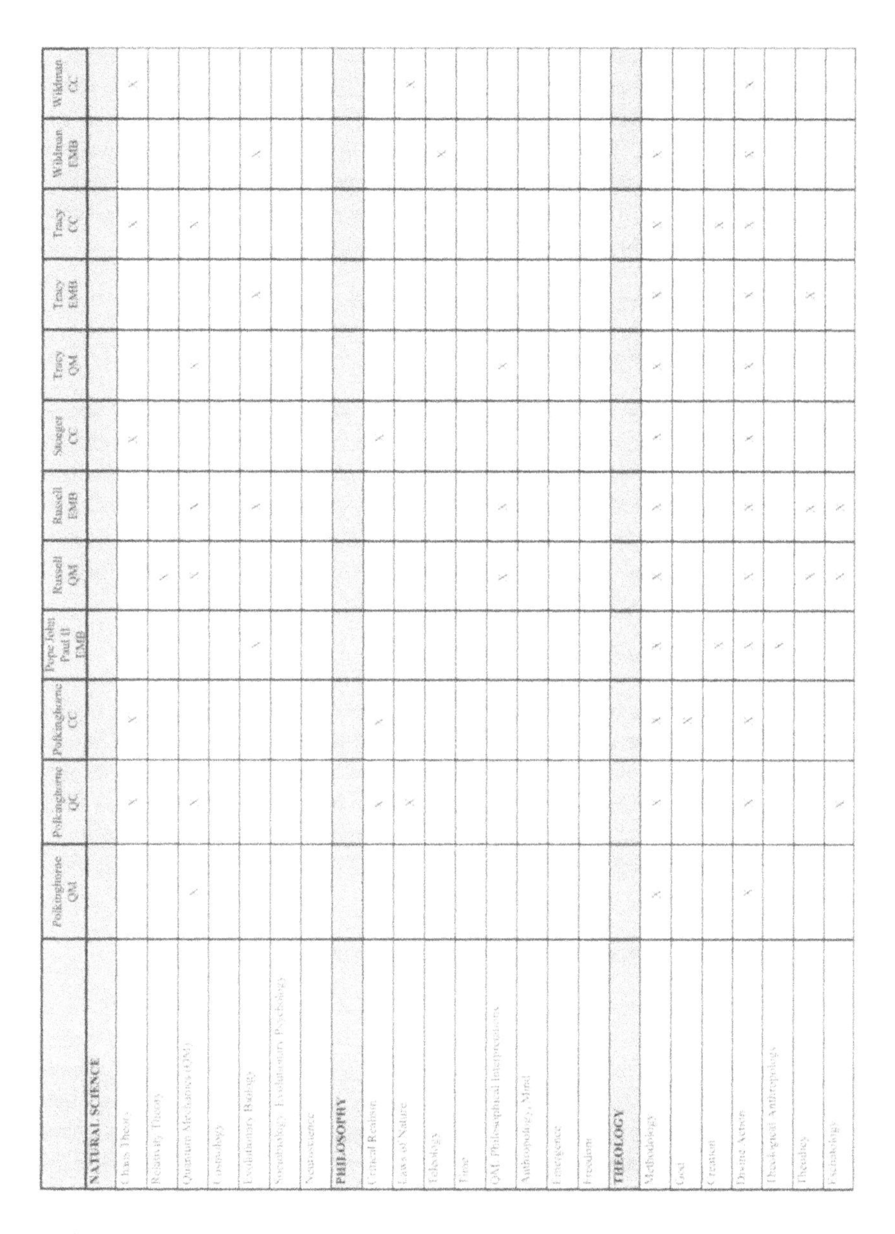

	Polkinghorne QM	Polkinghorne QC	Polkinghorne CC	Pope John Paul II EMB	Russell QM	Russell EMB	Stoeger CC	Tracy QM	Tracy EMB	Tracy CC	Wildman EMB	Wildman CC
NATURAL SCIENCE												
Chaos Theory		X	X				X			X		X
Relativity Theory					X							
Quantum Mechanics (QM)	X	X			X	X		X		X		
Cosmology												
Evolutionary Biology				X		X			X		X	
Sociobiology, Evolutionary Psychology												
Neuroscience												
PHILOSOPHY												
Critical Realism		X	X				X					
Laws of Nature		X										X
Teleology											X	
Time												
QM, Philosophical Interpretations					X	X		X				
Anthropology, Mind												
Emergence												
Freedom												
THEOLOGY												
Methodology	X	X	X	X	X	X	X	X	X	X	X	
God			X									
Creation				X						X		
Divine Action	X	X	X	X	X	X	X	X	X	X	X	X
Theological Anthropology				X								
Theodicy					X	X			X			
Eschatology		X			X	X						

4 Conclusions: Challenges and Progress in Theology and Science

In this final section I will briefly touch on the six areas in which I believe progress has been made through the VO/CTNS research series, including the initial publication (*PPT*) that served as a basis and warrant for the series. I will then suggest five areas of challenge generated by the series—either as unresolved issues in the series or as resulting from the progress of the series itself—and offer recommendations for future research.

4.1 Progress

This short overview paper is not the appropriate place for a detailed assessment of the ways the eight areas in philosophy and the seven areas in theology were developed by the authors in the series. (A sample of these developments can be found in Appendix E.[10]) However I will briefly touch on six areas in which I believe significant accomplishments were made and progress achieved in the series as a whole. Then I will suggest several ways in which progress has been achieved on the specific topic of divine action and science.

4.1.1 Areas of Progress in The Series as a Whole

4.1.1.1 New Methodology

The new methodology developed for and deployed in the VO/CTNS series included a) choosing a guiding theme for the entire series rooted in philosophical theology that could unify the theological interests of all its participants and bridge between theology and science, b) choosing participants with cross-disciplinary expertise, c) building in preconference interactions, d) agreeing to reading the conference papers in advance, and e) thorough postconference revisions of papers in light of conference discussions.

4.1.1.2 Landmark Publications

This series includes several pieces that have been extremely influential in the field. These include the statements by Pope John Paul II: on science and religion ("Message to George Coyne" in *PPT*), and on evolution ("Message to the Pontifical Academy of Sciences" in *EMB*). It also includes Ian Barbour's 4-fold typology on science and religion in *PPT*, preceding its publication in *Religion in an Age of Science,* which was to become a 'standard' for the field in the following decade.

4.1.1.3 Important Introductory Resources in Science and Philosophy

The series contains important introductory resources for future research. This includes key essays on science by Arbib, Ayala, Berry, Brothers, Cela-Conde, Chela-Flores, Chiao, Crutchfield et. al., Ellis and Stoeger, Hagoort, Heller, Isham, Jeannerod, Küppers, LeDoux, Shimony, Stoeger; essays on metaphorical language in science and theology by Happel, Hesse, Lash, McFague, Soskice; and essays on the philosophy of science and philosophical issues raised by science, by Arbib, Alston, Barbour, Butterfield, Clarke, Clayton, Cushing, Drees, Ellis, Happel, Heller, Hesse, Leslie, Meyering, Murphy, Peacocke, Polkinghorne, Redhead, Russell, Shimony, Soskice and Wildman/Russell.

---

[10] The text is excerpted and edited from the analytic introductions to the five volumes, four of which I wrote and one (*NP*) which was written by Nancey Murphy. It goes without saying that the choice of which to include reflects my own perspective and not necessarily those of the other editors in the series. More to the point, it was a difficult task both because I sincerely appreciate all of the chapters in the series and because I truly value the lasting collegiality, team effort and friendship with the authors.

4.1.1.4 Jointly-authored Essays and Coordinated, Separately-authored Essays on Interdisciplinary Research Topics in Theology, Philosophy and Science

This includes the joint essay by Isham and Polkinghorne on time in special relativity and its philosophical and theological significance, the joint essay by Wildman and Brothers on neuroscience and religious experience, the joint essay by Wildman and Russell on the philosophical and theological implications of chaos theory, and the joint essay by Cela-Conde and Marty on biology and culture. In addition there were coordinated essays on the theological significance of cosmological fine-tuning (i.e. the Anthropic Principle) by Ellis and Murphy, coordinated essays on evolutionary biology and human nature by Edwards and Hefner and coordinated essays on the philosophical and theological implications of quantum physics by Barbour, Clayton, Ellis, Murphy, Peacocke, Polkinghorne, Russell, Stoeger, and Tracy.

4.1.1.5 Novel Directions in Research in Theology and Science

This includes research on the ontological status of the laws of nature and the degree to which our scientific laws represent the laws of nature (Polkinghorne, Stoeger), on metaphor in science and in theology (Soskice, Barbour, Clifford, Happel, McMullin, Soskice), on time in nature and in theology (Drees, Happel, Isham and Polkinghorne, Lucas), on science and atheism (Buckley, Ellis), on science and models of God (Barbour, Edwards, Gilkey, McFague, Moltmann, Peters), on science and creation (Barbour, Ellis, Haught, Isham, Leslie, Murphy, Peacocke, Peters, Russell, Tracy), on science and the intelligibility of nature (Davies, Heller), science and human nature (Barbour, Clayton, Edwards, Ellis, Hefner, Murphy, Watts, Wildman and Brothers), on divine action and science (Alston, Barbour, Birch, Clayton, Edwards, Ellis, Happel, Murphy, Peacocke, Polkinghorne, Russell, Tracy) and on science and theodicy (Ellis, Russell, Tracy).

4.1.1.6 Major Impact on Scholars and the General Public

Sales of *PPT* and the series have been surprisingly high considering that "science and religion" is still a fairly specialized field among scholars. Over 3000 copies of *PPT* had been sold by the end of 2003, and it had been translated into Spanish and Arabic. Total sales for the five volumes in the series have topped 10,000 copies. Records taken by the CTNS Science and Religion Course Program indicate that over 250 courses internationally have included *PPT* or the volumes in the series. Finally, the CTNS website, which makes available summaries of all the chapters in the series, typically receives over 60,000 extensive visits per year.

4.1.2 Special Focus on Divine Action

I believe the series resulted in progress on the philosophical and theological topic of divine action in several ways. A sample of the chapters specifically on divine action is found in Appendix F.

4.1.2.1 On Terminology Regarding Divine Action

Over time we tended towards a shared meaning for key terms and concepts so that genuine differences and disagreements could be adequately illuminated by the common use of these terms and concepts. This in turn led to the possibility of solid conceptual progress on the diverse meanings of divine action in light of science. An early version of this commitment to shared meanings was published in the "Introduction" to the second volume in the series, *CC*, Section 3.4, pp. 9-13. Additional clarification came in key chapters throughout the series, with particularly helpful insights by George Ellis, Nancey Murphy, Bill Stoeger, and Tom Tracy, to which I also sought to contribute. Key terms include:

- laws of nature
- ontological indeterminism
- objective vs. subjective divine action
- direct (basic) vs. indirect divine action
- mediated vs. unmediated divine action
- compatibilist and incompatibilist views of divine action

4.1.2.2 On Distinguishing Between Six Approaches to Divine Action

Over time we also began to discover that a variety of distinct approaches were being taken by various scholars regarding divine action. Most followed one of these approaches, but in some cases scholars stipulated that eventually, at least, some combination of them would be needed as natural systems of increasing complexity and with increasing numbers of emergent properties and processes were considered. These approaches include four types of causality (termed top-down, whole-part, lateral, and bottom-up) as well as two broad metaphysical systems (process metaphysics and neo-Thomistic metaphysics/contemporary Catholic philosophy). An early attempt at listing these approaches was published in the same section in the "Introduction" to the second volume cited above. For details on the types of approaches to divine action see Appendix D.

4.1.2.3 Non-interventionist Objective Divine Action (NIODA) in Light of Science

In my own writings in the series I have suggested a goal for the divine action project which I believe represents what many of the other scholars in the series sought in their own ways. Drawing on the terminology noted above, I use the term "NIODA" as an acronym for this goal, namely "non-interventionist objective divine action." My goal is, then, an account of God's action in which certain events in nature mediate God's direct and objective action in a non-interventionist mode. In essence, NIODA would offer us, for the first time, an account of objective divine action that is not necessarily "miraculous" (in the Humean sense of divine acts which violate or suspend natural regularities/the laws of nature). Now in order for such divine action to be truly non-interventionist, nature at least at some level must be thought of as causally indeterministic. The focus of my research, then, is to search for and assess candidate theories in science for their capability of being given an indeterministic interpretation. In principle this could involve many theories at many different levels of complexity in nature. But even when we have one such scientific theory at one level which permits an indeterministic interpretation, we can claim that the direct, mediated effects of the objective acts of God occur within that domain of nature without intervention. The crucial role of science in thus offering the possibility for non-interventionist objective divine action is portrayed schematically (Figure 1), given ontological determinism or indeterminism in nature.

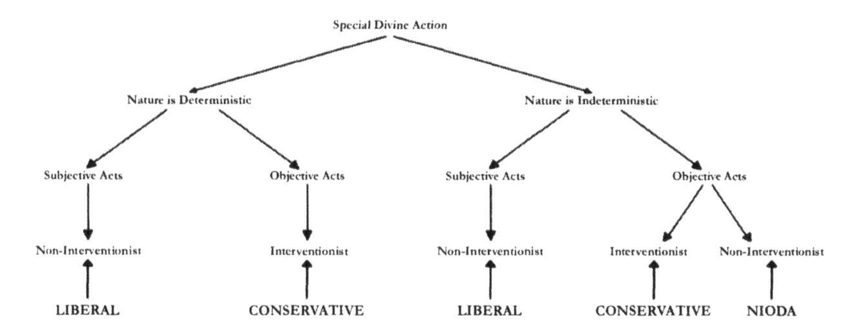

Figure 1: On the left half of the figure, nature, viewed through the lens of classical physics, is interpreted deterministically. This in turn leads to the historical split between liberal and conservative approaches to special divine action. For liberals, the notion of subjectively special divine action reduces, in essence, to a verbal redescription of what is in fact ordinary divine action. For conservatives, objectively special divine action requires interventionism and thus amounts to "miraculous" divine action (in the Humean sense). Note that determinism, as a philosophical interpretation of classical physics, forces the theological split between these approaches to divine action. On the right half of the figure, nature, understood through contemporary science, is interpreted indeterministically. Here we see that, while liberal and conservative approaches to divine action are still options, a third possibility arises for the first time: NIODA. NIODA combines the virtues of the liberal approach (non-interventionism) and the conservative approach (objective divine action) without their corresponding disadvantages. Note in particular that the indeterministic interpretation of nature allows us to separate out "miraculous" objective divine action from "non-miraculous" (non-interventionist) objective divine action, a move which has tremendous theological promise. The challenge is to find one or more areas in contemporary science that permit such an indeterministic ontology for nature. VO/CTNS scholars pursued a variety of areas in science in response to this challenge.

Results to date: I believe that quantum mechanics provides a particularly promising area for NIODA because it is clearly capable of supporting an indeterministic interpretation. I am not optimistic about chaos theory as it currently stands, since its only interpretation is deterministic, making objective divine action interventionist. Perhaps more complex theories of chaotic systems will one day be found which will, in turn, be open to an indeterministic interpretation, but these theories have yet to be discovered—and interpreted. I am not optimistic about top-down approaches which focus by analogy from open systems embedded in larger boundaried systems to the universe-as-a-whole and which depict divine action on the boundary of the embedding system because technical problems in scientific cosmology preclude us from viewing the universe as having a boundary (and because God's action on the boundary, if it existed, would still be interventionist).

Process theology clearly allows for non-interventionist divine action through the metaphysical conception of the intrinsic role of the divine subjective lure for each actual occasion, but that is only a starting point. One must still search the sciences to determine whether God's lure can actually be said to affect the outcome of these occasions in an unpredictable way and thus the debate over the ambiguous interpretations of science is still required. Neo-Thomism might be interpreted as including events which suggest objective divine action within the standard primary/secondary causal context but I am unconvinced that this can be done without violating the metaphysical distinction between primary and secondary causality and without the intervention of miracles.

4.2 Challenges

There are also a number of topics and issues that have emerged in the discussion which call for continued exploration. They constitute challenges, problems and insights whose sustained analysis is pivotal in making further progress. The importance of these topics and issues has been brought out by our work so far. They include previously recognized and newly formulated areas on the growing edge of theology/science research.

Actually new challenges are to be expected, even celebrated, because a mark of real progress is that initial problems come to be seen as partly confusions over terms and partly genuine issues to be addressed. When these issues are successfully addressed, this in turn leads to new insights into the depth and character of the overall problematic and to new questions requiring further attention. The VO/CTNS series is clearly successful in having responded to and having moved beyond many of the problems that the series initially faced in 1990. In doing so, it has exposed deeper issues and challenges for future research. The following is a brief itemization of some of these issues together with recommendations that they be addressed in the future.

4.2.1 Differences in the Doctrine of God

Most scholars referred to God in the language of generic monotheism. Some, however, made explicit reference to the Trinity (including Edwards, Moltmann, Peters, Russell). Still others worked explicitly with a doctrine of God as found within the framework of panentheism (both generic panentheism, i.e., Clayton, Peacocke, and process panentheism, i.e., Barbour). To what extent did these theological differences enhance or hinder progress in the theology/science research?

Recommendation: More explicit attention to similarities and differences in the doctrine of God should be made in future theology/science research.

4.2.2 The Relative Merits of Differing Metaphysical Systems

Some scholars adopted a specifically Whiteheadian metaphysics with variations (e.g., Birch, Barbour, Haught), others a neo-Thomistic/modern Catholic metaphysics with variations (e.g., Clifford, Edwards, Happel, McMullin, Stoeger). Most did not discuss metaphysics extensively. To what extent did this philosophical diversity enhance or hinder the conversations from making further progress? Moreover, while most scholars adopted some form of realism, at least in relation to science, some scholars (notably Drees) criticized this move in crucial ways. To what extent is a realist view of science or of theology helpful or problematic?

Recommendation: More explicit attention to the question of the need for an explicit metaphysics (or not) as well as to the assumption of a philosophy of realism (or its liability) should be given in future theology and science research.

4.2.3 Compatibilist/Incompatibilist Views in Divine Action

Some scholars (e.g., Happel, Peacocke, Soskice, Stoeger and Ward) seemed to assume a form of compatibilism regarding objective special divine action while others (e.g., Ellis, Polkinghorne, Murphy, Russell and Tracy) seemed to presuppose an incompatibilist view. My general concern is that the ambiguities in the way (in)compatibilism was being used and its relation to (in)determinism in nature actually complicated and even confused the conversations during the conference. This, in turn, may account in part for why some participants (notably Peacocke) tended to call quantum mechanical based forms of NIODA "interventionist."

Recommendation: Further clarify the meaning of (in)compatibilism and its relation to (in)determinism and (non)interventionism in future research.

4.2.4 The Ontological Status of the Laws of Nature

Some scholars (e.g., Peacocke, Ward, Soskice) seemed to presuppose a Platonic view of the laws of nature (e.g., they 'govern' the processes in nature). Most scholars, however (e.g., Russell, Stoeger, Tracy), seemed to presuppose that causal efficacy lies within nature as a gift of God and that the laws we formulate are descriptions of such efficacy in nature.

Recommendation: Further examination is needed of the concept of the "laws of nature" and their ontological status.

4.2.5 Criteria of Assessment for Proposals for NIODA

While there has been significant agreement, noted above, by scholars in the VO/CTNS series about the goal of obtaining a successful theory of non-interventionist objective divine action and its importance for theology as a whole, there has been significant disagreement about the best way to develop such a theory, i.e., which scientific theory to use, which philosophical interpretation of it is most persuasive, which model of the God/world causal relation should be used, etc. (For details see Appendix E). These areas of agreement and disagreement are discussed in many of the chapters of Parts II and III of the current volume.[11] There have also been important criticisms of the divine action project as a whole from scholars outside the series, notably from Nicholas Saunders.[12] Wildman's chapter in the current volume[13] includes a careful analysis and assessment of Saunders's arguments. The reader should note the criticisms of Wildman and Saunders published in *Theology and Science* from Clayton,[14] Polkinghorne,[15] Stoeger,[16] and Tracy,[17] as well as Wildman's response to them.[18] I include my criticism of Saunders and Wildman in the endnotes briefly.[19]

---

[11] For an earlier criticism of the way the concept of divine action was formulated in terms of direct vs. indirect and mediated from a neo-Thomistic perspective see Stephen Happel, "Divine Providence and Instrumentality: Metaphors for Time in Self-Organizing Systems and Divine Action," in *CC*, 416, esp. Section 4.6, 197-201.

[12] Nicholas Saunders, *Divine Action & Modern Science* (Cambridge: Cambridge University Press, 2002).

[13] Wildman's chapter in this volume was previously published in Wesley J. Wildman, "The Divine Action Project, 1988-2003," *Theology and Science* 2.1 (2004): 31-75.

[14] Philip Clayton, "Wildman's Kantian Skepticism: A Rubicon for the Divine Action Debate," *Theology and Science* 2.2 (2004): 186-190.

[15] John Polkinghorne, "Response to Wesley Wildman's 'The Divine Action Project,'" *Theology and Science* 2.2 (2004): 190-192.

[16] William R. Stoeger, S. J., "The Divine Action Project: Reflections on the Compatibilism/Incompatibilism Divide," *Theology and Science* 2.2 (2004): 192-196.

[17] Thomas Tracy, "Scientific Perspectives on Divine Action?: Mapping the Options," *Theology and Science* 2.2 (2004): 196-201.

[18] *Wesley Wildman,* "Further Reflections on 'The Divine Action Project," *Theology and Science* 3.1 (2005): 71-83.

[19] Saunders stipulates a test that any successful theory of non-interventionist objective divine action must meet, and the test is spelled out in terms of four distinct criteria. In my opinion, two of the four criteria of the test are *mutually contradictory*: that there is genuine openness in nature (i.e., ontological indeterminism) and that the laws of nature, viewed as ontological realities, determine individual events whether the laws are stochastic or deterministic. Because of this contradiction, Saunders's test fails to constitute be a valid test for assessing theories of divine action and Saunders's assessment of the failure of the proposals deployed by scholars in the VO/CTNS series based on his test should be set aside.

Wildman is also highly critical of the possibility of successful theories of non-interventionist objective divine action, but in this case his reasons are based on his agreement with Kant. According to Wildman, Kant showed that we must inevitably understand nature in terms of causal closure. Thus any theory of objective divine action

Recommendations: There are clearly a variety of issues here regarding what should count, in principle, for an acceptable theory of divine action. I suggest that the issues raised by Saunders can be laid to rest now, but the challenges raised by the other scholars noted here should be pursued vigorously as part of future VO/CTNS research.[20]

4.2.6 Natural Theodicy / Suffering in Nature

To the extent that the case for non-interventionist divine action in light of science has been strengthened by these volumes, so the problem raised by suffering in nature and God's relation to it (e.g., natural theodicy) is, arguably, exacerbated. (Note: Tom Tracy raises important objections to the claim that it is, in fact, exacerbated).[21] If God really does act in nature in ways that 'make a difference' in the course of natural history, what is the relation between such divine action and suffering in nature: Does God cause it? Does God allow it? Does God suffer with creation? What is the result of God's suffering with creation?

Recommendation: A new series by VO/CTNS on natural theodicy has already been launched to address these questions. The first conference, held at the Specola Vaticana in September 2005, focused on physics and cosmology.[22] Future conferences are being planned which then shift the scientific focus to evolutionary and molecular biology and, perhaps, to anthropology, the neurosciences and cognitive science, exploring the preconditions for the possibility of human moral evil in our biological, genetic and neurological roots.

---

will always be interventionist. My response is that quantum mechanics challenges Kant's insistence on causal determinism (in ways similar to how non-Euclidean geometry challenged his view of Euclidean geometry as a synthetic *a priori* judgment) and thus, contrary to Kant, quantum mechanics does allow for the possibility of ontological indeterminism in nature. For this reason I think Wildman's criticisms of the VO/CTNS proposals based on his agreement with Kant should also be set aside.

Note: Wildman offers an additional, and I think more serious, criticism of the divine action project based on what he understands to be the view of God underlying the proposals on divine action: namely, the problem of theodicy. Whether or not Wildman correctly represents that underlying view of God, the problem of theodicy is a serious one for any theory of objective divine action, non-interventionist or not. That is why it has already been raised and discussed frequently in the five volumes, particularly by Tracy, Ellis, Murphy and me. That is also why the problem of theodicy, whether or not it is genuinely exacerbated by the possibility of non-interventionist objective divine action, is a driving factor in the formulation of an overarching theme for a new series of VO/CTNS research (see the following section #7).

[20] I offer an extended analysis and critical assessment of the preceding issues in Robert John Russell, *Cosmology from Alpha to Omega: Theology and Science in Creative Mutual Interaction* (Philadelphia: Fortress Press, 2007), chaps. 4-6.

[21] Thomas F. Tracy, "Evolution, Divine Action, and the Problem of Evil," in *EMB*, 511-30.

[22] Nancey Murphy, Robert John Russell, and William R. Stoeger, S. J., eds., *Physics and Cosmology: Scientific Perspectives on the Problem of Natural Evil*, Vol. 1 (Vatican City State: Vatican Observatory Publications / Berkeley, Calif.: Center for Theology and the Natural Sciences, 2007).

4.2.7 Eschatology

Perhaps the most promising—and most challenging—theological response to natural theodicy is to move the conversation from the locus of creation theology where it is at present to that of redemption. If one claims that God's response to suffering in nature is to suffer with nature and in doing so to redeem nature, as many VO/CTNS scholars have suggested, this takes us directly to the various forms of the theology of the cross. Of course this, in turn, takes us to the resurrection of Jesus and this finally opens onto the question of eschatology—the coming of the new creation by God's transforming action modeled proleptically on the bodily resurrection of Jesus. Now the scope of "creation" in the theology/science discussions has always been the universe as a whole as understood by science. This, then, means that the scope of the "new creation" must also be the universe as a whole—not just human society/history (as in the varieties of liberation theology), or the earth's ecosystem (as in various forms of environmental ethics and ecofeminist theology), or even planet Earth itself (as reflected, say, in the writings of Teilhard de Chardin).

But how then are we to think about the transformation of the universe into the new creation in light of science? In my view this is the most serious challenge to, and most promising direction for, future research in Christian theology and science. Conversely without dealing explicitly with the "eschatology and science" question it is hard to see how the promissory note—that we can respond to natural theodicy by a theology of God's redemptive suffering with nature—can be cashed out. In the process, the importance of lifting up a Trinitarian doctrine of God mentioned previously becomes all the more urgent given the theological complexities raised by the cross and resurrection.[23]

Recommendation: The new series on natural theodicy should also take up the issue of Christian eschatology and science and frame it within an explicitly Trinitarian doctrine of God.

---

[23] Initial research includes the following: John Polkinghorne and Michael Welker, eds., *The End of the World and the Ends of God: Science and Theology on Eschatology* (Harrisburg, Pa.: Trinity Press International, 2000); John Polkinghorne, *The God of Hope and the End of the World* (New Haven, Conn.: Yale University Press, 2002); Ted Peters, Robert John Russell, and Michael Welker, eds., *Resurrection: Theological and Scientific Assessments* (Grand Rapids, Mich.: Eerdmans Publishing Company, 2002); Robert John Russell, "Eschatology and Physical Cosmology: A Preliminary Reflection," in *The Far Future Universe: Eschatology from a Cosmic Perspective*, George F. R. Ellis, ed. (Philadelphia: Templeton Foundation Press, 2002), 266-315; idem, "Cosmology and Eschatology," in *Oxford Handbook of Eschatology*, Jerry Walls, ed. (Oxford: Oxford University Press, in press).

5 Appendix A

Physics, Philosophy and Theology: A Common Quest for Understanding, Robert John Russell, William R. Stoeger, S. J., and George V. Coyne, S. J., eds.

Table of Contents

6 Appendix B

John Paul II on Science and Religion: The New View from Rome, Robert John Russell, William R. Stoeger, S. J., and George V. Coyne, S. J., eds.

Table of Contents

Karl Schmitz-Moorman, "Science and Theology in a Changing Vision of the World: Reading John Paul's Message in *Physics, Philosophy and Theology*"

Thomas F. Torrance, "The Message of John Paul II to Theologians and Scientists Commemorating the Third Centenary of Sir Isaac Newton's *Philosophiae Naturalis Principia Mathematica*"

Charles H. Townes, "A Response to the Message of John Paul II"

7 Appendix C: Key to the Authors of Chapters Cited in the Tables

Alston, W. "Divine Action, Human Freedom, and the Laws of Nature."

1. Arbib, Michael A. "Crusoe's Brain: Of Solitude and Society."

2. Arbib, Michael A. "Towards a Neuroscience of the Person."

1. Ayala, Francisco J. "Darwin's Devolution: Design without Designer."

2. Ayala, Francisco J. "The Evolution of Life: An Overview."

1. Barbour, Ian G. "Five Models of God and Evolution."

2. Barbour, Ian. "Neuroscience, Artificial Intelligence, and Human Nature: Theological and Philosophical Reflections."

Berry, Michael. "Chaos and the Semiclassical Limit of Quantum Mechanics (Is the Moon There When Somebody Looks?)"

Birch, Charles. "Neo-Darwinism, Self-organization, and Divine Action in Evolution."

Brothers, Leslie A. "A Neuroscientific Perspective on Human Sociality."

Butterfield, Jeremy. "Some Worlds of Quantum Theory."

1. Cela-Conde, Camilo J. "The Hominid Evolutionary Journey: A Summary."

2. Cela-Conde, Camilo J. and Gisele Marty. "Beyond Biological Evolution: Mind, Morals, and Culture."

Chela-Flores, Julian. "The Phenomenon of the Eukaryotic Cell."

Chiao, Raymond Y. "Quantum Nonlocalities: Experimental Evidence."

Clarke, Chris. "The Histories Interpretation of Quantum Theory and the Problem of Human/Divine Action."

1. Clayton, Philip. "Neuroscience, the Person, and God: An Emergentist Account."

2. Clayton, Philip. "Tracing the Lines: Constraint and Freedom in the Movement from Quantum Physics to Theology."

Clifford, Anne M. "Darwin's Revolution in the Origin of Species: A Hermeneutical Study of the Movement from Natural Theology to Natural Selection."

Coyne, George V., S.J. "Evolution and the Human Person: The Pope in Dialogue."

Crutchfield, James P., J. Doyne Farmer, Norman H. Packard, and Robert S. Shaw. "Chaos."

Cushing, James T. "Determinism versus Indeterminism in Quantum Mechanics: A 'Free' Choice."

1. Davies, Paul C. W. "The Intelligibility of Nature."

2. Davies, Paul. "Teleology without Teleology: Purpose through Emergent Complexity."

1. Drees, Willem B. "A Case against Temporal Critical Realism?: Consequences of Quantum Cosmology for Theology."

2. Drees, Willem B. "Evolutionary Naturalism and Religion."

3. Drees, Willem B. "Gaps for God?"

1. Edwards, Denis. "Original Sin and Saving Grace in Evolutionary Context."

2. Edwards, Denis. "The Discovery of Chaos and the Retrieval of the Trinity."

1. Ellis, George F. R. "Intimations of Transcendence: Relations of the Mind to God."

2. Ellis, George F. R. "Ordinary and Extraordinary Divine Action: The Nexus of Interaction."

3. Ellis, George F. R. "The Theology of the Anthropic Principle."

4. Ellis, George F. R. "The Thinking Underlying the New 'Scientific' World-views."

5. Ellis, George F. R. and William R. Stoeger. "Introduction to General Relativity and Cosmology."

6. Ellis, George F. R. "Quantum Theory and the Macroscopic World."

Gilkey, Langdon. "The God of Nature."

Green, Joel B. "Restoring the Human Person: New Testament Voices for a Wholistic and Social Anthropology."

Grib, Andrej A. "Quantum Cosmology, the Role of the Observer, Quantum Logic."

Hagoort, Peter. "The Uniquely Human Capacity for Language Communication: From POPE to [po:p] in Half a Second."

1. Happel, Stephen. "The Soul and Neuroscience: Possibilities for Divine Action."

2. Happel, Stephen. "Divine Providence and Instrumentality: Metaphors of Time in Self-Organizing Systems and Divine Action."

3. Happel, Stephen. "Metaphors and Time Asymmetry: Cosmologies in Physics and Christian Meanings."

Haught, John F. "Darwin's Gift to Theology."

Hefner, Philip. "Biocultural Evolution: A Clue to the Meaning of Nature."

1. Heller, Michael. "Chaos, Probability, and the Comprehensibility of the World."

2. Heller, Michael. "Generalizations: From Quantum Mechanics to God."

3. Heller, Michael. "On Theological Interpretations of Physical Creation Theories."

Isham, C.J. and J.C. Polkinghorne. "The Debate over the Block Universe."

1. Jeannerod, Marc. "Are There Limits to the Naturalization of Mental States?"

2. Jeannerod, Marc. "The Cognitive Way to Action."

Kerr, Fergus. "The Modern Philosophy of Self in Recent Theology."

Küppers, Bernd-Olaf. "Understanding Complexity."

1. LeDoux, Joseph E. "Emotions: A View through the Brain."

2. LeDoux, Joseph E. "Emotions: How I've Looked for Them in the Brain."

Lucas, John R. "The Temporality of God."

McMullin, Ernan. "Formalism and Ontology in Early Astronomy."

Meyering, Theo C. "Mind Matters: Physicalism and the Autonomy of the Person."

Moltmann, Jurgen. "Reflections on Chaos and God's Interaction with the World from a Trinitarian Perspective."

1. Murphy, Nancey. "Divine Action in the Natural Order: Buridan's Ass and Schrödinger's Cat."

2. Murphy, Nancey. "Evidence of Design in the Fine-Tuning of the Universe."

3. Murphy, Nancey. "Supervenience and the Downward Efficacy of the Mental: A Nonreductive Physicalist Account of Human Action."

4. Murphy, Nancey. "Supervenience and the Nonreducibility of Ethics to Biology."

1. Peacocke, Arthur. "Biological Evolution: A Positive Theological Appraisal."

2. Peacocke, Arthur. "Chance and Law in Irreversible Thermodynamics, Theoretical Biology, and Theology."

3. Peacocke, Arthur. "God's Interaction with the World: The Implications of Deterministic 'Chaos' and of Interconnected and Interdependent Complexity."

4. Peacocke, Arthur. "The Sound of Sheer Silence: How Does God Communicate with Humanity?"

1. Peters, Ted. "Playing God with Our Evolutionary Future."

2. Peters, Ted. "Resurrection of the Very Embodied Soul?"

3. Peters, Ted. "The Trinity In and Beyond Time."

1. Polkinghorne, John. "Physical Process, Quantum Events, and Divine Agency."

2. Polkinghorne, John. "The Laws of Nature and the Laws of Physics."

3. Polkinghorne, John. "The Metaphysics of Divine Action."

Pope John Paul II. "Message to the Vatican Observatory Conference on Evolutionary and Molecular Biology."

Redhead, Michael. "The Tangled Story of Nonlocality in Quantum Mechanics."

1. Russell, Robert John. "Divine Action and Quantum Mechanics: A Fresh Assessment."

2. Russell, Robert John. "Finite Creation without a Beginning: The Doctrine of Creation in Relation to Big Bang and Quantum Cosmologies."

3. Russell, Robert John. "Special Providence and Genetic Mutation: A New Defense of Theistic Evolution."

Shimony, Abner. "The Reality of the Quantum World."

1. Stoeger, William R., S. J. "The Immanent Directionality of the Evolutionary Process, and its Relationship to Teleology."

2. Stoeger, William R. "Contemporary Physics and the Ontological Status of the Laws of Nature."

3. Stoeger, William R. "Epistemological and Ontological Issues Arising from Quantum Theory."

4. Stoeger, William R., S. J. "The Mind-Brain Problem, the Laws of Nature, and Constitutive Relationships."

5. Stoeger, William R. "Describing God's Action in the World in Light of Scientific Knowledge of Reality."

1. Tracy, Thomas F. "Creation, Providence, and Quantum Chance."

2. Tracy, Thomas F. "Evolution, Divine Action, and the Problem of Evil."

3. Tracy, Thomas F. "Particular Providence and the God of the Gaps."

Ward, Keith. "God as a Principle of Cosmological Explanation."

Watts, Fraser. "Cognitive Neuroscience and Religious Consciousness."

1. Wildman, Wesley J. "Evaluating the Teleological Argument for Divine Action."

2. Wildman, Wesley J. and Leslie A. Brothers. "A Neuropsychological-Semiotic Model of Religious Experiences."

3. Wildman, Wesley J. and Robert John Russell. "Chaos: A Mathematical Introduction with Philosophical Reflections."

8 Appendix D: Typology of the Approaches to Divine Action

Six distinct approaches to the problem of divine action were pursued in the VO/CTNS series. Their main difference lies in the relation between where God's direct act is thought to take effect and where its indirect effects are experienced and understood as acts of God.

8.1 Top-down

This refers to God's action at a higher epistemic and phenomenological level than the level of the effects. So, for example, in the "mind/brain" problem, where language about mental states cannot be entirely reduced to—although it is constrained by—language about neuroscience, God might be thought of as acting at the level of mind (e.g., revelation) and thereby affecting the pattern of neuron firings. (The converse model of revelation—God affecting neuron firings to bring about mental inspiration—would be a form of "bottom-up" as discussed below.)

8.2 Whole-part

This type of causality or constraint refers to the way the boundary of a system affects the specific state of the system. One example is the formation of vortices in a bucket of water being heated. The vortices form because the shape of the bucket as well as the applied heat bring about large-scale patterns of movement in the water. Another example is the universe considered as a whole with the effects played out in local events in the universe (assuming that the universe can be said to have a boundary). In these cases, God may be thought of as affecting the boundary of the system, perhaps the boundary of the universe itself, and this action leads to specific states within the system/universe which we call objectively special, indirect divine acts.

8.3 Lateral

This refers to effects lying in the same epistemic level (e.g., physics) as their causes but at the end of a long causal chain. So the "butterfly" effect in chaos theory depicts small differences in the initial states of a chaotic physical system leading to large differences in later states of that same system. God, then, might act directly to set the initial conditions and thus bring about bulk states indirectly.

8.4 Bottom-up

This causality refers to the way the lower levels of organization affect the way more complex levels behave. Here God might act at the most elementary domains of an organism to achieve specific results which are manifest at the level of ordinary human experience. Quantum physics seems the most promising candidate for further inquiry into divine action through bottom-up causality.

Actually, most scholars want to combine most, or even all four, types of causality when it comes to human agency in the world and to God's action in human life and history. The challenge, however, is to conceive of God as acting in the processes of biological evolution or physical cosmology long before the arrival of any kind of complex biological organism (let alone humanity). Here bottom-up causality may be the only approach available.

It should be noted that these four approaches can be appropriated by scholars from a diversity of philosophical perspectives as can be seen in the chapters on divine action in the VO/CTNS series. However two additional approaches to divine action involve more explicit dependence on a specific overall philosophical system, even while using one or more of the preceding approaches:

8.5 Process Theology

This provides a metaphysical basis for a non-interventionist interpretation of divine action. Every actual occasion is influenced by God, who provides the "subjective lure," by efficient causality from the past ("prehension") and by the innate creativity of the occasion itself (its "mental pole" or "interiority"). Entities at all levels of organization are capable of experiencing God's action as the (non-interventionist) subjective lure without violating the regularities reflected in the laws of science.

8.6 Contemporary Catholic Theology

Much of contemporary Catholic theology has been influenced by a recovery of Thomistic views of divine action. Here a basic distinction is made between God acting as the primary cause of all events, creating them *ex nihilo* and holding them in existence, and God granting to all events a degree of natural or secondary causality (while still acting through these secondary causes), as reflected in the laws of nature discovered by science. In some cases, particularly where humankind is involved, God can also bring about special events of discernment and action without intervening in the ordinary flow of natural processes.

9 Appendix E: A Sample of Chapters in the VO/CTNS Series

9.1 Progress in Philosophy

9.1.1 Critical Realism; Laws of Nature & Ontological (In)determinism

In "Contemporary Physics and the Ontological Status of the Laws of Nature," Stoeger asks how we should think of the laws of nature. This is "an absolutely crucial question" underlying the entire discussion of science, philosophy and theology. In his essay, Stoeger defends the thesis that the laws, although revealing fundamental regularities in nature, are not the source of those regularities, much less of their physical necessity. They are descriptive and not prescriptive; they do not exist independently of the reality they describe. Stoeger thus rejects a "Platonic" interpretation of the laws of nature. Since they have no pre-existence with respect to nature, they do not ultimately explain why nature is as it is. Instead, the regularities which the laws of nature describe stem from the regularities of physical reality itself. Because the complexity of this reality subverts any attempt at a reductionist approach to science, a "theory of everything" is ruled out. Moreover, he questions the evidential basis for the claim that there may be other sets of actual or potential laws that might describe universes different from our own. This reduces the cogency of "many-worlds" arguments which hypothesize the existence of other universes as a means of explaining away the (supposed) fine-tuning of our own universe.

In "Chaos: A Mathematical Introduction with Philosophical Reflections," Wildman and Russell analyze the meaning of predictability in relation to chaotic systems based on the example of the "logistic equation." On the one hand, the logistic map is predictable in principle since the sequence of iterations is generated deterministically by the equation. Of course this is not surprising since chaos theory is, strictly speaking, a subset of classical, Newtonian mechanics. On the other hand, chaotic systems are "eventually unpredictable" in practice, since most values of the initial conditions cannot be specified precisely and this imprecision is eventually amplified into the divergent trajectories of the system. This leads to a definition of 'chaotic randomness' as a *tertium quid* between strict randomness (cf. quantum mechanics) and the complete absence of randomness (cf. classical non-chaotic mechanics).

Two philosophical conclusions are drawn from this. On the one hand, the deterministic whole-part relations between environment and system, the deterministic connnectedness implied in the butterfly effect, and the fact that much of the apparent randomness of vast portions of nature can now be brought under the umbrella of chaos, are best seen as supporting evidence for the hypothesis of metaphysical determinism: reality, at least at the classical level, is more roundly deterministic than we might otherwise have thought. On the other hand, there are profound epistemic and explanatory limitations on the testing of chaos theory due to the peculiar nature of chaotic randomness. In this sense, chaos theory places a new fundamental limit on how well the hypothesis of metaphysical determinism can be supported: When chaotic predictions break down, it may be due, not to the eventual unpredictability of the theory but to a genuine indeterminism at the classical level in chaotic systems which until now has not been evidenced and which calls for a new, more refined macroscopic physics.

9.1.2 Teleology in Nature

In "Darwin's Devolution: Design without Designer" Ayala develops a complex conception of teleology. An object or behavior is teleological when it gives evidence of design or appears to be directed toward certain ends. Features of organisms, such as the wings of a bird, are teleological when they are adaptations which originate by natural selection and when they function to increase the reproductive success of their carriers. Inanimate objects and processes, such as a salt molecule or a mountain, are not teleological since they are not directed towards specific ends. Teleological explanations, in turn, account for the existence of teleological features. Ayala then distinguishes between those actions or objects which are purposeful and those which are not. The former exhibit artificial or external teleology. Those resulting from actions which are not purposeful exhibit natural or internal teleology. Bounded natural teleology, in turn, describes an end-state reached in spite of environmental fluctuations, whereas unbounded teleology refers to an end-state that is not specifically predetermined, but results from one of several available alternatives. The adaptations of organisms are teleological in this indeterminate sense. Finally, teleological explanations are fully compatible with efficient causal explanations, and in some cases both are required.

9.1.3 Time in Relativity and its Philosophical Interpretations

In "The Debate over the Block Universe" Isham and Polkinghorne ask, is ours a world of timeless being (the "block universe") or of flowing time and true becoming? Proponents of the block universe appeal to special and general relativity to support a timeless view in which all space-time events have equal ontological status. The finite speed of light, the light cone structure, and the downfall of universal simultaneity and with it the physical status of "flowing time" in special relativity result in a heightened tendency to ontologize space-time. The additional arbitrariness in the choice of time coordinates in general relativity makes flowing time physically meaningless. Thus no fundamental meaning can be ascribed to the "present" as the moving barrier with the kind of unique and universal significance needed to unequivocally distinguish "past" from "future." Instead the flowing present is a mental construct, and four-dimensional space-time is an "eternally existing" structure. Opponents of the block universe begin by distinguishing between kinematics and dynamics. Special relativity imposes only kinematic constraints on the structure of space-time. The dynamics of quantum physics and chaos theory encourage a view of nature as open and temporal, thus allowing for both human and divine agency. The problem of the lack of universal simultaneity is lessened since simultaneity is an *a posteriori* construct. Philosophically disposed to critical realism, opponents are wary of the incipient reductionism of the block view. They resist the Boethian implications of relativity, and argue instead that divine omnipresence must be redefined in terms of a special frame of reference, perhaps one provided by the cosmic background radiation. In the end, is the debate merely philosophical or could it actually have scientific consequences? Proponents of the block universe challenge their opponents to decide between a mere reinterpretation of the existing theories of physics and the much stronger claim that these theories should be changed. If forthcoming, such changes ought to be testable empirically and would constitute a major achievement in the debate over time.

9.1.4 Quantum Mechanics and its Philosophical Interpretations

According to Cushing's argument in "Determinism Versus Indeterminism in Quantum Mechanics: A 'Free' Choice," the question of determinism versus indeterminism is "the fundamental issue" regarding the possibilities for particular divine action, and thus the importance of quantum mechanics. His central point is that "considerations of empirical adequacy and logical consistency alone" do not force one to chose the indeterministic view of quantum mechanics as found in the Copenhagen interpretation—David Bohm's interpretation offers an empirically valid deterministic alternative.

Bohmian mechanics is an objective (realist) and deterministic account in which the positions of the particles of the system function as "hidden variables" and must be included in a complete state description. As in the Copenhagen interpretation, the Schrödinger equation governs the evolution of the wave function, but an additional "guidance condition" governs the evolution of the particles' positions. With the inclusion of a quantum-equilibrium statistical distribution, Bohm's theory is empirically identical with standard quantum mechanics. Its ontology depicts particles following definite trajectories that are completely deterministic and observer-independent. The ontology, however, is nonlocal: instantaneous, long-range influences are included. Still Bohmian nonlocality is "benign," since the "no-signaling" theory of quantum mechanics prohibits sending messages faster than light. In order to discuss Bohmian ontology, Cushing points to "relational holism" since it seems to offer a better conceptual framework than one which distinguishes between separability and locality. It also suggests a world of temporal becoming since it includes a preferred frame for instantaneous action. Still this world is one in which everything, including the future, is determined. Such a world is reminiscent of Newton's idea of space as the divine sensorium. It certainly poses a challenge to our ideas of free will and divine action—as does the problem of evil.

9.1.5 Philosophical Anthropology, Philosophy of Mind and Free Will

Ian Barbour, in "Neuroscience, Artificial Intelligence, and Human Nature: Theological and Philosophical Reflections," argues that it is consistent with neuroscience and computer science on the one hand and a theological view of human nature on the other hand to understand a person as a multilevel psychosomatic unity who is both a biological organism and a responsible self. He considers the themes of embodiment, emotions, the social self, and consciousness. Barbour surveys biblical and theological accounts of the person that emphasize the integration of body and mind, reason and emotion, and individual and social groups. He then cites work by neuroscientists that highlights these same features, including Arbib's action-oriented schema theory, LeDoux's work on emotions, and Brothers' work on the neural bases of social interaction. The ways in which computers fall short of human capacities provide additional insight into human nature: to approach the level of human functioning, computers require analogues to embodiment, learning and socialization, and emotion. The question of the possibility of consciousness in a computer is particularly problematic. Barbour shows that the concepts of information, dynamic systems, hierarchical levels, and emergence are valuable for integrating insights from neuroscience and AI research with that of theology in a theory of human nature. Barbour argues that process philosophy provides a supportive metaphysical framework for understanding the concept of human nature that he has developed in this essay. Alfred North Whitehead's philosophy emphasizes processes or events rather than substances. These events are all of one kind (thus, monism) but are all dipolar—they have both an objective and a subjective phase. Thus, in attenuated form, experience can be attributed not only to humans and animals, but also to lower forms of life, and even to atoms. In its own way, process philosophy emphasizes the same themes that Barbour traced through theology, neuroscience, and AI research. So Barbour concludes that a dipolar monism based on process philosophy is supportive of a biblical view of the human as a multilevel unity, an embodied social self, and a responsible agent with capacities for reason and emotion.

9.1.6 Emergence, Supervenience and Downward Causality

In "Supervenience and the Downward Efficacy of the Mental: A Nonreductive Physicalist Account of Human Action," Murphy sets out to answer the question: If mental events are intrinsically related to (supervene on) neural events, how can it not be the case that the contents of mental events are ultimately governed by the laws of neurobiology? The main goal of her essay, then, is to explain why, in certain sorts of cases, complete causal reduction of the mental to the neurobiological fails. To do so, she first considers the concept of supervenience, offering a definition that runs counter to the "standard account." The concept of supervenience was introduced in ethics to describe the relation between moral and nonmoral (descriptive) properties; the former are not identical with the latter, but one is a "good" person in virtue of possessing certain nonmoral properties such as generosity. Supervenient properties are multiply realizable; that is, (in the moral case) there are a variety of lifestyles each of which constitutes one as a good person. Murphy criticizes typical attempts at formal definitions of "supervenience" for presuming that subvenient properties alone are sufficient to determine supervenient properties. She argues that many supervenient properties are codetermined by context—this move recognizes constitutive relationships not only at the subvenient level but also at the supervenient level itself or between the level in question and even higher levels of organization. Murphy thus concludes that it is this participation of entities in higher causal orders by virtue of their supervenient properties that accounts for the fact of downward causation.

Murphy then turns to the issue of mental causation: How do reasons get their grip on the causal transitions among neural states? The key to answering this question is the fact that neural networks are formed and reshaped (in part, at least) by feedback loops linking them with the environment; the environment selectively reinforces some neural connections but not others. Murphy points out that it is not only the physical environment that plays a downward causal role in configuring neural nets, but also the intellectual environment. It is the fact that mental states supervene, in Murphy's sense of the term, on brain-states—that is, that they are co-constituted by both brain-states and their intellectual context—that makes the occurrence of the brain-states themselves subject to selective pressures from the intellectual environment.

9.2 Progress in Theology

9.2.1 Methodology in Theology and Science

Russell's paper, "Finite Creation without a Beginning: The Doctrine of Creation in Relation to Big Bang and Quantum Cosmologies," is as much about methodology as it is about the specific implications of cosmology for the doctrine of creation. In the first section Russell focuses on how t=0 in Big Bang cosmology should be related to the doctrine of creation *ex nihilo*. Some have welcomed it as direct evidence of divine creation; others have dismissed it as irrelevant to the creation tradition. As Russell sees it, the argument on both sides has been shaped by the methodological assumptions made by Langdon Gilkey in his *Maker of Heaven and Earth* which dichotomize "ontological origination" (in which t=0 is irrelevant) and "historical/empirical origination" (in which t=0 carries the essential theological message). Instead, Russell believes that a new rapport between theology and cosmology can be achieved if historical/empirical origination can be indirectly related to ontological origination. He believes that the philosophical concept of finitude can provide this indirect relation between temporal finitude (historical origination, such as indicated scientifically by t=0), and ontological contingency (ontological origination, such as the doctrine of creation asserts about the universe). Taking a cue from the writings of Ian Barbour, Nancey Murphy and Philip Clayton, he frames his approach in terms of the methodology of a Lakatosian research program in theology. The creation tradition will form the core hypothesis of this program, with t=0 entering as confirming evidence through the use of a series of auxiliary hypotheses involving the concept of finitude deployed in increasingly empirical contexts of meaning. In the second section, he discusses the challenge to the preceding rapport brought about by inflationary Big Bang cosmology and quantum cosmology since, in both of these approaches, t=0 is abandoned. In response Russell suggests that we should distinguish between the theological claim that creation is temporally finite in the past and the further claim that this past is bounded by the event, t=0. Since the first claim by itself is sufficient for *creatio ex nihilo,* Russell says we can set aside arguments specifically over t=0 and yet retain the historical/empirical sense of the past temporal finitude of creation, an insight he refers to as "finite creation without a beginning." Following Lakatos, he then looks for novel predictions this might entail.

In "God as a Principle of Cosmological Explanation," Ward moves 'both ways' between theology and cosmology. He begins with a summary of the traditional doctrine of creation: God is a non-spatio-temporal being, transcending all that is created, including space-time, although immanent to all creation as its omnipresent Creator. Divine eternity is thus timeless, for God has neither internal nor external temporal relations. The act of creation is one of non-temporal causation. Whether there was a first moment is irrelevant to the doctrine. Ward admits that this view of God is congruent with the block universe interpretation of special relativity, but he is highly critical of it. Ward maintains that the doctrine of creation does not entail a timeless God. Although God transcends space-time as its cause, God is nevertheless temporal, since "...by creating space-time, God creates new temporal relations in the Divine being itself." Allowing God to have temporal relations makes it possible for God to act in new ways, make new decisions and bring into being in time an infinite number of new things. The inclusion of divine contingency along with divine necessity enriches the concept of omnipotence.

Ward then relates nomological models, which are dominant in physics and involve general principles and ultimate brute facts, to axiological models, which arise in the social sciences and describe the free realization of ultimate values. A nomological model realizes an aesthetic value, since the laws of nature are elegant and simple. An axiological model is ultimately factual, since values arise out of the natural capacities of sentient beings as described by physics and evolutionary biology. This inter-relationship is central to the Christian claim that "...goodness is rooted in the nature of things, and is not some sort of arbitrary decision or purely subjective expression of feeling." Quantum cosmologists attempt to offer a secular explanation of ultimate brute facts, but this minimizes the importance of freedom, creativity, and the realization of values. Theism can offer a comparable explanation of nature, but its advantage lies in its combination of nomological and axiological explanations. Theism is thus "the best possible intelligible explanation of the universe" and "the completion of that search for intelligibility which characterizes the scientific enterprise." He urges that we reconstruct the doctrine of creation in terms of creative emergence, i.e., the novel realization of intrinsic values grounded in the divine nature and emerging through the cooperative acts of rational creatures.

9.2.2 God

In "Tracing the Lines: Constraint and Freedom in the Movement from Quantum Physics to Theology," Clayton advances the argument for panentheism compared to classical theism. Both agree that the world is real and has its origin in an ultimate principle called spirit which is active in the world and is personal. Classical theism can become problematic if it places too great a distance between God and the world. Panentheism avoids this as it understands the world to be within God even while, as with classical theism, God is more than and distinct from the world. Here each physical event can be an expression of divine agency in a "top-down" manner which does not violate physical law. It also provides a metaphysics that coheres nicely with some of the interpretations of quantum physics which stress holism, veiled reality, interconnection, and interdependence.

In "The Trinity in and beyond Time," Peters develops a complex response to the perennial question of how an eternal God can act, and be acted upon, in a temporal universe. Classical theology made the problem particularly difficult by formulating the distinction between time and eternity as a "polar opposition." Peters' fundamental move is to presuppose a Trinitarian doctrine of God, thus including relationality and dynamism within the divine. By relating the economic and the immanent Trinity we take the temporality of the world into the divine life of God. To substantiate this move, Peters turns to the understanding of temporality in physics and cosmology. His overall aim is to show that the Trinitarian doctrine of God leads us to expect that the temporality of the world will be taken up eschatologically into God's eternity. The problem is that many current proposals conceive of eternity as timeless, and thus they all fail to solve the underlying problem posed by God's eternal experience of a temporal universe. Can we instead conceive of God as "enveloping time," transcending its beginning and its end and taking it up into the divine eternity? Hawking's cosmology might be helpful here. Although Hawking has given it an "anti-theological" interpretation, Chris Isham has shown how in this model God can be thought of as present to and active in all events of the universe even if there were no initial event. Peters then draws on the thought of Wolfhart Pannenberg, Robert Jenson and Jürgen Moltmann, to suggest that the doctrine of God might be required to explain the temporality of the world, including the arrow of time. Moreover the movement between economic and immanent Trinity, through creation, redemption, and consummation, could be seen as bringing the history of creation into the life of God ("Trinity in and beyond time").

9.2.3 Creation

Clifford writes on "Darwin's Revolution in the *Origin of Species*: A Hermeneutical Study of the Movement from Natural Theology to Natural Selection." She first examines Darwin's *Origin of Species* in relation to nineteenth-century British natural theology, which attempted to provide evidence from nature for God's sovereignty and purposeful design. Clifford warns us not to let the hegemony that Darwin's theory now enjoys undercut our interest in natural theology, partly because we would not otherwise fully appreciate what Darwin's revolution accomplished. In tracing Darwin's accomplishment she points to the way language in both science and theology, with its metaphorical character, shapes our claims about reality. Clifford then analyzes the role of metaphor in science by drawing on the writings of Janet Soskice, Paul Ricoeur, and Sallie McFague. She focuses on two of Darwin's key metaphors: "the origin of species" and "natural selection." Darwin's theory effectively shifted the meaning of "origins" by describing the emergence of new species while bracketing the question of the origin of life as such. He also transformed the meaning of species from fixed and discrete to fluid and possessing the capacity to evolve. Darwin's metaphor, "natural selection," combines meanings drawn from animal breeding by humans and from nature in the wild. It suggests that nature "chooses" and, though Darwin rejected vitalism, he has been read as deifying nature. Clifford points out that Darwin considered his theory compatible with belief in God, though his personal position seems to shift from belief to agnosticism.

According to Clifford, then, Darwin did not intend a warfare against Christianity, only against natural theology, and here only in the form of a highly rationalistic Christian theism coupled to a limited body of scientific data. He challenged Paley's watchmaker analogy for God. What might we find to replace it? McFague had proposed the metaphor of the universe as God's body, but Clifford suggests the metaphor of a mother giving birth. It brings together in dynamic tension the reproductive and evolutionary character of nature with the biblical doctrine of God as creator. It is panentheistic, rather than pantheistic, and is, according to Elizabeth Johnson, the "paradigm without equal," drawing on a wealth of biblical texts for God's relation to the world. Finally it is compatible with Darwin's rejection of God as designer, the immutability of species, and it takes up his concern to acknowledge the extent of suffering in nature.

In "The Theology of the Anthropic Principle" Ellis combines reflections on the Anthropic Principle with the theology of William Temple. He calls this a "Christian Anthropic Principle" (CAP) which seeks to account for the particular character of the universe in terms of the design of God who intends the evolution of creatures endowed with free will and the ability to worship the Creator. God's design, working through the laws of physics and chemistry, allows for the evolution of such modes of life in many places in the universe. "From this viewpoint, fine-tuning is no longer regarded as evidence for a Designer, but rather is seen as a consequence of the complexity of aim of a Designer whose existence we are assuming..."

This in turn entails five implications for the creation process. The universe must be orderly so that free will can function. God attains this goal through creating and sustaining the known physical laws which allow for the evolution of creatures with consciousness and free will. God has also given up the power to intervene directly in nature. The existence of free will makes pain and evil inevitable and requires that God's providence be impartial and that God must remain hidden from the world. God achieves both an impartial providence and hiddenness through the impartiality of the laws of nature. Yet revelation must be possible, so that God can disclose an ethical basis for life. Ellis turns to quantum indeterminacy to provide a basis for divine inspiration. While it is highly probable that life exists throughout the universe, he also claims that the number of individuals in the universe must be finite if God is to be able to exercise care for each. Thus the SETI project is of "tremendous religious significance" in testing this hypothesis. Finally, CAP leads us to the following questions: is our physical universe the only way to achieve the divine intention? how, more precisely, is the ultimate purpose imbedded in, and manifested by, the laws of physics? what proof can be given for CAP? To the latter question Ellis argues that the evidence for CAP is stronger than evidence for inflation or the quantum creation of the universe.

Peacocke's essay, "Biological Evolution—A Positive Theological Appraisal," begins with five broad features of biological evolution and the theological reflections they suggest. 1) *Continuity and emergence*. Although the seamless web of nature is explained by scientists using strictly natural causes, biological evolution is characterized by genuine emergence and a hierarchy of organization. Emergence, in turn, is God's action as the continuous, ongoing, and immanent Creator in and through the processes of nature. 2) *The mechanism of evolution*.

Although biologists agree on the central role of natural selection, some believe selection alone cannot account for the whole story. Peacocke describes eight approaches to the question which operate entirely within a naturalistic framework, assume a Darwinian perspective, and take chance to include both epistemic unpredictability and inherent unpredictability. Against Monod, Peacocke claims that chance connotes the many ways in which potential forms of organization are thoroughly explored in nature. This fact, for the theist, is one of the God-endowed features of the world reflecting the Creator's intentions.

3) *Trends, properties and functions which arise through, and are advantageous in, natural selection*. Drawing on G. G. Simpson and Karl Popper, Peacocke claims that there are "propensities" for properties such as complexity, which characterize the gradual evolution of complex organisms and contribute to the eventual existence of persons capable of relating to God. Thus the propensities for these properties can be regarded as the intention of God who continuously creates through the evolutionary processes. 4) *The ubiquity of pain, suffering and death in nature.* Pain and suffering are the inevitable consequence of possessing systems capable of information processing and storage. Death of the individual and the extinction of species are prerequisites for the creation of biological order. This, in turn, raises the problem of theodicy. Peacocke stresses that God suffers in and with the suffering of creatures. Moreover, God's suffering with Christ on the cross extends to the whole of nature. Sin consists in our alienation from God, a falling short of what God intends us to be. It arises because, through evolution, we gain self-consciousness and freedom, and with them, egotism and the possibility of their misuse.

5) *The theological significance of Jesus Christ in an evolutionary perspective.* Christ's resurrection shows that union with God cannot be broken even by death. His invitation to follow him calls us to be transformed by God's act of new creation within human history. But how is this possible for us now? This leads Peacocke to the problem of atonement. Here the suffering of God and the action of the Holy Spirit in us together must effect our "at-one-ment" with God and enable God to take us into the divine life.

9.2.4 Theological Anthropology

In writing on "Evolution and the Human Person: The Pope in Dialogue," Coyne presents an interpretive chapter for John Paul II's recent statements on evolution and the human person. Coyne sets the context by starting with the historical background of the Pope's statement which he describes in terms of three approaches to science and religion. During the seventeenth and eighteenth centuries, the Church attempted to appropriate modern science to establish a rational foundation for religious belief. Paradoxically, this led to the corruption of faith and contributed to the rise of modern atheism. The founding of the Vatican Observatory in 1891 signals the second approach. Here the Church attempted to combat anticlericalism by a vigorous, even triumphalistic, agenda. Finally, the twentieth century has seen the Church come to view science as offering rational support for theological doctrine. Coyne cites Pope Pius XII who, in 1952, took Big Bang cosmology as "bearing witness" to the contingency of the universe and to its creation by God.

John Paul II, however, has taken a new approach, calling for a genuine and open-ended dialogue in which science and religion, while distinct and marked by their own integrity, can contribute positively to each other. He stresses that, though evolution is an established scientific theory, philosophy and theology enter into its formulation, leading to several distinct and competing evolutionary world-views. Some of these—materialism, reductionism, and spiritualism—are "rejected outright." Instead a genuine dialogue begins as the papal message struggles with two views which may or may not be compatible: evolution according to science and the intervention by God to create the human soul. Although revelation is given an antecedent and primary role compared with scientific discovery, the religious message struggles to remain open, perhaps through a reinterpretation of what science tells us. One possibility would be the body-soul dualism taken by Pius XII. Instead John Paul shifts from an ontological to an epistemological interpretation of the appearance of what he then calls the "spiritual" in humanity. The message closes by indicating that the dialogue should continue.

Hefner's topic is "Biocultural Evolution: A Clue to the Meaning of Nature." He begins with the "two-natured" character of the human: the confluence of genetic and cultural information. These co-exist in the central nervous system (CNS) and have co-evolved and co-adapted, with the genetic making the cultural dimension possible. Though we are conditioned by our evolutionary development and our ecological situation, we are free to consider appropriate behaviors within an environmental and societal matrix of demands. The emergence of conditionedness and freedom are an evolutionary preparation for values and morality; the ought is built into evolution and need not be imported from external sources. Hefner speaks theologically of the human as "created co-creator." We encounter transcendence in several ways: as evolution and the ecosystem transcend themselves when they question their purpose through us; as we act in the non-human world and in culture; and as we open ourselves to our future. Thus the "project" of the human species is also nature's project, and the challenge for us is to discover its content. Human culture includes diverse strategies for living; its greatest challenge is science and technology. These provide the underlying conditions for our interrelated planetary community but also its pressure on the global ecosystem. In response Hefner calls for "a re-organization of consciousness" adequate to this crisis.

Christian theology can provide such perspective. The natural world is vested in meaning by its relation to God as creator *ex nihilo*. The doctrine of continuing creation emphasizes the way in which, at every moment of time, God creates in freedom and love, giving the world its evolutionary character, purpose, and meaning. Humanity, as created in the *imago Dei*, becomes a metaphor for the meaning of nature. Human sin represents the epistemic distance between the actual human condition and the primordial intentionality and love that God bequeaths the world. The key question for relating theology to science should be whether we believe that what governs the world at its depths is divine love. In Jesus Christ we discover both the normative image of God and the instantiated character of God's freedom, intentionality, and love. Hefner concludes with six claims regarding the created co-creator as a fully natural creature illuminating the capabilities of nature, the convergence of the human project and the project of nature, and the transcendence and freedom of both.

9.2.5 Theodicy

In "Original Sin and Saving Grace in Evolutionary Context," Edwards is concerned with rethinking the doctrine of sin and grace in light of biological evolution and with them the problem of theodicy in nature. After engaging the writings of Gerd Theissen and Sallie McFague, Edwards offers an extended critique of Hefner's work. He then turns to a Trinitarian doctrine of God in which love is revealed most radically in mutual, equal, and ecstatic friendship. Drawing on the writings of John Zizioulas, Walter Kasper, and Catherine Mowry LaCugna, Edwards suggests that if the essence of God is relational and if everything that is springs from persons-in-relation, then this points towards an ontology which he calls "being-in-relation." Moreover, such an ontology is partially congruent with evolutionary biology, including its stress on cooperative, coadaptive, symbiotic, and ecological relations. Still the struggle and pain of evolution leads Edwards to face the challenge of theodicy. He asserts that the Trinitarian God who creates through natural selection needs to be understood not only as relational but also as freely accepting the limitations found in loving relationships with creatures. The Incarnation and the Cross point to a conception of God related to natural selection through vulnerability and self-limitation. The God of natural selection is thus the liberating, healing, and inclusive God of Jesus. This God is engaged with and suffers with creation; at the same time, creatures participate in God's being and Trinitarian relationships.

Tracy writes on "Evolution, Divine Action, and the Problem of Evil." According to Tracy, any morally sufficient response to the problem of natural evil must identify the good for the sake of which evil is permitted, and it must explain the relation of evil to this good. One standard approach is to argue that God must permit some evils as a necessary condition for achieving various goods in creation. If we grant that the good cannot be achieved without permitting these evils, we may nonetheless object that the world contains far more of them than would be necessary to serve God's purposes. Tracy's response is that God must permit evil that does not serve as the means to a greater good if it occurs as a necessary by-product of preserving moral freedom and the integrity of the natural order. Precisely because these "pointless" evils do not generate particular goods, they will appear to us to be unnecessary. However, evils of this type must be permitted by God, and it will be up to us to prevent or ameliorate them. But just how much pointless evil is really required? Does the world instead contain gratuitous evils? The initial problem here is in assuming that we can calculate what could be considered the minimum amount of acceptable pointless evil, and thus that we could quantify and balance goods and evils. The deeper problem is in the assumption that the world really does include gratuitous evils. In fact we cannot even conclude that some evil is gratuitous merely because we cannot think of a reason for God's permitting it. Moreover we must recognize that we are in no position to see how each evil fits into the overall course of cosmic history, to comprehend all of the goods to which it may be relevant, or to recognize all of the consequences of eliminating it. In grappling with the reality of evil, we confront the limits of human comprehension and are forced to accept epistemic humility, as the Book of Job makes plain in God's speech from the whirlwind. We cannot expect to solve the problem of evil. Instead the central task for Christian faith in the face of evil is to proclaim and understand what God is doing to suffer with and to redeem creation.

9.2.6 Eschatology

Peters focuses on "Resurrection of the Very Embodied Soul?" According to Peters, the Christian understanding of eternal salvation is not threatened by the rejection of body/soul substance dualism. In fact, the rejection of dualism by both the cognitive neurosciences and the Christian tradition represents an important area of consonance between theology and science— namely, that human reality is embodied selfhood. Peters notes that, until recently, theologians have not been forced to distinguish between two ways of conceiving personal salvation: One, rooted primarily in the ancient Hebrew understanding, pictures the human person as entirely physical, as dying completely, and then undergoing a divinely effected resurrection. The other, a later view influenced by Greek metaphysics, pictures the human person as a composite of body and soul. When the body dies the soul survives independently until reunited with a body at the final resurrection. Although he favors the former tradition, in both of them the resurrection of the body is decisive for salvation.

How best then should we relate cognitive theory and theology? Peters first examines and rejects two "blind alleys": the notion of the "humanizing brain" developed by James Ashbrook and Carol Albright, and the artificial intelligence model of the human soul as disembodied information processing developed by Frank Tipler. In contrast to Tipler's view, Peters notes that belief in the resurrection, for Christian theology, does not depend on any natural process identifiable by science or philosophy, but on the witnessed resurrection of Jesus Christ at the first Easter. The Christian promise points toward an eschatological transformation—a new creation—to be wrought by God. Peters follows Wolfhart Pannenberg in connecting the resurrection to God's eschatological act wherein time is taken up into eternity, and wherein God provides for continuing personal identity even when our bodies disintegrate.

10 Appendix F: A Sample of the Chapters which Focus Specifically on Divine Action

In "Describing God's Action in the World in Light of Scientific Knowledge of Reality," the approach taken by Stoeger is to accept our knowledge of reality from the sciences, philosophy and other disciplines, including theology, and to integrate these results into a theory of God's action he refers to as "weakly critical realist." Included are elements of Aristotelianism and Thomism, particularly the notions of primary and secondary causality. Stoeger uses the term 'law' in two ways: first, to mean any pattern, regularity, process, or relationship as they are in themselves, and secondly our limited description of these patterns and relationships, etc. It does not necessarily imply determinism. While Stoeger acknowledges that there are certain aspects of divine action which we are able to understand better by letting science and theology critically interact, there are two aspects which seem thoroughly resistant to our understanding: the nexus between God and the secondary causes through which God acts, and nexus between God and the direct effects of divine action such as *creatio ex nihilo*.

If God acts through secondary causes it would seem to require the injection of information, and therefore energy, from outside the physical system. Though we cannot rule out such injections, they have never been observed and are unattractive from many points of view. Some scholars overcome this problem by allowing God to influence events at the quantum level. Stoeger admits that this is a solution but he finds it unattractive. While God is always acting through second causes, God's special action is almost always a function of God's invitation, or response, to persons. To locate such divine action at the quantum level removes it from the level of the personal. Another alternative is that God works within what is already given to make the recipient more receptive to what is available. Stoeger prefers the latter, though something of the former may be involved as well. The difficulty with higher regularities subsuming those at the lower-level is that we usually experience the lower level laws as constraining what can be done on a higher level while not being supplanted by them. Nevertheless there is a great deal left under-determined by the lower-level constraints within which agents, including God, can function.

Following the philosophical conclusions raised in the chapter discussed above, Wildman and Russell then ask what relevance chaos theory has for theology. On the one hand, the authors stress that chaos theory does not permit a philosophical interpretation of nature as open to the non-interventionist action of God. On the other hand, chaos theory can be seen to challenge polemical determinists: due to the new fundamental limitation on the testability of chaos theory, one can never fully exclude the possibility that the breakdown in predictability actually arises from the inapplicability of classical physics to the ordinary macroscopic world. Perhaps classical physics as we now have it, including chaos theory, complexity and self-organization, will be replaced by a better model of the world at the classical level which points towards ontological indeterminism at the classical level (e.g., through what John Polkinghorne calls "downwards emergence") and, in turn, which allows for non-interventionist divine causality in the macroscopic world without an appeal to quantum physics.

The first part of Edwards' chapter, "The Discovery of Chaos and the Retrieval of the Trinity," explores the general concept of divine action from the perspective of what many are calling a "retrieved" Trinitarian theology. In the West, Trinitarian theology as inherited from Augustine and Aquinas emphasized an individual and psychological model of the Trinity rather a communitarian one. The newer Trinitarian theology, built instead on the writings of Richard of St. Victor and Bonaventure, offers a theology of divine action which understands the Trinity as a communion of mutual relationships which are dynamic, ecstatic, and fecund. Edwards argues that the universe is God's Trinitarian self-expression, that there are "proper" roles for the Trinitarian persons in creation, and that divine interaction with creation is characterized by the vulnerability and liberating power of love.

The second part of the paper focuses on particular divine actions, such as the incarnation, the Holy Spirit, and divine providence. Edwards's reflections can be summarized via six statements: (1) The Trinitarian God works in and through the processes of the universe, through laws and boundary conditions, through regularities and chance, through chaotic systems and the capacity for self-organization. (2) This Trinitarian God allows for, respects, and is responsive to the freedom of human persons and the contingency of natural processes, but is not necessarily to be denied a knowledge of future contingent events. (3) We must take into account not only the divine action of continuous creation, but also particular or special divine acts. (4) If God is acting creatively and responsively at all times and also in particular ways, this seems to demand action at the level of the whole system as well as at the everyday level of events, and at the quantum level. (5) Particular divine acts are always experienced as mediated through created realities. (6) The unpredictability, openness, and flexibility discovered by contemporary science is significant for talk of particular divine action because it provides the basis for a worldview in which divine action and scientific explanation are understood as mutually compatible, but it is not possible or appropriate to attempt to identify the "causal joint" between divine action and created causality.

Moltmann writes on "Reflections on Chaos and God's Interaction with the World from a Trinitarian Perspective." He first describes five models of the God-world relation: (1) According to the Thomistic model, God is the *causa prima* of the world. God also acts through the *causae secundae* which serve as God's instruments. (2) The interaction model postulates a degree of reciprocal influence between God and the world. This model can include the Thomistic model, but not vice versa. (3) The whole-part model, taken from biological systems theory, emphasizes that the whole is more than and different from its parts. In complex and chaotic systems this difference shows up in the form of top-down causality. The whole-part model is more inclusive than the previous models and sheds light on God's indirect effect upon the world as a whole. (4) The model of life processes emphasizes the open character of biological systems. The present state of a living system is constituted by its fixed past and its open future. More generally, the world process is open to God as its transcendental future. (5) Finally, Moltmann considers two central theological models: creation and incarnation. Here God creates by a process of self-limitation (or *tzimtzum*). The limitation on God's omnipresence creates a habitation for the world; the limitation on God's omniscience provides the world with an open future. Moltmann believes this model is the most inclusive of the five.

Moltmann is critical of the interaction model, seeing it as a theistic model in which God is the absolute Subject who may intervene at will in nature. In the modern period it was replaced by two even more problematic models: deism and pantheism. In their place Moltmann commends to us a Trinitarian model in which God both transcends and is immanent in the world. According to this model God acts upon the world through God's presence in and perichoresis with all things. Next Moltmann discusses eschatology, the new creation of all things. Here the future is not a state of completion but a process of continuing openness, one in which all finite creatures will participate in God's unending and open eternity even as God participates in their temporality. The openness of chaotic, complex, and evolutionary systems is suggestive of this vision, and seems inconsistent with a future conceived of as completed. Finally Moltmann asks whether the universe as a whole should be thought of as an open system. The growth of possibilities for such systems, their undetermined character, and their dependence on an influx of energy suggest that the universe itself might be open to energy and thus "a system open to God."

In "The Metaphysics of Divine Action" Polkinghorne claims that any discussion of agency requires the adoption of a metaphysical view of reality. Most physicists, he claims, interpret Heisenberg's uncertainty principle as implying an actual indeterminacy in the physical world, rather than an ignorance of its detailed workings. However, there are problems about amplifying their effects and it leads to "an episodic account" of divine agency. Polkinghorne prefers an approach based on interpreting the unpredictabilities of chaotic dynamics as indicating an ontological openness to the future whereby "active information" becomes a model for human and divine agency. He interprets sensitivity to small triggers as indicators of the vulnerability of chaotic systems to environmental factors, with the consequence that such systems have to be discussed holistically. The resulting metaphysical conjecture Polkinghorne calls a "complementary dual-aspect monism" in which mind and matter are opposite poles or phases of the created world. This scheme is anti-reductionist, stressing a contextualist approach in which the behavior of parts depends on the whole in which they participate. Polkinghorne then discusses some of the consequences of this point of view, including the insight that God's knowledge of the world of becoming will be truly temporal in character.

Clayton's chapter, "Tracing the Lines: Constraint and Freedom in the Movement from Quantum Physics to Theology," gives two reasons why constructive theology should engage in dialogue with quantum physics: it cannot afford the fideistic position that results from disengaging with science, and it should seek a more hypothetical, fallible, and revisionist method that opens itself up to the engagement without becoming fully relativistic. Should physics, then, provide constraints on how God might act? Clayton's response is that it does if divine agency occurs in the physical world in conformity with physical law, and this becomes Clayton's "wager." Clayton then explores three quantum mechanical constraints on divine action. The first is the role of the observer. Minimalists focus on macroscopic measurements by an observer who is never within the quantum mechanical system being studied. Maximalists introduce subjectivity and consciousness in explaining a quantum experiment, despite the resistance of many physicists. Here Clayton finds another crucial issue at work: reductionism assumed by minimalists versus emergence and even dualism assumed by maximalists. The second issue is the "many-worlds" interpretation as compared with the irreducible role of subjectivity in nature. Both of these interpretations are deeply influenced by metaphysics. Clayton then turns to his third issue, indeterminism and free will. He reminds us how the early defenders of the Copenhagen view saw the free choice of an experimenter as playing an irreducible role in the outcome of the experiment. Clayton then claims that ontological indeterminism is a necessary condition for an incompatibilist view of free will.

After assessing the writings of Bernard d'Espagnat, Fritjof Capra, David Bohm, and Ken Wilber, Clayton turns to classical theism and panentheism. Theism in general asserts that the world has its origin in the divine Spirit as an active and personal force in the world. Classical theism has attractive features but it can become problematic when it places too great a distance between God and the world. To Clayton, panentheism avoids this difficulty, particularly as it understands the world to be within God even while God is more than and distinct from the world. Here each physical event can be an expression of divine agency in a "top-down" manner which does not violate physical law. It also provides a metaphysics that coheres nicely with some of the interpretations of quantum physics previously discussed, particularly those which stress holism, veiled reality, interconnection, and interdependence.

In "Divine Action in the Natural Order: Buridan's Ass and Schrödinger's Cat," Murphy argues that the problem of divine action requires a revised metaphysical theory of matter and natural causality. Her proposal is that we view the causal powers of created entities as inherently incomplete. No event occurs without divine participation but, apart from creation *ex nihilo*, God never acts except by means of cooperation with created agents. Her paper attempts to show how this account can be reconciled with contemporary science, focusing on divine action at the quantum level. First Murphy argues that any satisfactory theory of divine action must allow for objectively special divine acts, it must not undercut our scientific picture of the law-like regularity of natural processes, and it must include a "bottom-up" approach: if God is to be active in all events, then God must be involved in the most basic of natural events. Current science suggests that this most basic level is quantum phenomena. It is a bonus for theology that we find a measure of indeterminacy at this level since it allows for an account of divine action wherein God has no need to overrule natural tendencies or processes. This cooperation rather than coercion is in keeping with God's pattern of respecting the integrity of other higher-level creatures, especially human creatures. She then spells out the consequences of this proposal regarding the character of natural laws and regarding God's action at the macroscopic level. One of these is that the laws of nature must be descriptive, rather than prescriptive; they represent our human perceptions of the regularity of God's action. She then replies to some of the objections that have been raised against theories of divine action based on quantum indeterminacy and explains how the essay's proposal meets the criteria of adequacy set out in the beginning.

Tracy writes on "Particular Providence and the God of the Gaps." Here he takes up a persistent modern problem for theories of divine action: Do they require "gaps" in the causal order and therefore incompleteness in scientific explanations? Tracy considers three strategies which reply to this question. The first avoids conflict between scientific and theological claims by insisting that God does not act in history but rather enacts history as a whole. Tracy considers its paradigmatic formulation by Friedrich Schleiermacher as well as a contemporary version developed by Gordon Kaufman. Unfortunately these proposals leave us with a series of questions about how God can enact history without acting in history. The second strategy affirms that God does act in the world to affect the course of events, but holds that this does not require any gaps in the causal structures of nature. Brian Hebblethwaite contends that God acts in and through the causal powers of creatures because the whole network of created agencies is "pliable, or flexible, to the providential hand of God." John Compton suggested that, just as we routinely describe certain movements of the human body both as a series of physical events and as intentional actions, so we can describe events in the world both as part of a causally complete natural order and as acts of God. Tracy shows that these two approaches are undone by internal inconsistencies. The third strategy grants that particular divine action requires causal incompleteness in the natural order. Chaos theory does not provide the needed openness; instead it presupposes an unbroken causal determinism. Quantum mechanics, however, can be interpreted in terms of causal indeterminism. Tracy contends that indeterminism will only be theologically interesting if the determination of such events by God can make a macroscopic difference. If so, then God could affect the course of events without disrupting the structures of nature.

In "Divine Action and Quantum Mechanics: A Fresh Assessment" Russell argues that, if one interprets quantum mechanics philosophically as pointing to ontological indeterminism, then one can construct a robust bottom-up, noninterventionist, approach to objective, mediated, direct divine action (NIODA). In this approach, God's mediated indirect acts at the macroscopic level, understood as both general and special providence, arise in part from God's mediated direct action at the quantum level. God sustains the deterministic time-development of elementary processes described by the Schrödinger equation, and God also brings about irreversible interactions that are not described by the Schrödinger equation (i.e., "measurements") which can occasionally lead to specific macroscopic effects. Thus divine action ultimately results in the regularities of our everyday world, which we attribute to general providence, and in specific macroscopic events in our everyday world, which we view as special providence.

Russell begins with methodological clarifications: His thesis does not explain how God acts or constitute an argument that God acts, but merely shows that this theory of divine action is consistent with natural science. It is neither an epistemic nor an ontological "God of the gaps" argument. It does not reduce God to a natural cause. It does not propose that God alters the wavefunction. A bottom-up approach such as this, which seems the best way to discuss God's action during the billions of years from the early universe to the evolution of primitive organisms on Earth, can be combined with top-down approaches when sentient life is considered. Although quantum mechanics can be given multiple philosophical interpretations, so can every scientific theory and this poses a problem for all theological engagements with science. The key is to take a "what if" strategy, exploring the implications of one particular interpretation to its fullest without incurring the foundationalist problems of natural theology. Next Russell turns to the "measurement problem" whose scope, he argues, is much wider than usually acknowledged. It includes: micro-macro (e.g., the absorption of a photon by the retina), micro-meso (e.g., the capture of an electron by an interstellar dust particle), and irreversible micro-micro (e.g., proton-proton scattering in the presence of heavy nuclei) interactions, though it does not include reversible micro-micro interactions (e.g., proton-proton scattering in free space). The term "quantum event" can be defined as referring to all such irreversible interactions

Turning to theological issues, Russell weighs arguments for viewing special divine action as either "ubiquitous" or "episodic" and concludes that "pervasive" is more helpful. He proposes that the spatial and temporal characteristics of the wavefunction point to divine action as both global and local. He discusses scientific and theological challenges raised by special relativity, suggesting that we need a richer theological conceptuality of "time and eternity." He then discusses four crucial theological issues. First, does God act providentially in all quantum events, or only in some? Russell prefers the first option, though there are advantages and disadvantages to both. Second, if divine action and mind/brain top-down causality are both operative in acts of human volition, how do we avoid what Russell calls "somatic over-determination"? Russell suggests one possibility: that God acts in all quantum events until the rise of life and consciousness, after which God limits God's action, leaving room for top-down, mind/brain causality. Third, why does God not act to minimize suffering, disease, death, and extinction in nature, i.e., the problem of natural theodicy? Russell proposes that we can give a more persuasive response to theodicy if we move from creation theology to eschatology via a Trinitarian theology of redemption, particularly as developed by Wolfhart Pannenberg. This, in turn, leads to Russell's fourth issue: the meaning and intelligibility of eschatology based on the bodily resurrection of Jesus in light of physics and cosmology.

Writing on "Ordinary and Extraordinary Divine Action: The Nexus of Interaction," Ellis elaborates the conclusions reached by Tracy, Murphy, Russell and others concerning the role of quantum indeterminacy in a contemporary understanding of divine action. He claims that some account of special divine action is necessary if the Christian tradition is to make sense. However, there are two important constraints to be reckoned with: divine action must not conflict with a scientific understanding of nature and it must respond to the challenge of theodicy. Ellis's analysis focuses on the nature of bottom-up and top-down causation in hierarchical systems, predicated upon the assumption that chaotic dynamics does not provide the required openness in physical systems and that top-down causation alone does not provide for an adequate account of divine action. He distinguishes between generic top-down causation, in which boundary conditions produce a global effect upon all the entities in a system, and specific top-down causation, which involves local interactions with elements of the lower-level system. Special divine actions would seem to entail the latter. However, specific top-down causation seems to require, in turn, that there be an intrinsic openness or indeterminacy at the very lowest level of the hierarchy of complexity. Thus, a study of the possibilities for divine action via top-down causation leads inevitably to a consideration of divine action at the quantum level.

Ellis takes God's action to be largely through the ordinary processes of nature which result in the emergence of higher levels of order, including, finally, free human beings. Special divine action focuses on the theology of revelation, God's offering us intimations of God's will for our lives. Ellis speculates that quantum events in the brain (directed by God) might be amplified to produce revelatory thoughts, images, and emotions. If God has adequate reason to restrict divine action to a combination of ordinary action and revelation then the problem of evil does not take on the same dimensions as it does when it is assumed that God might freely intervene in any sort of process at any time.

Barbour develops "Five Models of God and Evolution." In the first part of his paper, Barbour describes the evolution of Darwinism over the past century. He then outlines four philosophical issues which characterize the interpretation of evolution. Self-organization is the expression of built-in potentialities and constraints in complex hierarchically-organized systems. This may help to account for the directionality of evolutionary history without denying the role of law and chance. Indeterminacy is a pervasive characteristic of the biological world. Unpredictability sometimes only reflects human ignorance, but in the interpretation of quantum theory, indeterminacy is a feature of the microscopic world and its effects can be amplified by non-linear biological systems. Top-down causality arises when higher-level events impose boundary conditions on lower levels without violating lower-level laws. Barbour places top-down causality within the broader framework of holism. Communication of information is another important concept in many fields of science, from the functioning of DNA to metabolic and immune systems and human language. In each case, a message is effective only in a context of interpretation and response.

Each of these has been used as a non-interventionist model of God's relation to the world. If God is the designer of a self-organizing process as Paul Davies suggests, it would imply that God respects the world's integrity and human freedom. Theodicy is a more tractable problem if suffering and death are inescapable features of an evolutionary process for which God is not directly responsible. But do we end up with the absentee God of deism? The neo-Thomist view of God as primary cause working through secondary causes as defended by Bill Stoeger tries to escape this conclusion, but Barbour thinks it undermines human freedom. Alternatively, God as providential determiner of indeterminacies could actualize one of the potentialities present in a quantum probability distribution. But does God then control all quantum indeterminacies or only some of them?—the possible options are discussed by Ellis, Murphy, Russell, and Tracy. God as top-down cause might represent divine action on "the world as a whole," as Arthur Peacocke maintains. But this is problematic since the universe does not have a spatial boundary and the concept of "the-world-as-a-whole" is inconsistent with relativity theory. Grace Jentzen and Sallie McFague view the world as God's body but Barbour is concerned that this model also breaks down when applied to the cosmos.

Process theology offers a fifth model of God's action in the world via the interiority of all integrated events viewed as moments of experience. Rudimentary forms of perception, memory, and response are present in lower organisms;

sentience, purposiveness, and anticipation are found in vertebrates while consciousness occurs only at the highest levels of complex organisms. The process model resembles but differs from each of the four models above. God as designer of self-organizing systems is a source of order, but the God of process thought is also a source of novelty. God acts in indeterminacies at the quantum level, but also within integrated entities at higher levels. God acts as top-down cause, not through the cosmic whole but within each integrated system which is part of a hierarchy of interconnected levels. Communication of information can occur through events at any level, not primarily through quantum events at the bottom or the cosmic whole at the top. God is persuasive, with power intermediate between the omnipotent God of classical theism and the absentee God of deism. God is present in the unfolding of every event, but God never exclusively determines the outcome. This is consistent with the theme of God's self-limitation in contemporary theology and with the feminist advocacy of power as empowerment. Process theology has much in common with the biblical understanding of the Holy Spirit as God's activity in the world. Barbour concludes by considering some objections to process thought concerning panexperientialism, the charge that it is a "gaps" approach, and the abstract character of philosophical categories in the context of theology.

SCIENTIFIC ISSUES: GROUND COVERED AND HORIZONS UNFOLDING

George Ellis

The key issue that was helped scientifically by the VO/CTNS series of meetings was looking at integration of the different sciences and considering their relation to each other. The series as a whole provided a very useful overview of the three major areas of physics application to the world and the universe:[1] the large (the cosmos), the small (quantum physics), and the complex (life and the brain), as well as their interactions with each other. These causal interconnections are diagrammed in Figure 1. Furthermore because the series looked at foundational issues, it identified a number of topics of fundamental importance that cut across these specific application areas: issues of ontology and epistemology, issues of reductionism and emergence, and the foundational importance of information. The links of each application to these themes are also diagrammed in Figure 1. Finally, and in many ways most importantly, the approach taken emphasized the complex and interacting nature of the causal nexus (indicated by the enclosing box in Figure 1).

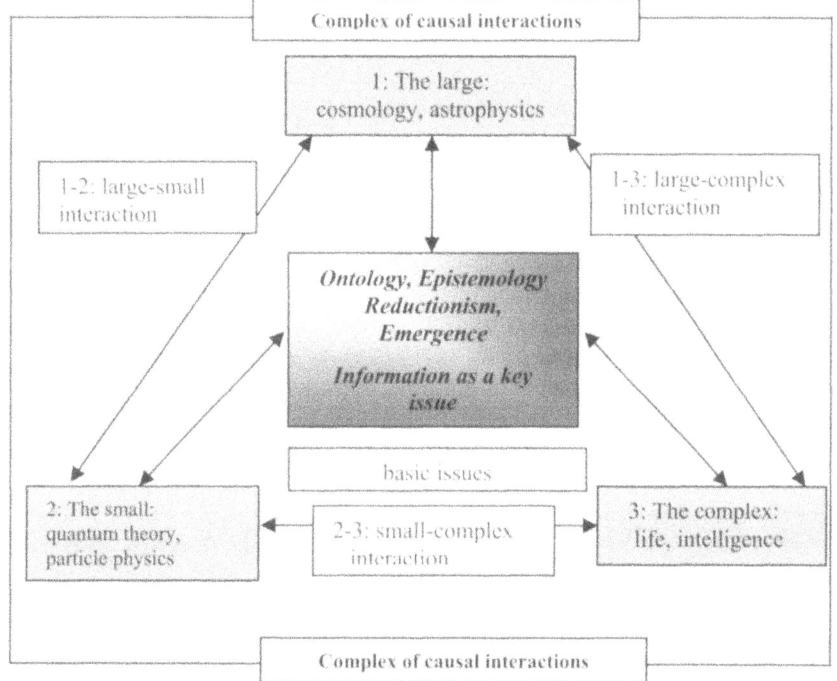

Figure 1. The nexus[2] of causal interactions between the large, the small, and the complex, and their foundations.

---

[1] Roger Penrose, *The Large, the Small, and the Human Mind* (New York: Cambridge University Press, 1997).

[2] **nexus**: a complicated series of connections between different things. *Oxford Advanced Learner's Dictionary*, 6th ed. (Oxford: Oxford University Press, 2000).

There are strong causal links in action at any time ('laws of nature'), and there are many of them:[3] fundamental and effective laws of physics, constitutional equations, initial conditions and boundary conditions, bottom-up and top-down interactions between causal levels in the hierarchy of structure. This makes quite clear that any attempt to claim that only one factor is *the* cause of some physical or biological effect will always be highly misleading and indeed a profoundly inadequate basis on which to make philosophical or metaphysical claims (as some attempt). Thus, the primary usefulness of the series was the refusal to be boxed in by the boundaries of any specific scientific discipline. This emphasis has helped useful developments arise from the meetings, for example the various further meetings that have taken up the issue of emergence[4] and my own cross-disciplinary look at the natural and life sciences.[5]

In what follows, I will look in turn at each subject and its relation to the foundations, and then at the interactions between them. This may be a useful way to understand the integrated set of causal interactions that are in fact in operation in the real universe all the time. In doing so I will not so much be looking at what was presented at the VO/CTNS meetings,[6] rather I will be looking at what flows out of those discussions when viewed in a broad framework.

1 The Large: Cosmology

We have a good description of the universe at recent times, and cosmology has made great strides there.[7] But it always rests on earlier foundations, so the issue of the origin of the universe is the key point here. We can describe the causal interactions structuring the universe and work out their consequences, but what led to their specific nature, and what led to the universe being the way it is by imposing particular boundary or initial conditions?

The many proposals for quantum creation of the universe[8] have multiplied since the Quantum Cosmology and the Laws of Nature meeting (*inter alia*, pre-big bang theory, Randall-Sundrum brane universes, the ekpyrotic universe, cyclic universes, universes based on the holographic principle have been proposed) but the fundamental issue remains the same: there is no way to test our theories about the origin of the universe directly, and indirect tests

---

[3] J. Trefil, *Cassell's Laws of Nature: An A-Z of Laws and Principles Governing the Workings of Our Universe* (London: Cassell, 2002).

[4] N. H. Gregersen, ed., *From Complexity to Life: On the Emergence of Life and Meaning* (New York: Oxford University Press, 2003); P. Clayton and P. C. W. Davies, eds., *The Re-emergence of Emergence* (New York: Cambridge University Press, 2006).

[5] G. F. R. Ellis, "An Integrative View of Science and Cosmology," (2002): http://www.mth.uct.ac.za/~ellis/cos0.html.

[6] See http://www.ctns.org/books.html for summaries of most of the resulting articles.

[7] J. Silk, *A Short History of the Universe* (New York: W. H. Freeman, 1997); J. A. Peacocke, *Cosmological Physics* (Cambridge: Cambridge University Press, 1999).

[8] C. J. Isham, "Quantum Theories of the Creation of the Universe," in *QC*, 51-89.

(through their effects on the perturbation spectrum, for example) are very weak and non-unique.

We run into the fundamental limitations of observational and experimental testing in cosmology: unless we live in a small universe (where we have already seen round the universe since the time of decoupling), most of the universe (indeed maybe even a zero fraction of what there is) is unobservable, and the physical laws underlying the development of the really early universe are untestable (the energies involved are too high for us to reach them in particle accelerators).[9] However, this problem is marginally improved with the realization that despite some claimed theorems to the contrary, there do indeed exist singularity-free universe models, which need not even attain the quantum cosmology era.[10] If we lived in such a universe it is indeed possible that we might eventually be able to test all the physics involved, because quantum gravity never comes into play. But that may or may not be the case: most would say it is unlikely.

The most sustained recent attempt to deal with the issue of the origin of the universe is via the idea of a *multiverse*: we live in one universe amongst many, and indeed the claim is made that in the ensemble of universes, all that is possible happens: all possible universes occur.[11] To those who realize the seriousness of the anthropic fine-tuning of the universe (a relation between the large and the complex, considered again below), this is the most likely approach. However, here we directly run into the tension between epistemology and ontology: although this is the favored approach of a number of cosmologists, it cannot be regarded as a scientifically provable proposition, despite attempts by some to make this case.[12] Yet it might be true. Curiously, in cosmology the tendency is to resolve the epistemology/ontology tension in the opposite way to that in quantum theory: the fashionable way in cosmology is to assume existence despite the lack of proof, whereas in quantum theory the inability to show there is a causal connection is taken to prove there is none: uncertainty is taken to be ontological rather than epistemological.

---

[9] G. F. R. Ellis, "Before the Beginning: Emerging Questions and Uncertainties," *Astrophysics and Space Science* 269-279 (1999): 693-720.

[10] G. F. R. Ellis and R. M. Maartens, "The Emergent Universe: Inflationary Cosmology with No Singularity and No Quantum Gravity Era," (2003): http://arXiv.org/abs/gr-qc/0211082.

[11] D. W. Sciama, "Is the Universe Unique?," in *Die Kosmologie der Gegenwart*, G. Borner and J. Ehlers, eds. (Serie Piper, 1993); M. Tegmark, "Is the Theory of Everything Merely the Ultimate Ensemble Theory?," *Annals of Physics* 270 (1998): 1-51, http://arXiv.org/gr-qc/9704009; M. J. Rees, *Just Six Numbers: The Deep Forces that Shape the Universe* (New York: Basic Books, 1999); idem, *Our Cosmic Habitat* (Princeton, N.J.: Princeton University Press, 2001); idem, "Concluding Perspective," (2001): http://arXiv.org/astro-ph/0101268; M. Tegmark, "Parallel Universes," (2003): http://astro-ph/0302131 and *Scientific American* (May 2003): 41-51.

[12] See G. F. R. Ellis, U. Kirchner, and W. R. Stoeger, S. J., "Defining Multiverses," (2003): http://arXiv.org/abs/astro-ph/0305292.

One of the most profound issues here is the problem of realized infinities:[13] if all that is possible happens, we have multiple infinities of realized universes in the ensemble. In particular, the uncountable continuum of numbers plays a devastating role here: is it really conceivable that universes occur with *all* values of both the cosmological constant and the gravitational constant, or *all* values of the Hubble constant at the instant when the density parameter takes the value 0.97?

To make sense of the multiverse proposal, much more work needs to be done to define the space of all possible models considered, to justify why that space should have the structure it is assumed to have, and to justify why specific distribution functions of universes should be realized on that space.[14] But even given all that, this proposal will remain a metaphysical rather than scientific one: the existence of some specific ensemble will remain an issue of faith rather than proof. There will never be any way to prove that one specific ensemble exists rather than any other one.

The foundational role of information in cosmology is realized in recent intriguing discussion of the so-called Holographic principle—focusing on bounds on the information that can be given on a null surface in order to specify what happens in a spacetime region bounded by that surface.[15] The cosmological implications are still being explored,[16] but in the present context one point is clear: this is a focus on the significance of information at the macro-level. Indeed, one can Fourier analyze the data at any partitioning surface for the universe at some arbitrary time t_0 (for example, the surface of last scattering of radiation where the microwave background radiation decoupled) determining the co-moving spectrum of perturbations about the Robertson-Walker background on that surface. At the linear perturbation level, the data at each scale at an arbitrary time t_0 is independent of that at other scales, and determines what happens at later times at that co-moving scale.[17] The data at each scale can be regarded as causally effective and also is independent of the data at other scales; thus, the perturbations at each macro scale are irreducible aspects of the hierarchy of structure (they are not reducible to a summation of perturbations at lesser length scales) and so should be considered to be ontologically real. They do not emerge from smaller scale structures; they exist in their own right.

From this viewpoint, the key issue in cosmology is, where does the information come from that sets initial conditions for the universe? Now it is

---

[13] R. Bousso, "The Holographic Principle," *Reviews of Modern Physics* 74 (2002): 825, http://arXiv/org/abs/hepth/0203101.

[14] Ibid.

[15] Ibid.

[16] C. J. Hogan, "Holographic Discreteness of Inflationary Perturbations," *Physical Review* (2002): D66 023521, http://arXiv.org/abs/astro-ph/0201020; idem, "Holographic Bound on Information in Inflationary Perturbations," (2004): http://arXiv.org/abs/astro-ph/0406447.

[17] The relevant perturbation evolution equations are ordinary differential equations, once the Fourier analysis has been performed; see e.g. G. F. R. Ellis, J. Hwang, and M. Bruni, "Covariant and Gauge-Independent Perfect Fluid Robertson-Walker Perturbations," *Physical Review D*, Vol. 40, No. 6 (1989): 1819-26.

true that if inflation took place,[18] then most of this information is forgotten and new information is inserted (via quantum uncertainty) that orders the later structure that appears—a cosmological form of autonomous emergence of inhomogeneous structure. But the point is that inflation only takes place if you put in the right initial information in the first place (too much anisotropy will kill it off, for instance,[19] as will the wrong laws of physics or physical constants); only specific kinds of initial information lead to inflation. Additionally, some pre-inflationary (`Transplanckian') information may remain identifiable in the later perturbation spectrum long after inflation has occurred:[20] basically very small scale information gets transformed to large scale information.

Despite our good understanding of the development of structure in the expanding universe, a fundamental mystery remains: the complexity of astronomical structure is self-generated by gravitational instability acting on seed perturbations, and this spontaneous structure growth is the basis on which all other complexity in the universe comes into existence. On the face of it, this violates all usual statements about entropy and the growth of structure, which ignore the effects of gravity.[21] When gravity is appreciable in an inhomogeneous situation, either entropy decreases because structure formation takes place (the Second Law does not hold in these conditions), or some unusual definition of entropy is appropriate where increased entropy does not correspond to increasing structural disorder.

The point here is that we still have no good understanding of the nature of gravitational entropy and its relation to the difficult problem of the arrow of time,[22] despite some intriguing speculations, for example Roger Penrose's suggestion that it is related to the Weyl tensor.[23] If correct, this proposal implies that gravitational entropy increases only for particular initial conditions for the universe rather than being a fundamental physical law: that is, it is an emergent phenomenon based in top-down action from boundary conditions in the far past (and perhaps the far future as well).

---

[18] A. Guth, *The Inflationary Universe: The Quest for a New Theory of Cosmic Origins* (Boston: Addison Wesley, 1997).

[19] A. Rothman and G. F. R. Ellis, "Can Inflation Occur in Anisotropic Cosmologies?," *Physical Letters B*, Vol. 180, No. 1-2 (1986): 19-24.

[20] J. Martin and R. H. Brandenberger, "The Trans-Planckian Problem of Inflationary Cosmology," (2000): http://arXiv.org/abd/hep-th/0005209; L. Bergstrom and U. H. Danielsson, "Can MAP and Planck Map Planck Physics?," (2003): http://arXiv.org/abs/hep-th/0211006.

[21] G. F. R. Ellis, "Comment on 'Entropy and the Second Law: A Pedagogical Alternative,' By Ralph Baierlein," *American Journal of Physics* 63 (1995): 472; I. Stewart, "The Second Law of Gravitics and the Fourth Law of Thermodynamics," in *From Complexity to Life*, Gregersen, ed., 114.

[22] D. Zeh, *The Physics of Time Asymmetry* (Berlin: Springer, 1989); J. J. Halliwell, J. Perez-Mercader, and W. H. Zurek, *The Physics of Time Asymmetry* (Cambridge: Cambridge University Press, 1996).

[23] Roger Penrose, *The Emperor's New Mind* (Oxford: Oxford University Press, 1989).

Elucidation of the nature of gravitational entropy remains a major outstanding problem of classical physics.

2 The Small: Quantum Theory

In the case of quantum theory,[24] the measurement problem (when to use a Schrödinger evolution and when reduction of the wave function) is again related to the arrow of time because of the time asymmetry of the measurement process.[25] The choice of this arrow cannot be resolved on the basis of local physics alone, for there is no preferred direction of time in the local spacetime structure; so, in the case of quantum physics too, this arrow (apparently inviolable: we have no evidence of backwards arrows of time occurring locally) is presumably an emergent phenomenon based in top-down action from boundary conditions in the far past and future.

The ontology of quantum mechanics must be approached with caution.[26] It is through the collapse of the wave function that uncertainty enters microphysics, and the standard view is that this is an ontological rather than epistemological uncertainty. The Bell theorems discussed in depth in the VO/CTNS meeting show that there is no hidden variable if physics is non-local; but many EPR-style entanglement experiments show that physics is indeed nonlocal[27] ("quantum correlations...truly reflect the nonlocal nature of quantum mechanics"),[28] so hidden variables are quite possible without a major change in this regard. If they were present, the uncertainty would be epistemological rather than ontological—we cannot determine its cause if we have no effective handle to determine the nature or value of the hidden variable. Thus a broad investigation will keep open the hidden variable possibility. A further point is that we cannot even prove the outcome of quantum theory is random, because according to Chaitin's crucial analysis (extending Gödel's theorem to probability and computation), we can never actually prove that any sequence of numbers is random.[29] Randomness is a reasonable supposition rather than proven proposition.

The celebrated issue of entanglement raises crucial issues of ontology: how do single separate real entities emerge from the collective? This underlies

---

[24] Richard P. Feynman, *QED: The Strange Theory of Light and Matter* (Princeton, N.J.: Princeton University Press, 1988); Michael E. Peskin and Daniel V. Schroeder, *An Introduction to Quantum Field Theory* (Reading, Mass.: Perseus Books, 1999).

[25] Penrose, *The Emperor's New Mind.*

[26] E. McMullin, "Formalism and Ontology in Early Astronomy," in *QM*, 55-78; A. Shimony, "The Reality of the Quantum World," in *QM*, 3-16; J. Butterfield, "Some Worlds of Quantum Theory," in *QM*, 111-40.

[27] R. Y. Chiao, "Quantum Nonlocalities: Experimental Evidence," in *QM*, 17-40.

[28] P. Blanchard, D. Giulini, E. Joos, C. Kiefer, and I. Stamatescu, eds., *Decoherence: Theoretical, Experimental and Conceptual Problem* (Berlin: Springer, 2000), preface.

[29] G. J. Chaitin, "Randomness and Mathematical Proof," in *From Complexity to Life,* Gregersen, ed., 19.

a major attack on reductionism by Laughlin,[30] claiming that most particle properties may be emergent phenomena, thus echoing Bohm's extolling of the implicate order.[31] Either view implies that simplistic reductionism is fatally flawed at its foundations. Entanglement is rapidly reduced by decoherence ("a quantum mechanical process that dynamically describes the apparent loss of quantum coherence due to coupling of the system under observation to other degrees of freedom, which escape direct observation")[32] induced by interaction with the environment, which some claim solves the measurement problem, but Butterfield has made it clear that this is not the case.[33] However decoherence does underlie emergence of a classical order, and is thus an example of a crucially important top-down action from the environment to local physical systems (which is fundamental to the possibility of the emergence of life). Again, the arrow of time is writ large into this effect; how it gets there is not clear.

One of the lessons one might suggest emerges from physics is that its interlocking structure means that one will always run into trouble if one looks at the parts rather than the whole. An ongoing problem for quantum physics is the divergences that are dealt with at a practical level by renormalization procedures but remain a worrying feature. However, loop quantum gravity[34] suggests that spacetime might be quantized at the Planck level; then spatial Riemannian geometry is discrete, leading to a discrete volume spectrum.[35] Perhaps this should not be ignored. It is conceivable that this spacetime and spatial quantization could solve some of the divergences of quantum field theory (and hence perhaps be related to the value of the cosmological constant?), in addition to solving the problem of the initial singularity in cosmology.[36]

Finally, it seems to be widely accepted that information is a key concept in quantum theory, and that reformulation of quantum theory with information as a central concept is a key step forward (this has particularly resulted from investigations into quantum computation and the centrality in that theory of the concept of a *qubit*). The aim of showing that physics actually emerges from this concept (Wheeler's idea of 'it from bit') has still to be realized.[37] But in

---

[30] R. B. Laughlin, "Fractional Quantisation," *Reviews of Modern Physics* 71 (1999): 863-74.

[31] D. Bohm and B. J. Hiley, *The Undivided Universe* (London: Routledge, 1993).

[32] Blanchard et al., eds., *Decoherence*, preface. Note that this is a kind of hidden variable explanation.

[33] Butterfield, "Some Worlds of Quantum Theory."

[34] L. Smolin, *Three Roads to Quantum Gravity* (New York: Basic Books, 2001); C. Rovelli, "Notes for a Brief History of Quantum Gravity," (2001): http://www.arXiv.org/abs/gr-qc/0006061.

[35] M. Bojowald, *Class Qu Grav* 17 (2000): 1509.

[36] M. Bojowald, "Absence of Singularity in Loop Quantum Cosmology," (2001): http://arXiv.org/abs/gr-qc/0120069.

[37] J. D. Barrow, P. C. W. Davies, and C. L. Harper, eds., *Science and Ultimate Reality: Quantum Theory, Cosmology and Complexity* (Cambridge: Cambridge University Press, 2004).

any case, the reality of the causal effect of information seems clear from these analyses. Physical interactions affect what happens, but its meaning is best understood in terms of the information encoded. That information has an ontological reality apart from the physical system through which it is embodied, it can be encoded in different ways and transmitted from point to point. It is not encompassed in any particular physical representation.

3 The Complex

The issue of the emergence of complexity has historical, developmental, and functional dimensions. There are three specific aspects I wish to consider: developmental and functional issues, historical issues (evolution), and the particular case of the brain.

3.1 Developmental and Functional Issues

One of the great virtues of the VO/CTNS series of meetings was how they highlighted the levels of complexity in the hierarchical structure of the physical and biological world and their interaction with each other, emphasizing the basis of emergence in the key ideas of top-down causation[38] and self-organization.[39] Additionally, the key role of information has become clear,[40] biochemically based in the recognition of information at the supra-molecular level[41] and electrochemically in signal processing by neurons,[42] enabling pattern recognition by self-organizing neural networks.[43] It may be suggested that the essential way this information is effective is by setting goals

---

[38] A. Peacocke, *An Introduction to the Physical Chemistry of Biological Organisation* (Oxford: Oxford University Press, 1989); N. Murphy, "Supervenience and the Non-reducibility of Ethics to Biology," in *EMB*, 463-90; G. F. R. Ellis, "Quantum Theory and the Macroscopic World," in *QM*, 259-92; A. Peacocke, "The Sound of Sheer Silence: How Does God Communicate with Humanity?," in *NP*, 245-8; N. Murphy, "Supervenience and the Downward Efficacy of the Mental: A Non-Reductive Physicalist Account of Human Action," in *NP*, 147-64.

[39] B. O. Kuppers, "Understanding Complexity," in *CC*, 93-106; S. A. Kaufman, *The Origins of Order: Self-Organisation and Selection in Evolution* (Oxford: Oxford University Press, 1993).

[40] B. O. Kuppers, *Information and the Origin of Life* (Cambridge, Mass.,: MIT Press, 1990); C. A. Pickover, ed., *Visualizing Biological Information* (Singapore: World Scientific, 1995); G. F. R. Ellis, "On the Nature of Emergent Reality," in *The Re-emergence of Emergence*, Clayton and Davies, eds., pages.

[41] J. M. Lehn, *Supramolecular Chemistry* (Weinheim: VCH Verlag, 1995).

[42] A. Scott, *Stairway to the Mind* (New York: Copernicus/Springer, 1995); J. LeDoux, *Synaptic Self: How Our Brains Become Who We Are* (New York: Viking, 2002).

[43] G. A. Carpenter and S. Grossberg, *Pattern Recognition by Self Organizing Neural Networks* (Cambridge, Mass.: MIT Press, 1991).

in the hierarchical structure of feedback systems that occur in truly complex systems.[44]

This is a key feature in the emergence of true complexity,[45] and one can characterize higher levels of organized complexity, firstly, by the existence of such feedback systems, and secondly, by the origin of their goals when they exist (in-built through heredity, affected by memory, or determined consciously). In conjunction with top-down action, this feature enables the functional emergence of autonomous levels of structure[46] with self-consistent phenomenological laws of behavior (in physics terms, effective theories)[47] and associated languages to describe them (this clearly occurs in the case of the software in computer systems[48] and in the genetic code).[49] Thus each level of structure must be assigned an ontological reality: each exists as a feature of reality in its own right, even though it is structured of many constituent parts or modules. A table is real, as well as the atoms and molecules of which it is made.

This complexity comes into being in each individual through the processes of developmental biology, based in the top-down action of positional information which controls how the genetic information in DNA is read;[50] much is still to be clarified in this regard, for even such simple issues as how size is encoded in DNA, and of course proteomics is now a major area of research.[51] But DNA is only part of the solution: cells beget cells without specific instruction from the DNA.[52] Not all biological information is in the gene; some of it is simply inherited in complex self-replicating structures. Clarifying how the non-genetic information is coded and reproduced is an important issue (for example it involves the extraordinary coordinated action

---

[44] Ellis, "On the Nature of Emergent Reality," see http://www.mth.uct.ac.za/~ellis/emerge.doc; S. Beer, *Brain of the Firm* (New York: Wiley, 1972).

[45] N. A. Campbell, *Biology* (Menlo Park, Calif.: Benjamin Cummings, 1990); C. Starr and R. Taggart, *Biology: The Unity and Diversity of Life* (Boston: Brooks/Cole, 2001); D. Randall, W. Burggren, and K. French, *Animal Physiology: Mechanisms and Adaptations* (New York: W. H. Freeman, 2002); R. Rhoades and R. Pflanzer, *Human Physiology* (Philadelphia: Saunders College Publishing, 1989).

[46] Campbell, *Biology*.

[47] S. Hartmann, "Effective Field Theories, Reductionism, and Scientific Explanation," *Studies in History and Philosophy of Modern Physics* 32 (2001): 267-304.

[48] A. S. Tanenbaum, *Structured Computer Organization* (Englewood Cliffs, N.J.: Prentice Hall, 1990).

[49] G. F. R. Ellis, (2002): http://www.mth.uct.ac.za/~ellis/cos4.html, Table 4.6.

[50] S. F. Gilbert, *Developmental Biology* (Sunderland, Mass.: Sinauer, 1991); L. Wolpert, *Principles of Development* (Oxford: Oxford University Press, 1998).

[51] C. O'Donovan, R. Apweiler, and A. Bairoch, "The Human Proteomics Initiative (HPI)," *Trends Biotechnol.* 19 (2001): 178.

[52] E. Fox Keller, *The Century of the Gene* (Cambridge, Mass.: Harvard University Press, 2001); F. H. Harold, *The Way of the Cell: Molecules, Organisms and the Order of Life* (Oxford: Oxford University Press, 2001).

of microtubules and actin in cell division). The attempt to reduce biology to genes fails.

3.2 Historical Issues: Evolution

These structures arise through the historical process of evolutionary development.[53] One important aspect here is the potential timescale problem if one attempts to search the space of molecular possibilities randomly, for this space is characterized by immense numbers[54] (greater than 10^{121}) rather than astronomical numbers (of the order of 10^{80}). This is potentially resolved by the existence of autocatalytic hypercycles,[55] which provide preferred paths through the space of molecular possibilities. Then of course the question is why chemistry is so constructed (on the basis of the underlying physics) that these paths lead to viable living organisms. This is one of the deep questions underlying the existence of life: it is an aspect of the anthropic question, picked up again below.

There are two further related deep issues. First, can one truly generate new qualities *ex nihilo* through evolutionary processes leading to emergences, or must there be in some sense pre-cursor states preceding their existence in the fabric of reality?[56] Clearly this is a contentious issue, that deserves more searching examination. The second is an absolutely crucial feature of the relation between biology and the social sciences: the question of whether evolutionary biology has the ability to fully explain morality, as claimed by many evolutionary psychologists and evolutionary game theorists.[57] The counter claim strongly made at our meetings (and which I support) is that this does not explain ethics and morality, it explains them away;[58] thus these give partial but not total explanations. This will remain highly contested terrain in the future, related to the one mentioned above because the issue is whether or not ethics and morality can arise completely *ex nihilo*, or whether there must in some sense be precursors to their existence before they can come into being.

---

[53] F. J. Ayala, "The Evolution of Life: An Overview;" in *EMB*, 21-58; idem, "Darwin's Devolution: Design without Designer," in *EMB*, 101-16; C. J. Cela-Conde, "The Hominid Evolutionary Journey: A Summary," in *EMB*, 59-78; J. Chela-Flores, "The Phenomenon of the Eukaryotic Cell," in *EMB*, 79-98.

[54] W. M. Elsasser, *Reflections on a Theory of Organisms: Holism in Biology* (Baltimore: Johns Hopkins, 1998).

[55] M. Eigen and P. Schuster, *The Hypercycle: A Principle of Natural Self-Organisation* (Berlin: Springer, 1979); B. Kuppers, *Molecular Theory of Evolution: Outline of a Physico-Chemical Theory of the Origin of Life* (Berlin: Springer, 1985).

[56] I make the latter case in my article "Natures of Existence (Temporal and Eternal)," in *The Far-Future Universe*, G. F. R. Ellis, ed. (Philadelphia: Templeton Foundation Press, 2003), 316-354.

[57] L. D. Katz, ed., *Evolutionary Origins of Morality: Cross-Disciplinary Perspectives* (Bowling Green, Ohio: Imprint Academic, 2000).

[58] C. J. Cela-Conde and G. Marty, "Beyond Biological Evolution: Mind, Morals, and Culture," in *EMB*, 445-62; Murphy, "Supervenience and the Nonreducibility of Ethics to Biology."

3.3 The Mind and Brain

Enormous progress in neurobiology is clarifying neural correlates to mental processes and identifying the associated relevant brain regions.[59] Studies are pursuing the unknown neural features underlying the crucial human use of language[60] and symbolic systems,[61] and the 'hard problem' of consciousness[62] remains as elusive as ever.

This is a huge area, and there are just three issues I will pursue. First, of critical importance is the reality of consciousness and its causal effects. It is of course extraordinary that anyone who takes the trouble to write a book or scientific paper should deny this reality, thus undermining the validity of everything they write; nevertheless, many do. Thus the defense of the reality of consciousness[63] is a crucial need in the academic enterprise. Associated with this is the defense of free will, which (as Anton Zeilinger pointed out at the Wheeler Birthday Meeting) is a central premise at the foundation of the possibility of all the sciences. I will assume that reality here, and comment on a further aspect: this means an ontological reality must be assigned to thoughts, emotions, and social human constructions[64] because of their ability to have causal effects on the physical world. Thus a simple-minded physicalist approach to reality cannot provide a causally complete understanding of the world around us.

Secondly, the discussions on the brain and neuroscience omitted what I have since come to regard as one of the key features of brain development and functioning, namely Gerald Edelman's concept of Neural Darwinism. The key point here is that, after developmental processes establish a great variety of connection patterns between neurons,

> a process of synaptic selection occurs within the repertoires of neuronal groups as a result of behavioral experience...these changes occur because certain synapses within and between groups of locally coupled neurons are strengthened and others weakened without changes in the anatomy. This selectional process is constrained

---

[59] R. Carter, *Mapping the Mind* (Berkeley, Calif.: University of California Press, 1999).

[60] P. Hagoort, "The Uniquely Human Capacity for Language Communication: From POPE to [po:p] in Half a Second," in *NP*, 45-56; Michael Tomasello, *Constructing a Language: A Usage-Based Theory of Language Acquisition* (Cambridge, Mass.: Harvard University Press, 2003).

[61] T. Deacon, *The Symbolic Species: The Co-Evolution of Language and the Human Brain* (London: Penguin, 1997).

[62] D. J. Chalmers, *The Conscious Mind: In Search of a Fundamental Theory* (Oxford: Oxford University Press, 1996).

[63] M. Donald, *A Mind So Rare: The Evolution of Human Consciousness* (New York: W. W. Norton, 2001).

[64] K. Popper and J. Eccles, *The Self and its Brain: An Argument for Interactionism* (Berlin: Springer, 1977); Penrose, *The Large, the Small, and the Human Mind*; Ellis, "Quantum Theory and the Macroscopic World"; idem, "True Complexity and its Associated Ontology," in *Science and Ultimate Reality*, Barrow et al., eds., 606-36.

> by brain signals that arise as a result of the activity of diffusely projecting value systems, a constraint that is continually modified by successful output.[65]

The unit of selection is neuronal groups.[66] The same idea is proposed *inter alia* by Deacon,[67] LeDoux,[68] and Schore[69] (who relates it to the idea of 'parcellation').[70]

This argument extends the Darwinian type of understanding from the evolutionary processes that historically led to the existence of the brain to also underpinning both brain developmental processes and brain functioning. This is in accord with the way that such processes are now understood to underlie the functioning of the immune system through the process of clonal selection proposed by Burnet.[71] Thus such principles are already known to occur in human physiological functioning, giving the same benefits as discussed here: putting in place a mechanism that can deal efficiently with conditions already encountered, but that can also deal adequately with situations that have never before been encountered by the organism. Through this mechanism, "In a very literal sense, each developing brain region adapts to the body in which it finds itself."[72] Developmental and functional issues merge with evolution: it is just a question of different timescales.

Third, while there was discussion of significance of emotions at the meeting,[73] this is a very fast moving area and the discussion did not tackle some of the specifics now available, such as the systems characterized in detail by Panksepp.[74] The emotional system interacts with the higher cortex, while the immune system interacts with the emotional system.[75] This establishes the

---

[65] G. M. Edelman and G. Tononi, *Consciousness: How Matter Becomes Imagination* (London: Penguin, 2001), 84.

[66] G. M. Edelman, *Neural Darwinism: The Theory of Group Neuronal Selection* (Oxford: Oxford University Press, 1989), 43-69; idem, *Brilliant Air, Brilliant Fire: On the Matter of Mind* (New York: Basic Books, 1992), 95-99.

[67] Deacon, *The Symbolic Species*, 202, 205.

[68] LeDoux, *Synaptic Self.*

[69] A. N. Schore, *Affect Regulation and the Origin of the Self: The Neurobiology of Emotional Development* (Hillsdale, N.J.: Lawrence Erlbaum, 1994), 162, 253, 257.

[70] Ibid., 19, 250, 258.

[71] F. M. Burnet, *The Clonal Selection Theory of Acquired Immunity* (Cambridge: Cambridge University Press, 1959).

[72] Deacon, *The Symbolic Species*, 205.

[73] J. LeDoux, "Emotions: A View through the Brain," in *NP*, 41-4.

[74] J. Panksepp, *Affective Neuroscience: The Foundations of Human and Animal Emotions* (Oxford: Oxford University Press, 1998); J. Panksepp, "The Neuro-Evolutionary Cusp between Emotions and Cognitions: Implications for Understanding Consciousness and the Emergence of a Unified Mind Science," *Evolution and Cognition* 7 (2001): 141; see also M. Solms and O. Turnbull, *The Brain and the Inner World: An Introduction to the Neuroscience of Subjective Experience* (New York: Other Press, 2002).

[75] E. Sternberg, *The Balance Within: The Science Connecting Health and Emotions* (New York: W. H. Freeman, 2000).

causal pathways through which Neural Darwinism acts.[76] They enable development of the mind in conjunction with its social environment[77] and thereby, for example, underlie the important development of social modeling.[78] Further development of these themes should be rewarding.

4 Interactions between Large and Small

The standard model of cosmology assumes that the local action of gravity everywhere determines the large-scale evolution of the universe, a typical case of bottom-up action. The great development of cosmology as a physical science occurred because of the realization that nuclear physics could be applied in the early universe, leading to the understanding of nucleosynthesis of light elements,[79] and that in principle particle physics could be applied at even earlier times, leading to the theory of inflation.[80] The latter provides an extraordinary example of bottom-up action; sub-microscopic quantum fluctuations become the seeds of growth of clusters of galaxies, and lead to observable anisotropies in the cosmic background radiation. However, there is a considerable problem here: essentially the same underlying theory (the theory of the quantum vacuum) predicts a value of the cosmological constant—and hence the present day acceleration rate of the expansion of the universe—120 orders of magnitude larger than observed.[81] This has been called the worst prediction in the history of theoretical physics, and no agreed resolution is in sight. Clearly, this is still work in progress.

Nucleosynthesis, in contrast, is an example of top-down action: the overall density of matter in the universe determines the outcome of the micro-processes of nuclear reactions, setting initial conditions for the evolution of matter in the later universe. This results in the fact that observations of primordial element abundances can be used to determine the average density of baryons in the universe.

---

[76] G. F. R Ellis and J. Toronchuk, "Neural Development: Affective and Immune System Influences" in *Consciousness and Emotion: Agency, Conscious Choice and Selective Perception* R. D. Ellis and N. Newton, eds. (Philadelphia: John Benjamins, 2005), 81-119.

[77] M. A. Arbib, "Crusoe's Brain: Of Solitude and Society," in *NP*, 419-48; D. J. Siegel, *The Developing Mind: How Relationships and the Brain Interact to Shape Who We Are* (New York: Guilford Press, 2001); M. Donald, *Origins of the Modern Mind* (Cambridge, Mass.: Harvard University Press, 1991); idem, *A Mind So Rare*; I. Stewart and J. Cohen, *Figments of Reality: The Evolution of the Curious Mind* (Cambridge: Cambridge University Press, 1997).

[78] L. A. Brothers, "A Neuroscientific Perspective on Human Sociality," in *NP*, 67-74.

[79] S. Weinberg, *The First Three Minutes* (New York: Basic Books, 1977); Joseph Silk, *The Big Bang* (New York: W. H. Freeman, 1989).

[80] Guth, *The Inflationary Universe*; Silk, *A Short History of the Universe*.

[81] S. Weinberg, "The Cosmological Constant," *Reviews of Modern Physics* 61 (1989): 1.

There are many other examples of top-down action from cosmology to local physics:[82] some conjectured, and some that seem crucial. The most famous conjectured such relation is Mach's principle,[83] suggesting that the origin of inertia is an effect caused by the sum of all the most distant matter in the universe. Although this idea played a crucial role in the development of general relativity theory, it has been hard to pin it down in practice and even today it is controversial, some claiming it is fully incorporated into standard general relativity theory, with others denying this.[84] However, there is one intriguing link to present day quantum gravity theories: while loop quantum gravity[85] gives good solutions for black holes and cosmology, it has difficulties in producing flat spacetime as a solution to the quantum gravity equations. This is exactly what Einstein would have hoped for: from a Machian perspective, there should be no spacetime when no matter is present. From this viewpoint, this is a strength rather than a weakness of the theory.

Another proposal for top-down action is Dirac's conjecture[86] that as the universe is changing, the fundamental constants of physics should be changing with time also. In the past, variation of the gravitational constant was a very active area of research, but the data shows any such change is very small and not of astrophysical interest. At present, it is a possible variation of the fine structure constant that is generating some excitement (it has been claimed that astronomical observations support this idea) and may turn out to be important; it certainly needs investigation. However, some of the recently much-publicized proposals that the speed of light might vary are not on a solid basis.[87]

There are three cases in which top-down causation from the universe to local physics may indeed be claimed to be important. The first is something that does *not* happen: the universe does not interfere with local physics! There are all sorts of ways such interference could happen: we could be bathed in high-intensity cosmic rays or gravitational waves, or constantly bombarded by black holes. Life would not be possible if any of these happened.

The question here is when is it possible for there to be 'isolated' local systems in the universe, i.e. systems able to get on with their own affairs without interference from the external world so they can be regarded as separate physical entities. This is true in the real universe because on large scales there are only small deviations from spatial inhomogeneity and

---

[82] G. F. R. Ellis, "Cosmology and Local Physics," *New Astronomy Reviews* 46 (2002): 645-58, http://arXiv.org/abs/gr-qc/0102017.

[83] D. W. S. Sciama, *The Unity of the Universe* (London: Faber and Faber, 1959); idem, *The Physical Foundations of General Relativity* (New York: Doubleday, 1969).

[84] J. Barbour and H. Pfirster, eds., *Mach's Principle: From Newton's Bucket to Quantum Gravity* (Boston: Birkhauser, 1995).

[85] Smolin, *Three Roads to Quantum Gravity*; Rovelli, "Notes for a Brief History of Quantum Gravity."

[86] P. A. M. Dirac, "New Basis for Cosmology," *Proceedings of the Royal Society of London* 165 (1938): 199.

[87] J. P. Uzan and G. F. R. Ellis, "'c' is the speed of light, isn't it?," *American Journal of Physics*, Vol. 73, No. 3 (2005): 240-7.

anisotropy[88] which implies that the effect of disturbances by gravitational waves is minimal (in fact we cannot even detect them). Thus this is a top-down effect of considerable importance—the universe says, I will not interfere!

This is not the same as saying there is no interaction—there is indeed an interaction, because gravity is a long-range force (and the sum effect from all matter in an infinite universe could in principle diverge). Thus there is an interaction, but of such a nature as to allow local isolated systems to exist (mathematically expressed in limitations on the size of the Weyl tensor, which locally represents the free gravitational field).[89] The contrast with quantum theory is interesting; in the quantum case, interaction with the environment leads to decoherence, enabling isolated systems to exist at small scales; in the cosmic case, it is lack of interference from the environment that allows isolated systems to exist (and this is probably a requirement in order that Newtonian laws of gravity hold at a local level).[90]

A specific famous example of such lack of interference, worth commenting on in its own right, is Olbers's paradox:[91] the fact that the night sky is dark, rather than being at the surface temperature of the sun as predicted by simple static cosmological models, which circumstance would certainly destroy any possibility of life on earth. In fact, because of the standard thermal history of the expanding universe resulting from its hot big bang evolution,[92] the temperature of the night sky is the 3K of the cosmic background radiation, which allows the sky to be a heat sink for our waste radiation, hence enabling the thermodynamics of the biosphere to function.

The third key top-down effect is the determination of the arrow of time associated with the irreversibility of local physics,[93] commented on in sections 1 and 2. As explained there, known local micro-physics cannot supply that arrow, even though there is time asymmetry already at the micro level. It seems to be the result of top-down action from the universe,[94] although the precise way that this happens is still not clear—perhaps through the time-asymmetric quantum measurement process, or decoherence? But if so, how is the global arrow impressed on these processes? (in physics calculations, it is just put in by hand). This is a key effect underlying the universal local increase

---

[88] The universe is 'almost Robertson-Walker.' See W. Stoeger, S. J., R. Maartens, and G. F. R. Ellis, "Proving Almost-Homogeneity of the Universe: An Almost-Ehlers, Geren and Sachs Theorem," *Astrophysical Journal* 443 (1995): 1-5.

[89] Penrose, *The Emperor's New Mind.*

[90] Ellis, "Cosmology and Local Physics."

[91] H. Bondi, *Cosmology* (Cambridge: Cambridge University Press, 1960); E. R. Harrison, *Darkness at Night: A Riddle of the Universe* (Cambridge, Mass.: Harvard University Press, 1987); idem, *Cosmology: The Science of the Universe* (Cambridge: Cambridge University Press, 2000).

[92] Weinberg, *The First Three Minutes*; Silk, *The Big Bang.*

[93] Zeh, *The Physics of Time Asymmetry*; Halliwell et al., *The Physics of Time Asymmetry.*

[94] G. F. R. Ellis and D. W. Sciama, "Global and Non-Global Problems in Cosmology," in *General Relativity*, L. O'Raifeartaigh, ed. (Oxford: Oxford University Press, 1972), 35-59; Penrose, *The Emperor's New Mind.*

of entropy when gravity can be ignored, and hence the one-way flow of time in macro-physics and daily life. It is thus of crucial importance for evolution and consciousness.

5 Interactions between the Small and the Complex

The way in which the complex emerges from structured combinations of simple components through bottom-up and top-down interactions between the levels in the hierarchy and the purposeful use of information has already been discussed in section 3. Ultimately, this is based in the underlying physics: quantum field theory and the standard model of particle physics.[95] The overall key question is why physics is of such a nature as to provide the foundations for complexity,[96] when arbitrarily chosen laws of physics would almost certainly not do so; they would not allow organic chemistry to function, for example. This is a key aspect of the anthropic question:[97] why is the universe of such a nature as to allow life to exist? This will be pursued further in the next section. Three other specific issues will be pursued here.

The first crucial issue is the existence of free will. As stated in section 3.3, I take it that free will exists both because of the obvious data from daily life, and because if it is not true, then attempting to pursue science is a meaningless exercise, since we are then in no position to 'attempt' anything or make any rational judgements—things just happen in a pre-ordained way! But given this assumption, the key question is: How is free will possible, given the underlying strict control of the behavior of matter by physical laws? At the meetings, this was discussed in depth, and related strongly to the idea of supervenience in a very useful way.[98] However, whether this by itself is able to deal completely with the issue is unclear to me. Can a fully causal theory of micro-action together with the ideas of re-entry of neural circuits,[99] top-down causation, and supervenience supply the required answer? Or, might one need here to consider the possibility that free will has something to do with quantum theory and quantum uncertainty in the brain?

---

[95] J. Allday, *Quarks, Leptons and the Big Bang* (Bristol: Institute of Physics, 1999); Peskin and Schroeder, *An Introduction to Quantum Field Theory*.

[96] P. Anderson, "More is Different," *Science* 177 (1972): 377; S. Schweber, "Physics, Community, and the Crisis in Physical Theory," *Physics Today* (November 1993): 34; C. J. Hogan, "Why the Universe is Just So," *Reviews of Modern Physics* 72 (2000): 1149.

[97] J. D. Barrow and F. J. Tipler, *The Anthropic Cosmological Principle* (Oxford: Oxford University Press, 1986); J. Leslie, *Universes* (London: Routledge, 1989); Y. V. Balashov, "Resource letter AP-1: The Anthropic Principle," *American Journal of Physics* 59 (1991): 1069-76; J. Gribbin and M. Rees, *Cosmic Coincidences: Dark Matter, Mankind and Anthropic Cosmology* (London: Black Swan, 1991); Rees, *Our Cosmic Habitat.*

[98] T. C. Meyering, "Mind Matters: Physicalism and the Autonomy of the Person," in *NP*, 165-177; Murphy, "Supervenience and the Downward Efficacy of the Mental"; P. Clayton, "Neuroscience, the Person, and God: An Emergentist Account," in *NP*, 181-214.

[99] Edelman, *Neural Darwinism*, 4-8; idem, *Brilliant Air, Brilliant Fire*, 81-98; Edelman and Tononi, *Consciousness*, 79-92.

I am aware that this is a proposal that is unwelcome to many, and additionally that many regard it as unnecessary: they believe that free will can emerge from the aforementioned features without any need to invoke quantum theory. I still need to understand the latter claim: it seems paradoxical. But in any case, once one has raised it, the issue stands in its own right: is quantum theory directly significant in the operation of the brain, whether or not it is the answer to the issue of free will?

Now one must be clear here that of course quantum theory underlies the physics and chemistry that enables the brain to function (giving the stability of matter, underlying the periodic table, enabling covalent bonding for instance), and in that sense underlies brain function. But most neurophysicists do not deal with quantum calculations; they rather use effective theories of chemical bonding and ion transport, for example. The question is whether any essential quantum features such as entanglement, quantum tunnelling, holographic information storage, the nature of quantum measurement, or the uncertainty principle play key roles in brain function. There is a small but intriguing set of writings suggesting this may indeed be the case: quantum effects may indeed be significant in brain function.[100] The topic is ignored in the main neuroscience literature, but that may well be simply because neuroscientists are not trained in quantum physics, and hence naturally avoid the topic.[101]

The second issue, then, is the question: Is there any reason to expect the possibility of quantum effects playing a role in the brain, apart from the question of free will? From the integrative viewpoint of this paper, there certainly is, if one accepts the standard views on the evolutionary origin of the brain and additionally accepts the propaganda fed to us by Dawkins, Dennett, and others on the almost limitless powers of evolution. The point then is that, given the claimed powers of evolution, if there is any way that the brain could utilize quantum processes to enhance its computing power, then evolution should have found it. If quantum computing is possible, then we should expect the brain to utilize it, unless there is some physiological reason why it cannot be realized in living systems. Now it may be that decoherence processes prevent this from happening, but one would need to be very certain of this before dismissing totally the possibility of biological quantum computing.[102] A related important question is whether microtubules may play an important role

---

[100] C. A. Meier, ed., *Atom and Archetype: The Pauli/Jung Letters 1932-1958* (Princeton, N.J.: Princeton University Press, 1992); Penrose, *The Emperor's New Mind*; idem, *Shadows of the Mind: A Search for the Missing Science of Consciousness* (Oxford: Oxford University Press, 1996); H. P. Stapp, *Mind, Matter, and Quantum Mechanics* (Berlin: Springer, 1993); J. C. Eccles, *How the Self Controls Its Brain* (Berlin: Springer, 1994); J. M. Schwartz and S. Begley, *The Mind and the Brain: Neuroplasticity and the Power of Mental Force* (New York: Reagan Books, 2002).

[101] However, see http://www.u.arizona.edu/~chalmers/biblio/6.html#6.3 for a listing of over 60 relevant writings.

[102] L. P. Rosa and J. Faber, "Quantum Models of the Mind: Are They Compatible with Environmental Decoherence?," *Physical Review E*, Vol. 70, No. 3 (2004): 031902-7.

in information processing.[103] The answer is not clear; but this is clearly an issue of importance, for it would require a major rethinking about aspects of neurology and the way the brain works if true. Thus this is clearly an issue worth pursuing in depth.

Returning to the question of free will, how could quantum uncertainty help here? Would that not just introduce a random element into thought rather than an option of free choice? No, this is not necessarily the case, provided one takes the crucial step of denying the usual assumption that quantum uncertainty is ontological, and rather regard it as epistemological—in effect we have to regard outcomes at the quantum level as uncertain simply because we are unable to measure or control all the variables involved in specific outcomes. That is, one can advance a form of hidden variable theory as providing all the freedom one might want for free will to take place fully within the ambit of present day physics. According to the standard theorems in quantum theory, that means this action has to be nonlocal; that is, the concept of mind would be non-local in physical terms. That could well be acceptable. This could possibly tie into those proposals that see the observer as having significant effects on the quantum measurement process, as suggested by Wheeler and others.

One is here perhaps reaching the further shores of speculation, and I am not specifically making any definite proposal here. Rather, I am saying that open-mindedness about the issue may be sensible. What I can say for certain is that many speculations in the current theoretical physics and cosmology literature are at least as uncertain as this—and they are published in reputable journals.

The third and final issue is a return to the arrow of time in the present context—the disjunction between irreversible macrophysics that governs complex systems, and the reversible microphysics that underlies it. According to Alwyn Scott,[104] time irreversibility enters the Huxley-Hodgkin (HH) equation for nerve impulse propagation but not the underlying Schrödinger and Maxwell equations, for whom both directions of time are equally good. How a unique direction of time is enforced in the solutions characterized by the effective (HH) equations when it is not present in either of the underlying equations nor in the local spacetime structure remains mysterious, as is the fact that the same direction of time is chosen by all 10^{11} neurons in an individual and indeed all the neurons in different individuals in a population. As stated before, this probably arises by some kind of downward action from the cosmos as a whole to the assembly of molecules comprising the neurons but not apparent in the behaviors of the individual protons, neutrons and electrons that comprise these molecules. How this comes about is still not clear, but perhaps is related to quantum decoherence and/or measurement (collapse of the wave function). One thing one should be completely clear about, however, is the following: any argument based on increasing probabilities as time goes by will not do—for it always works both ways in time if one does not take the cosmological time asymmetry into account.

---

[103] J. Faber, R. Portugal, and L. P. Rosa, "Information Processing in Brain Microtubules," (2004): http://ArXiv.org/abs/q-bio.SC/0404007.

[104] Scott, *Stairway to the Mind*, 52.

6 Interactions between the Complex and the Large

The very profound link between the large and the complex is that the universe provides the context within which life can exist—the habitat for life.[105] Thus, this raises in its fundamental form the anthropic issue already mentioned several times: why is this habitat congenial to life, when most universes imaginable are not?[106]

Now this congeniality happens in two ways. First, the universe provides a large-scale habitat which is mostly hostile (most parts of the universe cannot host life easily: they are either too hot or too cold) but has some biophilic regions based on the existence of stars surrounded by planets. One of the major present debates is whether this is a very rare or a frequent occurrence. The odds in favor of this occurring often are being improved by the discovery of more and more extra-terrestrial planets, although the ones we can actually detect are probably rather bio-unfriendly—if we can detect their presence at a great distance they are very large, and so have very large gravitational fields that would crush the forms of life we have on earth; they also are probably more like the giant planets in the solar system (Jupiter and Saturn), with atmospheres of methane and ethane, than the Earth. Nevertheless, as there are about 10^{11} stars in each of the about 10^{11} galaxies in the observable part of the universe, even a very low probability of planets around a sun will lead to a large number of planets, of which quite a number might be expected to be biophilic.

Indeed, if this were not so—if, as claimed for example by Ernst Mayr,[107] we live on the only inhabited planet in the universe—we would in a profound sense have returned to a Copernican universe view: in terms of the relation between the large and the complex, we would be located at the unique center of the universe. At least as a working hypothesis, one may plausibly assume this is not so: that in a universe which is approximately spatially homogeneous in the large (for whatever reason, it obeys the cosmological principle[108] at present times) life will occur elsewhere as well as on earth.

In any case, the existence of planets certainly depends on the existence of galaxies made of stars, and these will not occur in all universes—for example if the cosmological constant were too large then no structures would form in the expanding universe, as gravitational attraction would be overwhelmed by cosmic repulsion. Furthermore, if the universe had too short a lifetime—say 100,000 years—then the 'deep time' required for biological evolution to take place would not be available—the universe might come to an end before life

---

105 Rees, *Our Cosmic Habitat.*

106 Barrow and Tipler, *The Anthropic Cosmological Principle*; Leslie, *Universes*; Balashov, "Resource letter AP-1."

107 E. Mayr, *Towards a New Philosophy of Biology* (Cambridge, Mass.: Harvard University Press, 1988).

108 H. Bondi, *Cosmology* (Cambridge: Cambridge University Press, 1960); E. R. Harrison, *Cosmology: The Science of the Universe* (Cambridge: Cambridge University Press, 2000).

could have come into existence. And as mentioned above, if the universe interfered with local systems too much, or the background radiation temperature never dropped below 3000K, then life would not be possible. Hence, in the ensemble of all conceivable universes, only a small fraction provide the global environment—galaxies containing planets—within which life can exist.

The second issue of congeniality arises in terms of the nature of the laws of physics—if we consider the family of all possible universes, in how many are the laws of physics of such a kind as to enable the existence of hierarchically structured complex systems? Does that physics allow atoms and molecules to exist, for example? It is conjectured by some cosmologists that the laws of physics might be different in different parts of the universe—for example, the constants of nature or even the number of spatial dimension could be different in different expanding domains in chaotic inflationary universe[109] or high-density regions in the universe could collapse to form black holes and then by some unknown process re-expand into a new expanding universe domain with new physical constants. Indeed, Lee Smolin has conjectured that this could happen in such a way that a Darwinian selection process will take place with the universe becoming more bio-friendly as time proceeds due to a process of natural selection leading to preferential reproduction of universe domains that are more likely to produce collapse to form black holes.[110]

Now there is no observational evidence whatever that either of these proposals is correct, firstly because the supposed underlying physics has not been shown to be true (indeed, the supposed inflation field has not even been uniquely identified in physical terms), and secondly because there cannot be observational evidence for the other expanding domains (with different physics) that are supposed to exist, because they are beyond the observational horizon: we can receive no information at all from them. No improvement in detector technology in the future will change this fact.[111] Nevertheless, these examples strongly make the point that different universes might have different laws of physics, and we must take that into account in considering cosmological possibilities,[112] particularly because there may be a top-down effect on the nature of the laws of physics from the global nature of the universe.

It is then important to note that the actual laws of physics are very exceptional: most conceivable laws of physics will not support life. A series of 'anthropic coincidences' have to be satisfied if complex structures are to

---

[109] A. D. Linde, *Particle Physics and Inflationary Cosmology* (Harwood, Chur, Switzerland, 1990).

[110] L. Smolin, *The Life of the Cosmos* (Oxford: Oxford University Press, 1999).

[111] However it could be possible to show chaotic inflation is not true by showing the universe has positively curved spatial sections with such a small radius of curvature we have seen round it already, in which case there is only one expanding universe domain—and the recent WMAP data on the cosmic background radiation may marginally support the view that this is indeed the case, see Joseph Silk, (2003): http://www.arXiv.org/astro-ph/abs/0303127.

[112] Ellis, "Before the Beginning."

form:[113] the number of spatial dimensions must be 3, the mass difference between the proton and the neutron must lie in a small range, the ratio between the strong and electromagnetic force strengths is highly constricted, and so on.[114] Why does the physics that is instantiated in the real universe lie in this tiny corner of the possibility space (and hence allow us to exist and observe the universe)? The answer necessarily has to be provided by metaphysics rather than physics, in the sense that we cannot test any proposed solution by any experiment or observation: in the real universe, we simply find instantiated the particular laws of physics that in fact apply, and have no way to change them.

I will not pursue the metaphysical options here,[115] except to refer again to the issue of an ensemble of universes or multiverse, discussed in section 1. This is now increasingly suggested as the scientific solution to the problem. However, it is completely unverifiable because there is no causal connection whatsoever between our universe and the postulated other universes in a disconnected ensemble of really existing universes, and no observational connection of any kind is possible to the other expanding universe domains in a chaotic inflationary model, so this is a metaphysical solution rather than a scientific one in the usual sense. Belief in its truth is just that, an article of faith rather than a proposition that can be verified. Nevertheless, the idea clearly has a strong scientific appeal. It is worth developing the idea further: actually investigating the idea in more detail, looking in detail at how it can be implemented, and so on.[116]

A particularly important kind of top-down action from the cosmos to complex systems is the causal link that governs the choice of the local arrow of time everywhere in the universe, as discussed in the previous section. It seems likely, for example, that the cosmos puts boundary conditions on solutions to Maxwell's equations that determine the local electro-magnetic arrow of time at the microphysical level,[117] which in turn determines this arrow at the macro level. How this all happens is a crucial question.

The relation between complex systems and the macro-universe may throw some light on the controversy regarding the block universe and the nature of time.[118] As indicated in the previous section, there is good reason to assume free will is real and hence there is an ontological reality to the freedom this allows in physical existence. If we draw a block universe diagram (see Figure 2), the past of the event "x" on the shown world line at the present time t_0 is already determined, but the future is not. One can only draw this diagram,

---

[113] Barrow and Tipler, *The Anthropic Cosmological Principle*; Leslie, *Universes*; Balashov, "Resource letter AP-1"; Gribbin and Rees, *Cosmic Coincidences*; Rees, *Our Cosmic Habitat.*

[114] Tegmark, "Is the Theory of Everything Merely the Ultimate Ensemble Theory?"; Ellis, (2002): http://www.mth.uct.ac.za/~ellis/cos7.html, Table 7.1.

[115] See G. F. R. Ellis, "The Theology of the Anthropic Principle," in *QC*, 363-400.

[116] Ellis et al., "Defining Multiverses."

[117] Ellis and Sciama, "Global and Non-Global Problems in Cosmology," 35-59; Penrose, *The Emperor's New Mind.*

[118] C. J. Isham and J. C. Polkinghorne, "The Debate over the Block Universe," in *QC*, 139-48.

showing what happens for the entire block bounded by t_1 in the past and t_2 in the future, when one knows everything that happens between those bounding times. Taking into account the free will of the 'observer,' who in this context should rather be called 'actor,' the data at t_1 does not suffice to determine what happens at t_2. Rather, on each world line what happens at time t_0 is only determined by what the actor decides to do at that time.

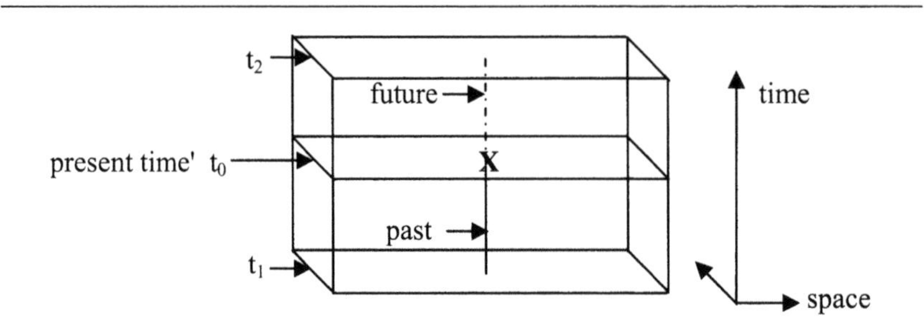

Figure 2: Block diagram of the universe. Time runs vertically up and space horizontally (2 dimensions are shown in this representation.)

Thus if we consider a whole family of actors filling the space at time t_0, where that time surface is arbitrarily chosen in spacetime and then used to determine a conventional correspondence between events on each actor's world line (it has no physical significance), then for each observer what happens in the observer's history prior to the surface $t=t_0$ is known and in the past; what happens after the surface $t=t_0$ lies to the future and is unknown. The whole block universe picture can only be drawn by knowledge available to the family of observers when all have reached the time surface $t=t_2$. The assumption that such a diagram can represent in one view the complete time evolution of a system both before and after the present time is only tenable if we restrict our considerations to the time-reversible and in principle predictable equations of fundamental physics. Taking into account true complexity, and particularly free will, the situation is different: this diagram can be seen to be a theoretical construction that cannot be completed until the time at the surface $t=t_2$ is reached on all the world lines contained in the diagram. But then what we are calling 'the present time' will correspond to the surface $t=t_2$ rather than $t=t_0$.

7 The Interacting Whole

This paper has considered links between the major aspects of sciences, and the complex interactions that occur between these aspects. The overall key point that emerges is that causality is always multi-dimensional. Partial ('nothing but') explanations ignore the taken-for-granted dimensions of causation that are not the current focus of attention. Analyzing the components and partial interactions of the whole in a reductive way is an essential part of attaining understanding; but claiming this partial view is the same as the whole is a crucial error when interpreting fundamental aspects of the nature of reality or making philosophical deductions. Associated with this multi-dimensional nature of causality is a distinct ontological reality that can be assigned to each aspect of science considered here: its applications to the large, the small, and the complex. In each case, the associated kinds of information must be taken to be real (else we have a causally incomplete scheme). This must be taken seriously in any attempt at a philosophical analysis of the implications of science.

To complete the picture, I now consider briefly the limitations of physics (the basic underlying science), as well as the limitations of the scientific enterprise as a whole.

7.1 The Limitations of Physics

Emergent effects are determined by a combination of chance (historical contingency) and necessity (physical laws), but there is additionally a role of conscious choice in the case of humans. As emphasized above, this is a key feature in our analyses of the world around us—without conscious choice, the attempt to understand does not make sense. We can only believe the outcome of our arguments and analyses if we have the ability to relate them to evidence, logic, and coherence.

The issue facing scientists, then, is this: are we trying to construct a causally complete theory of interactions that affect events in the physical world? If so, then humans must be included in the causal system (e.g. taking into account the fact that physicists carry out experiments) and the emergence of all the higher order phenomena mentioned above must be also taken into account. If not, we will necessarily have a causally incomplete theory, and must not pretend it can satisfactorily link human behavior to physics and chemistry.

The challenge to physics is that the higher levels are demonstrably causally effective, in particular, consciousness is causally effective; but conscious plans and intentions and emotions are not describable in present day physical terms. The key point is that human intentions and goals are not just convenient auxiliary variables that summarise physical microstates; rather, they are essential variables in many causal processes. Without them we cannot adequately model causation involving human beings (for example we cannot predict whether a pair of spectacles or a jumbo jet will be likely to emerge from a mental process). They are irreducible higher-level quantities that are clearly causally effective in the physical world. Thus physics has two choices: either

1. Extending its scope of description to encapsulate such higher level causal effects, for example, including new variables representing thoughts and intentions and so enabling it to model the effects of consciousness and its ability to be causally effective in the real physical world, or

2. Deciding that these kinds of issues are outside the province of physics, which properly deals only with inanimate objects and their interactions. In that case physics must give up the claim to

give a causally complete description of interactions that affect the real physical world.

Whichever option is adopted, the concept of a 'theory of everything' as usually understood by physicists (a unified theory of fundamental forces and interactions such as String Theory)[119] must be acknowledged to be unable to give a complete account of all causally effective aspects of the physical world, which includes the biological world. At minimum, physics has to be related somehow to the world of thoughts and feelings before it can make any claim to provide causal completeness. The challenge is to see if physics can sensibly be extended towards this goal.

7.2 The Limitations of Science

The domain of science has here, in essence following Penrose,[120] been related to the large, the small, and the complex. However, despite this wide scope, there are aspects of existence that science *per se* cannot handle (although it may help us understand their context and implications)—they lie outside the domain of science. This is indicated in Figure 3.

In particular, science cannot handle ethics or aesthetics, because in each case there is no scientific experiment that can determine the relevant results or quantities. There are not even scientific units defined to handle these cases (the beauty of a picture or the degree of evil of an act, for example). However, as is implied by the analysis above, these topics are causally effective—the striving of artists to attain beauty results in tangible artifacts, for example.

Science also cannot handle metaphysics—that is, determining the nature of fundamental reality. In particular, the ontology of the laws of nature—where are they writ and how they are effective—is unsolved.[121] Because there are no experiments that can resolve this issue, investigating this lies beyond the scope of science—it is the domain of metaphysics. However, science can handle with great effectiveness investigating the effective nature of the laws of physics—indeed this is its forte.

---

[119] B. Greene, *The Elegant Universe: Superstrings, Hidden Dimensions and the Quest for the Ultimate Theory* (London: Jonathan Cape, 1999).

[120] Penrose, *The Large, the Small, and the Human Mind.*

[121] W. R. Stoeger, "Contemporary Physics and the Ontological Status of the Laws of Nature," in *QC*, 207-31; idem, "The Mind-Brain Problem, the Laws of Nature, and Constitutive Relationships," in *NP*, 129-46.

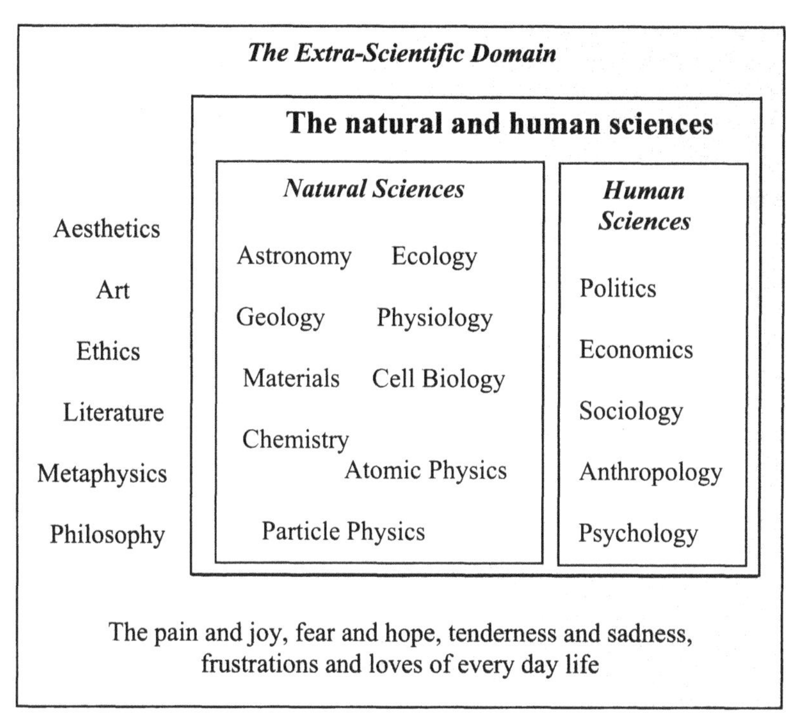

Figure 3: A brief summary of the natural and human sciences, as well as some features of the extra-scientific domain.

7.3 Concluding Comments

In his book *The Character of Physical Law,*[122] Richard Feynman summarizes the hierarchy of structure, starting with the fundamental laws of physics and their application to protons, neutrons, and electrons, going on to atoms and heat, and including waves, storms, stars, as well as frogs and concepts like 'man,' 'history,' 'political expediency,' 'evil,' 'beauty,' and 'hope.' He then says the following: (pp. 125-126):

> Which end is nearer to God, if I may use a religious metaphor. Beauty and hope, or the fundamental laws? I think that the right way, of course, is to say that what we have to look at is the whole structural interconnection of the thing; and that all the sciences, and not just the sciences but all the efforts of intellectual kinds, are an endeavour to see the connections of the hierarchies, to connect beauty to history, to connect history to man's psychology, man's psychology to the working of the brain, the brain to the neural impulse, the neural impulse to chemistry, and so forth, up and down, both ways. And today we cannot, and it is no use making

[122] R. Feynman, *The Character of Physical Law* (Cambridge, Mass.: MIT Press, 1990), 124-25.

> believe we can, draw carefully a line all the way from one end of this thing to the other, because we have only just begun to see that there is this relative hierarchy.
>
> And I do not think either end is nearer to God. To stand at either end, and to walk off that end of the pier only, hoping that out in that direction is the complete understanding, is a mistake. And to stand with evil and beauty and hope, or with fundamental laws, hoping that way to get a deep understanding of the whole world, with that aspect alone, is a mistake. It is not sensible for the ones who specialize at one end, and the ones who specialize at the other, to have such disregard for each other...The great mass of workers in between, connecting one step to another, are improving all the time our understanding of the world, both from working at the ends and from working in the middle, and in that way we are gradually understanding this tremendous world of interconnecting hierarchies.[123]

The VO/CTNS series have played a useful role in this regard. Perhaps the most obvious missing scientific dimension that should be explored in a follow-up is the relation of all this to the social and human sciences. However, that will be a difficult task, because those subjects are so diverse and so disunited in themselves.

---

[123] Ibid., 125-26.

II. PHILOSOPHICAL ANALYSIS OF SPECIFIC ISSUES IN THE SERIES

Philip Clayton
Nancey Murphy
Wesley J. Wildman

TOWARD A THEORY OF DIVINE ACTION THAT HAS TRACTION

Philip Clayton

The goal of this paper is to sketch the framework for, and to defend, a theory of divine action that maintains "traction" with the sciences.

1 Clarifying the Central Positions on Divine Action

The discussion of science and divine action is framed by three logical possibilities:

1. There is no divine action in the world.
2. All actions are God's actions; certain events only seem to be caused by created objects and persons.
3. Some events are caused by God and others are caused by finite agents.

Given the philosophical and theological traditions considered by the authors in the VO/CTNS series—a large set but not a complete one—(3) actually subdivides into only a few major options. Consider the following five:

> (a) The primary cause/secondary cause distinction, traceable back in particular to Thomas Aquinas: God is the primary cause of every event that occurs in the world. Without God's initial creation of the world, God's ongoing preservation of the finite order, and the general consent of God, no finite events whatsoever would exist. Moreover, God has to allow every new outcome; in this sense divine concurrence remains a necessary condition for each event. Still, alongside the primary (divine) cause, a secondary cause is always active. For example, whatever one says about the role of God in the event, if you kick the ball and it flies between the goalposts, clearly you have played a causal role in scoring a goal for your team. Whether or not God plays a causal role in lightning, the release of static electricity certainly does. Science studies the chain of secondary causes; theology studies the logic of primary causality.
>
> (b) Process theism: Drawing on the work of Alfred North Whitehead, process theists also understand every event to involve a combination of divine input and input from a finite agent. Here, however, the overarching model is mentalistic or experiential rather than mechanical. For Whitehead, finite reality consists of a series of "actual occasions" or moments of experience. For any given occasion, past events serve as input or data, which the occasion then synthesizes in its own unique, creative manner. Along with the other data available to it, God also offers to the occasion an "initial aim," which applies the general divine ideals to that particular situation and its needs. Even though God never coerces an outcome, the divine input is still offered to every occasion.

(c) Evangelical theism (or orthodox theism, or classical philosophical theism): This view begins with the creation of the universe by God out of nothing, and hence with divine omnipotence. Once God as first cause has brought the world into existence, four different types of events can be distinguished. First, an independent natural world exists, which functions according to natural law. Antecedent states of affairs, combined with natural laws, account for physical effects. Second, there are individual human agents who have souls or minds. Their free decisions account for certain actions in the world, as when you decide to pick up a pencil and then we observe your body to carry out this action. (The status of animals is ambiguous on this [often Cartesian] account; they are sometimes classified as lower-level souls, and sometimes as merely mechanical causes like other physical causes.) Third, on certain occasions at least there is a direct causal influence of God on human minds, sometimes with their willing concurrence and sometimes against their will. ("Saul, why do you persecute me? It is hard for you to kick against the goads," Acts 26:14, NIV). This divine influence can compel mental outcomes, e.g., causing an agent to have a thought or carry out an action apart from the agreement of his will. Finally, there can be direct actions by God in the physical world, in which a physical outcome is the result of divine input independent of any prior physical state of affairs. When God causes a miracle, God suspends the physical laws and their normal effects.[1]

(d) History as a whole as a single act of God: This is the position most strongly associated with the work of Maurice Wiles.[2] On this view, the entire course of history represents one ongoing act of God. There are no individual events in the world that are caused by a direct divine influence, such that the outcome is different than it would have been based on the previous state of affairs in the world. Nonetheless, it is true to say that the whole thing qualifies as an act of God.

(e) Panentheistic divine action. I defer treatment of this constructive proposal to the end of the essay.

The literature in this field—including the five VO/CTNS volumes—is replete with metaphorical and figurative statements about divine action. As Paul Ricoeur (and, following him, Sallie McFague) have shown, metaphors are not irrelevant to the outcome, and indeed the content, of theological debates.[3]

---

[1] C. S. Lewis, *Miracles: A Preliminary Study* (San Francisco: Harper San Francisco, 1996); C. John Collins, *The God of Miracles: An Exegetical Examination of God's Action in the World* (Wheaton, Ill.: Crossway Books, 2000).

[2] See Maurice Wiles, *God's Action in the World: The Bampton Lectures for 1986* (London: SCM Press, 1986).

[3] See Paul Ricoeur, *The Rule of Metaphor: Multi-Disciplinary Studies of the Creation of Meaning in Language* (Toronto, Canada: University of Toronto Press, 1993); Sallie McFague, *Metaphorical Theology: Models of God in Religious Language*

Nonetheless, the only way to make progress on the exceedingly difficult topic of divine action in a scientific world is to require the translation of metaphorical and figurative language into the most specific and least ambiguous position possible—even if much is lost in the translation. For example, if an advocate says, "I do not intend to assert that God *literally* does anything; I intend all language about divine action to be figurative," [4] then we should classify her position as a version of (1) above—even if her theology includes talk about the acts of God.

I do not intend to address the pros and cons of these three positions here. For purposes of gaining an overview of the divine action debate it is much more important to analyze the *logic* of the debate: its structural features, its inherent difficulties, and the long-term prospects for making progress in choosing among the options. One achievement of the series is that there is far more data today for carrying out this analysis than there was in 1987.

1.1 Different Positions Give Rise to Different Criteria of Success

I suggest that each of the five major options under (3) above gives rise to a different strategy for dealing with scientific data. Much of the ambiguity, and some of the failures, in the five volumes of the series are explained by the fact that one's fundamental approach to the science-theology debate will be affected by one's particular understanding of what divine action *is*. Failing to see this, advocates of one view commonly become highly frustrated with advocates of another; authors tend to believe that their opponents are avoiding what they take to be "the real issues" thereby making things too easy for themselves.

- Advocates of the primary cause/secondary cause distinction tend to be happy to pursue a complete scientific description of the causal history that produces any specific effect, whether in the physical, biological, or psychological realms. They are unworried if the causal history for some event is seen, scientifically speaking, as sufficient for producing that event—for, after all, scientific causes are "merely" secondary causes and in no way (on this view) obfuscate the role of God as the primary cause.

- Advocates of process theism frustrate their opponents with their belief that the causal histories and explanations given by science do not really represent what is occurring. Every development in the world is actually the result of creative syntheses by actual occasions. The actual occasions that constitute the trajectory of a particle in Newtonian physics are, admittedly, extremely uncreative; tests of multiple particles indicate identical results to a very high degree of

(Philadelphia: Fortress Press, 1982); idem, *Models of God: Theology for an Ecological, Nuclear Age* (Philadelphia: Fortress Press, 1987).

[4] I address the logic of this position in, "Can Liberals Still Believe that God (Literally) Does Anything?," *CTNS Bulletin* 20.3 (2000): 3-10.

(mathematical) precision. Hence, the process theologian is unperturbed by any scientific description or prediction of events in the natural world. In fact, she will be much more opposed to the idea of direct divine interventions (option c) than she will be worried about any concrete empirical results in science. Should the scientist however claim that no purpose or agency or experience is involved in some physical or biological system, the process thinker will accuse him of advocating a metaphysics unjustified by the scientific data themselves.

- Advocates of evangelical theism (or orthodox theism, or classical philosophical theism) are generally happy with standard scientific descriptions and explanations for most occurrences in the natural world. For "normal" natural events they stand perhaps closest of all to the scientist and her worldview: there really is a physical world that (generally) follows natural laws without the need for divine causes or the "panexperientialism" of process thought. It is just that there are also crucial exceptions to this pattern. Not only can we not exclude the possibility that God intervenes at his discretion to directly cause a particular outcome; we have biblical reasons to think that the Lord, in fact, often does so. The paradigmatic case of divine intervention is the resurrection of Jesus Christ, which by definition is not the result of a natural process. Perhaps scientists will be able to detect that this (or another) supernatural event has occurred, as in evangelical arguments for the historicity of the bodily resurrection of Jesus.[5] But they will never able to explain this (or any other) miracle without including the direct agency of God.
- To the advocates of history as a whole as a single act of God, all three of the previous positions are mistaken in looking for some direct role for God in the world. The theologian today merely needs to show that the existence of the finite world as a whole can meaningfully be spoken of as a single act of God. To do anything beyond this is to create an unnecessary clash with science. Nor, they argue, is one likely to be able to achieve anything more.
- I will have more to say about panentheistic divine action below.

1.2 What Is the "Fall-Back" Position on Divine Action?

Arguably, the "history as a whole as a single act of God" position represents the default or "fall-back" position in the divine action debate as it runs through the various VO/CTNS volumes. That is, it (or a variant on it) is the position

---

[5] See Gary R. Habermas and Anthony Flew, *Did Jesus Really Rise from the Dead?: the Resurrection Debate* (San Francisco: Harper & Row, 1987); cf. also Wolfhart Pannenberg, *Jesus—God and Man*, 2nd ed. (Philadelphia: Westminster Press, 1977).

that writers fall back to when they become convinced that their earlier positions are untenable.

Consider the case of Thomas Tracy. Over the course of the volumes, Tracy seems to have become more skeptical of his robust theory of quantum divine action in *CC*. His essay in *EMB* is slightly more cautious, and his contribution to *QM* seems to signal an important change in position. Tracy there notes that "the argument for moving in this direction"—by which he means "the idea of direct divine action at points of causal openness in the world"—is "not as strong as it may at first appear."[6] Now, he seems almost agnostic on the question: "Whether nature is arranged in such a way that this amplification of quantum effects can occur apart from human contrivance" is "a question of empirical fact, and it is an unsettled one."[7] He still refers to "the possibility that quantum transitions might serve as triggers for chaotic processes," but he does so, as he says, "even more hesitantly,"[8] since this hypothesis "is accompanied by a number of fundamental scientific uncertainties."[9] The theory of divine action that Tracy had earlier defended is "an enticing possibility, but it remains to be seen whether it will become more than that."[10]

In his final essay in the series, Tracy appears to be more attracted to the idea of a single divine action, as in option (d) above. He notes that "the key consideration," or decision for one to make, is "whether the idea of divine action in response to human actions *requires* that God act in ways that affect the course of events in the world once the world's history is underway."[11] It is no longer clear to Tracy, in other words, that one needs to say that God acts at any temporal moment subsequent to the initial creation. In fact, Tracy summarizes his argument in the essay with the thesis, "responsive divine action does *not* require that God act directly to alter the course of events in the world."[12]

Given his new attraction to what I read as a version of the "single divine action" theory, Tracy seems somewhat less invested in the quantum mechanics debate to which he had earlier contributed. He writes, "theologians have less at stake than it might first appear"—presumably Tracy means: than it had first appeared to him—"in the question of whether the science of quantum mechanics (or of chaos theory, or of higher-level emergent systems) provide openings in nature's causal structures through which God can act without intervening."[13] Tracy's comments represent particularly clear evidence that the default or fall-back position in the debate over divine action lies in the "single divine act" position.

---

[6] Thomas F. Tracy, "Creation, Providence, and Quantum Chance," in *QM*, 248.

[7] Ibid., 256.

[8] Ibid.

[9] Ibid., 257.

[10] Ibid.

[11] Ibid., emphasis added.

[12] Ibid., emphasis added.

[13] Ibid.

2 Nine Core Assumptions of the Divine Action Series

The VO/CTNS volumes have made certain assumptions about the relationship between science and metaphysics. It is the nature of assumptions to exclude positions that might otherwise have been important contenders. Still, without making these assumptions, it would not have been possible to engage in the sort of advanced, coherent research project which these volumes have represented. An outsider, looking back over the various essays, might infer that the following nine principles were being used as background assumptions for the VO/CTNS project:

1. *Science gives rise inevitably to certain metaphysical questions.* These questions are meaningful and can be discussed in a rational manner.
2. *Logical positivism is a deeply wrongheaded position.* That is, none of the authors accept that the only meaningful statements are those that can be made about empirical matters (or can be directly derived from them). Of course, some authors are more cautious about speculative statements that are further removed from the empirical or experimental data, whereas other authors do not shy away from robust metaphysical and theological speculation. Not surprisingly, the scientists who have written the various essays detailing the scientific results in each field are generally the most skeptical about metaphysics: a number seem to doubt that any rational progress can be made in metaphysics or theology.[14]
3. *It is possible to raise meaningful questions about scientific theories that are not themselves scientific questions.* In doing so, one must be guided by the details of the theories. Yet the scientific details do not by themselves provide answers to these open questions. In at least some cases, other scientific disciplines help to provide the answers, as when knowledge of the history and sociology of science helps one to evaluate a particular scientific theory, or when an insight from the new field of information biology helps to explain cell functioning. In other cases, however, the only way to resolve the questions that arise is to turn to a type of argument more typical of classical philosophical discussions.
4. *Advances in science, and differences among scientific theories, provide data that is relevant to addressing the question of divine action.* For example, the authors in the series seem to assume that answers to the question of divine action will be different depending on whether one accepts Aristotelian natural philosophy, or Newtonian physics, or a physics grounded in contemporary quantum mechanics. They assume that the challenges for talking about God's influencing the history of the emergence of life on this planet will be different (and presumably more difficult!) after

---

[14] See James T. Cushing, "Determinism versus Indeterminism in Quantum Mechanics: A 'Free' Choice," in *QM*, 99-110.

Darwin and the Neo-Darwinian synthesis. And they assume that contemporary developments in the neurosciences affect, directly or indirectly, what can meaningfully be said about God's influence on human thought.

5. By accepting (4), the authors presumably also accept that *scientific theories can either support or undercut—that is, they can be used either to verify or falsify—metaphysical theories*. A much smaller number of the authors, apparently, accepts the converse: that metaphysical theories can in turn support or undercut (verify or falsify) scientific theories. Many authors do accept, however, that metaphysical or, specifically, theological theories are relevant for assessing scientific theories. A larger number accepts that metaphysical or theological positions might at least assist in interpreting scientific theories.

6. *The degree of connection between science and metaphysics or theology varies depending on the question being raised, and perhaps on which scientific discipline one has in mind*. In some cases the philosophical discussions in these volumes remain quite close to the scientific fields. In these cases the authors seek to derive as many of their answers as possible from the sciences themselves. One thinks for example of the chapters by Jeremy Butterfield, James Cushing, and Michael Redhead in *QM*. In other cases the authors derive their philosophical or theological conclusions from other sources altogether.[15]

7. The common subtitle for the entire series is "Scientific Perspectives on Divine Action." The use of this subtitle would seem to imply that all authors writing for the volumes accept that the question of God's action in the world is a meaningful question and can be debated in a rational fashion. This in turn entails that they accept that one can rationally debate *any philosophical questions about science that are as broad, metaphysically speaking, as the discussion of divine action*. For example, debates about scientific realism or the nature of mind would not seem further removed from science than the question of divine action. Moreover, if one is prepared to debate the question of divine action in a rational manner, wouldn't it follow that one is no less prepared to debate the question of the existence of God?

   In fact, however, assumption (7) is clearly not shared by many of the authors in the five volumes! Many who have written for the series—usually scientists—are skeptical, implicitly or explicitly, about whether the question of divine action can be discussed meaningfully, and they would presumably be no less cautious about other metaphysical questions of equal breadth. Nonetheless, the series as a whole, if it is to offer any guidance

[15] See Stephen Happel, "The Soul and Neuroscience: Possibilities for Divine Action," in *NP*, 281-304.

on the topic of "scientific perspectives on divine action," must presuppose that enough metaphysical knowledge is possible for one to address the question of divine action in a rational and meaningful manner—even if the discussion does not lead in the end to a clear resolution.

8. *There is vast disagreement on the epistemic rights of theologians.* Some authors believe that theology can draw on its own sources of knowledge, above or beyond metaphysical reflection. Others sharply resist this assumption. This turns out to be one of the biggest disagreements that keeps arising across the series, and it is a disagreement of immense significance. Wouldn't a productive debate on competing theories of divine action have to presuppose some shared parameters about what moves theologians are or are not allowed to make? For example, the authors seem to assume that the question of divine action and science cannot be resolved by an individual's appeal to his or her own personal religious experience. One wants to know, however, what license, if any, the theologian is to be given in making her case that would not generally be accepted in standard metaphysical disputes. We return to this urgent question below.

9. A series on divine action, if it is to succeed, would seem to need to formulate some *core parameters for exploring the relationship between theology and science*. Yet in fact the authors of the series are deeply divided on the question of whether theology and science directly connect, or whether theology-science discussions must be mediated through other disciplines—and, if they must be so mediated, which disciplines should play this role. It is to this task of specifying at least some core parameters that I now turn.

3 Goal: A Theory of Divine Action That Maintains "Traction" with the Sciences

3.1 Defending the Value of Traction

All theories of divine action are examples of theological language. Theological language has "traction" if and only if it makes claims that can be impacted, either positively or adversely, by the results of philosophical critique, historical-critical research, or scientific knowledge. Confirming traction is traditionally known as *apologetics* (from 1 Peter 3:15). A number of articles in the VO/CTNS series attempt to establish confirming traction.[16] Obviously

---

[16] See Nancey Murphy, "Evidence of Design in the Fine-Tuning of the Universe," in *QC*, 401-28. Outside the series a leading advocate of positive traction in recent years has been Michael Dembski. See Dembski, *The Design Inference: Eliminating Chance Through Small Probabilities* (New York: Cambridge University Press, 1998); idem, *Intelligent Design: The Bridge between Science and Theology* (Downers Grove, Ill.: InterVarsity Press, 1999); idem, *No Free Lunch: Why Specified Complexity Cannot Be Purchased without Intelligence* (Lanham, Mass.: Rowan and Littlefield, 2002); idem, ed., *Signs of Intelligence: Understanding Intelligent Design* (Grand Rapids, Mich.:

theologians are happy to accept confirming traction when it is offered. To value disconfirming traction comes less easily. One values disconfirming traction when one attempts to relate theology to science in such a way that the two could conflict *and* one acknowledges that this conflict, when it occurs, is a bad thing. For reasons explained by the philosopher of science Karl Popper, it is in principle easier to establish disconfirming than confirming traction.

Many participants in the VO/CTNS series do not agree that maximizing traction is an important goal. Theology that has minimal traction, or that is tractionless, is theology that does not connect with science, history or philosophy in such a way that (e.g.) scientific theories, scientific method, philosophical critique, and/or historical-critical criticism can become relevant to assessments of the rational credibility of its assertions. Such theologies may connect with science in many other ways—they may refer to scientific examples or explain theological beliefs with reference to their implications (positive or negative) for science. To whatever extent a theology fails to value and pursue traction, to that extent it becomes fideistic.

Note that traction (and hence fideism) is a matter of degree. Take theories of divine action (TDAs) as examples of theological language. One discovers at least three levels of traction, which I designate by subscripts:

- Traction_1: a given TDA is *derived from* science, history or philosophy
- Traction_2: a given TDA, which one has other reasons to assert (e.g., it is implied by the religious tradition with which she is associated), is tested by science, history or philosophy. "Tested" means that something it says, or something that is entailed by it, faces a real possibility of being rationally counterindicated—"falsified" in this specific sense—by results in these fields.
- Traction_3: a given TDA is shown to be consistent with science (or historical criticism, or philosophy) as we know it.

Traction may not be an absolute value. But it seems to me that theologians should value traction if one or more of the following conditions holds, and that there is an inconsistency in accepting the conditions without valuing traction:

- If they view the rational credibility of Christian theology as under challenge today, and especially if they believe that theology as a discipline in the university is in crisis.

Brazos Press, 2001). The Intelligent Design argument is clearly summarized by Del Ratzsch, *Nature, Design and Science: The Status of Design in Natural Science* (Albany, N.Y.: SUNY Press, 2001), and a (partisan) history is given in Thomas Woodward, *Doubts about Darwin: A History of Intelligent Design* (Grand Rapids, Mich.: Baker Books, 2003). For a related argument concerning evolutionary biology see Michael Behe, *Darwin's Black Box: The Biochemical Challenge to Evolution* (New York: Free Press, 1996).

- If they view theology as at least partly overlapping with metaphysics, and they believe that metaphysical reflection is in crisis.
- If they accept a model of "public theology" as opposed to theology as the private discourse of faith.
- If they see theology as allied with metaphysics in one way or another. For theology's loss of rational credibility has to do with the fact that it is often seen as "tennis without a net." That is, nothing really controls whether a theological claim is likely actually to be true. Yet rational criticizability is a core value in metaphysical argumentation (even if it is not always attained).

Since I think that most, if not all, of these conditions hold, I place a high value on traction in the theology-science debate.

3.2 Traction and the Burden of Proof

I have argued that specific theological statements can have traction. But can Christian theism as such be given traction? It depends on which of the three tractions one has in mind. Although I cannot make the case here, I assume that traditional Christian theism does not have $traction_1$; that is, I assume the failure of a natural theology for traditional Christian theism.

But what about $traction_2$? Could Christian theism be formulated in such a way that it gives rise to a TDA that, for its part, achieves traction in this sense? (Recall that a TDA with $traction_2$ would assert or entail at least some propositions which can be rationally counterindicated ("falsified") by results in science, history, or philosophical criticism.) It seems to me that it could. That is, there are some TDAs derived from Christian theism that, if they are successful, would have $traction_2$. Consider these four:

1. The traditional belief in Christian miracles.[17] The claim that some event is a miracle in this sense would be falsified by evidence that the objects in question acted according to their natural powers (read: in accord with natural law as we know it).

2. Some claims for the historicity of the resurrection of Jesus. One thinks of the sort of positive apologetics found in Gary Habermas's *Did Jesus Really Rise from the Dead?* or in Wolfhart Pannenberg's *Jesus—God and Man.*[18]

3. Some positions in the "Intelligent Design" (ID) movement. We have already considered the work of Michael Dembski, for example, and I cited Michael Behe's *Darwin's Black Box.*

---

[17] As defined by Thomas Aquinas in *Summa Contra Gentiles* III, 100-3.

[18] Habermas and Flew, *Did Jesus Rise from the Dead?*; Pannenberg, *Jesus—God and Man.*

4. The proposal that non-interventionist objectively special divine action occurs at the level of quantum indeterminacy (as proposed in the VO/CTNS series by Russell, Ellis, Murphy and Tracy). This proposal would be falsified if, for example, the criticisms raised by Nicholas Saunders turn out to be valid. Wesley Wildman has nicely spelled out the conditions under which this claim would be fully testable.[19]

This is not the place to debate the merits of these four proposals. Here I have sought merely to show that traction is possible and that it is of value (given certain conditions, which I believe hold).

What if I am wrong and *no* TDA has $traction_{1 \text{ or } 2}$? Would it be enough if one's TDA had $traction_3$, that is, if it was consistent with science as we know it? The question is urgent, since many authors in the VO/CTNS series seem in fact to have had $traction_3$ as their only (traction-related) goal.

That some TDAs have $traction_3$ seems obvious. In fact, it is all too easy to come with examples of $traction_3$: all theories of primary and secondary causality; Wilesian theories of history as a single act of God; all theories according to which the acts of God in history are "hidden" from human knowers but revealed only to "the eyes of faith"; all acts of God as understood by Karl Barth; (1) and (2) in Tom Tracy's typology, though probably not (3); all TDAs that appeal to mystery or to the absolute disanalogy between God and humanity; and any of a wide variety of "compatibilist" theories of divine action. I thus conclude that $traction_3$ does not constitute a serious test for a TDA. $Traction_3$, I fear, amounts to "tennis without a net": no matter what shot one hits, one can find some way to call it "in." Although not completely without value, $traction_3$ is not significant enough as a goal to merit the primary attention of theologians.

One wants to know, then: where, in the end, does the burden of proof lie? For many in the world who self-classify as Christian theists, I suppose, the lack of serious traction ($traction_{1 \text{ or } 2}$) does not matter at all. Perhaps their belief in the truth of Christianity is based on direct religious experience, or on the authority of scripture or the Church. Or they do not accept the model of public theology.

But for public theologians, I suggest, *there are multiple reasons not to make the claim that God acts in the world if this claim has no* $traction_{1 \text{ or } 2}$. First, the success of science supports an epistemic presumption in favor of naturalism. Second, research in comparative religious studies, such as that stemming from the Boston-based Crosscultural Comparative Religious Ideas Project, uncovers a large number of religious worldviews that are viable in the sense of meeting the (low) standard of $traction_3$.[20] Isn't a suspension of belief more rationally justified in the face of this diversity of great religious-metaphysical traditions? Third, one's own preferences for Christian theism can

[19] Wesley Wildman, "The Divine Action Project, 1988-2003," *Theology and Science*, Vol. 2, No.1 (2004): 31-76. Wildman's own belief is that a TDA that met all his conditions could be constructed but would be incompatible with quantum physics.

[20] See Robert Neville, ed., *The Human Condition* (Albany, N.Y.: SUNY Press, 2001); idem, *Ultimate Realities* (Albany, N.Y.: SUNY Press, 2001); idem, *Religious Truth* (Albany, N.Y.: SUNY Press, 2001).

be sufficiently accounted for by one's own cultural context. Fourth, if one has to decide based on metaphysical arguments alone, I suggest that powerful metaphysical arguments in both Western and Eastern traditions support the notion of "God" as the ground of being or Ultimate Reality rather than as "a" person. Fifth, other metaphysical positions, such as Spinozism, are more coherent with quantum physics[21] and possibly with the neurosciences[22] than is Christian theism. Finally, process metaphysics offers a much more systematic metaphysical framework for interpreting "experience as a whole" (including science) than do most versions of Christian theology.[23]

Thus I conclude that every effort should be made to develop a TDA with $\text{traction}_{1 \text{ or } 2}$ since, if this project fails, the grounds may be insufficient to justify adhering to a Christian-theist TDA. In the absence of $\text{traction}_{1 \text{ or } 2}$, one lacks sufficient reason to sustain the belief that there is a God who carries out special divine actions in the world.

4 The Core Problem: The Relationship between Science, Metaphysics and Theology

We have seen that there are deep disagreements among the authors in the series. Unfortunately, the disagreements include topics on which there would have to be agreement, at least among a majority of participants, if the series is to be viewed as the expression of a single research program. Instead, the evidence suggests that the VO/CTNS series in fact implicitly contains a rather wide variety of research programs, each one of which, if pursued, would lead to a different sort of debate and different conclusions. Each author's essay implicitly emphasizes the themes and presents the arguments that would advance her particular research program on the topic.

In light of these divergencies, the chances of achieving consensus on the core question of the relationship between science, metaphysics and theology are probably vanishingly small. Yet *not* to attempt to formulate a shared position on this topic is to risk futility and guarantee equivocation. In hopes of reaching some conclusions on this topic, I approach the debate through a dialogue with one of the other participants in this conference, Owen Thomas.

4.1 Theology and Metaphysics

Theology is not identical to metaphysics, but it always presupposes a metaphysical position. Every theology that makes cognitive claims about its object is *ipso facto* metaphysical. Yet not every metaphysical statement is theological.

What then is the "something more" of theology? I have defined Christian theology as "level-two discourse concerning level-one beliefs, attitudes, and

---

[21] See the work of Bernard d'Espagnat.

[22] This is the claim made by Jonathan Bennett in his Spinoza commentary. The claim has been echoed recently by Antonio Damasio, *Looking for Spinoza: Joy, Sorrow, and the Feeling Brain* (Orlando, Fla.: Harcourt Inc., 2003).

[23] See David Ray Griffin, *Reenchantment Without Supernaturalism: A Process Philosophy of Religion* (Ithaca, N.Y.: Cornell University Press, 2001).

practices of the Christian community."[24] This definition and the discussion in which it occurs was developed against the backdrop of Pannenberg's definition of theology as "the science of God,"[25] although it diverges from Pannenberg's position in important respects. Note also Ian Barbour's definition of theology:

> Theology is the systematic and self-critical reflection of a paradigm community concerning its beliefs. The theologian traces the ways in which the memory of historical exemplars has shaped the life and thought of the community. He explores the relationships among its central models and doctrines and the implications of its view of nature, man and God.[26]

Owen Thomas cites Peter van Inwagen's definition of metaphysics as "the study of ultimate reality." The central metaphysical questions are "What are the most general features of the world and what sorts of things does it contain? ...Why does [such] a world exist? ... What is *our* place in the world?"[27] Although it might seem obvious, one should add that metaphysicians accept the obligation to formulate their positions in words and to provide arguments for the views they advance—commitments that poets and mystics need not make.

In the previous definitions I do not simply equate theology and metaphysics, as if every theological statement belongs within metaphysics. In fact, theologians sometimes stand closer to poets and mystics, rejecting the demand to formulate positions and provide arguments. I see no point in legislating that this cannot be done. Nonetheless, the theologian who engages in an academic debate on "scientific perspectives on divine action" of the sort represented by these five volumes must needs find herself drawn into the domains of theology that overlap more fully with metaphysics. Indeed, there is good reason to accept the demands of metaphysical argumentation: the VO/CTNS volumes mean to engage the natural sciences, not to dictate to them. The only way to be successful at this task is to eschew appeals to authority, whether scriptural or ecclesial, or to what can be known by faith alone. Some theologians may chafe under the restrictions. Still, for purposes of this series and its goals, I suggest, the theologian should accept the methodological requirements of metaphysics.

It follows that every theological position in this debate will be an instance of a metaphysical position; everything that holds of the relation of metaphysics to natural science will also hold of the relationship of theology to natural science.

---

[24] Clayton, *Explanation from Physics to Theology* (New Haven, Conn.: Yale University Press, 1989), 167. Because of important differences across the religions, one cannot just substitute "Jewish" or "Buddhist" for "Christian" in this definition and get the same result.

[25] Wolfhart Pannenberg, *Theology and the Philosophy of Science* (Philadelphia: Westminster Press, 1977), 297ff.

[26] Ian Barbour, *Myths, Models and Paradigms: A Comparative Study in Science and Religion* (New York: Harper & Row, 1974), 176.

[27] Peter van Inwagen, *Metaphysics* (Boulder, Colo.: Westview Press, 1993), 4.

4.2 Science and Metaphysics

In an unpublished essay, Owen Thomas emphasizes the importance of verification and falsification in distinguishing between science and metaphysics. It is intrinsic to a scientific theory to be testable, whereas metaphysical theories—which generally address "experience as a whole"—are often too general to be falsified by any specific test. As the physicist Dan Black pointed out in a recent conference in Minneapolis, it is standard to find groups of 50 or more scientists working together on a single research project, whereas it is a rare thing to find even two metaphysicians collaborating on a single project!

Nonetheless, this epistemic difference must not be allowed to expand into grounds for asserting the in-principle incommensurability of science and metaphysics. Many if not most authors in the VO/CTNS series implicitly—that is, as judged by their actual method of proceeding—resist the conclusion "that no proper scientific theory can support or undercut any proper metaphysical theory and vice versa."[28] Contrast this view with the passage from Polkinghorne quoted twice by Thomas: Metaphysics must be "consonant with its physical base," even though metaphysics is no more "*determined* by science than the foundations of a house completely determine the character of the building. *In each case there is constraint but not entailment.*"[29] One cannot help but notice the parallel with genetics and human behavior: genes clearly constrain human behavior and possibilities, providing dispositional tendencies for the organism; but they do not entail or determine all the behaviors of the organism over its lifespan.

Why would one wish for a closer partnership between science and metaphysics, and hence between science and theology, than Thomas allows? The reasons are legion. Such a partnership will reduce the appearance that metaphysics is akin to playing tennis without a net, that nothing really constrains its conclusions. It will attenuate the abstractness of metaphysical arguments if they are required to touch down in the real world of scientific research more often. Doing so, in addition to affecting the style of metaphysical debate, will also avoid decades-long (if not centuries-long!) periods during which metaphysicians have developed systems that stood in conflict with already-established scientific knowledge of the world. In the case of theology, connections with science help to combat anti-intellectualism and reduce appeals to authority and tendencies toward fundamentalism. As an extrinsic benefit, a scientifically based theology will win a greater hearing in intellectual circles around the world today.

Can one prove in advance that this goal of a closer partnership is actually achievable—that, for example, metaphysical theories can actually be responsive to science? No, the proof can only come at the end of the process of seeking to understand their interrelationships. Only as one binds the two fields

---

[28] Owen Thomas, "Metaphysics and Natural Science," Unpublished paper delivered at the Vatican Observatory /CTNS Project Capstone Conference (Castel Gandolfo, Italy: Sept 8-12, 2003), 3.

[29] John Polkinghorne, "The Metaphysics of Divine Action," in *CC*, 147, emphasis added.

more closely through detailed, sustained work on a narrower set of topics over an extended period of time will one know the extent of the mutual constraint. The VO/CTNS series, with its sustained attention to the question of divine action in light of scientific results, has done more than any single series to chart this territory. Yet without agreement on this question it will be difficult to make any further progress.

Perhaps it is easiest to state the two errors that authors most need to avoid. On the one hand, one should resist the tendency to harden the real epistemic differences between science and metaphysics into incommensurability, such that metaphysics is not affected by scientific developments and science remains unaffected by one's metaphysical position. On the other hand, one should avoid an equation of science and metaphysics, such as that proposed by Charles Birch: "I do not see the various discourses of science, religion, and philosophy as clearly distinct. They flow into one another."[30] Birch uses this blurring of the line between them—for which, I fear, his own essay is in part responsible—to suggest that "from its beginning evolution is the evolution of subjects."[31] As Owen Thomas points out, fusing the scientific and metaphysical perspectives as Birch does raises the question "whether the biological data are most successfully interpreted on the basis of materialism or some other metaphysical view."[32] He rightly wonders, "Now is this a scientific or a metaphysical question?"

The answer, then, is "both!" One can acknowledge that biological data require interpretation in two different ways: on the basis of the naturalistic assumptions that are intrinsic to all empirical science, and in light of a specifically metaphysical view. The neo-Darwinian synthesis that underlies contemporary biology gives the best available explanation of physical structures and biological functions. But the process metaphysics that Birch defends gives a better answer to the question, "What metaphysical postulates might help to explain the appearance of purpose in the evolution of mind, rationality and moral awareness?" Moreover, the strengths of these two different approaches emerge most strongly if they are developed, for a time anyway, in relative independence, with neither being subsumed into the other.

4.3 Scientific Metaphysics

Eventually, however, one begins to become dissatisfied by the two isolated sets of theories with their different functions and criteria. Eventually but inevitably, the mind asks, which is *ultimately* the better explanation? If the two domains stand separately as incommensurable discussions—Wittgensteinian language games in which "never the twain shall meet"—then no justified answer can be given concerning their interrelationships. Only if there is a *tertium quid*, a type of rational discourse that draws on both fields, is there any hope of making comparative evaluations. Indeed, without this sort of comparative evaluation there is no chance of achieving the "creative mutual

[30] Charles Birch, "Neo-Darwinism, Self-Organization, and Divine Action in Evolution," in *EMB*, 226.

[31] Ibid., 236.

[32] Owen Thomas, "Metaphysics and Natural Science," 9.

interaction" of science and theology that lies at the core of Robert Russell's vision.

I borrow the name for this *tertium quid* from the same article by Riffert and Cobb that Thomas cites,[33] "scientific metaphysics." Riffert and Cobb show how Whitehead's system was motivated by the desire to overcome the rift that had developed between science and metaphysics by the beginning of the 20th century. The Whiteheadian vision, as they present it, is a powerful corrective to the Great Divorce that still characterizes much of metaphysics and even more of theology. At the same time, it's not clear that these two authors do adequate justice to the severity of the difficulties standing in the way of scientific metaphysics today. When one takes these difficulties seriously, one if forced to acknowledge a break—at present, though not necessarily in principle—between science and metaphysics. Given the success of the empirical sciences as independent disciplines, it is no longer possible to treat them as merely "specialized metaphysics," or to say, as Whitehead does, that metaphysics offers the genus "for which the special schemes of the sciences are the species."[34]

Specifically, the Whiteheadian view does not do adequate justice to the epistemic superiority of the sciences. Here, "epistemic superiority" refers to the existence of decision procedures that allow for rational decisions between competing options and that generally lead to widespread consensus within a scientific discipline. Around the world scholars agree on successes and failures in science, whereas one often has difficulty finding agreement with one's closest metaphysical allies on the basic tenets of one's (allegedly) shared position! Given the epistemic superiority of the sciences, it's not possible to justify an actual full symmetry of knowledge claims in science and metaphysics. Metaphysicians and theologians *work toward* such a symmetry, but it is not a present possession. As a result, we must interpret the conceptual unity posited by every great metaphysical system *as a regulative ideal for knowledge* in the Kantian sense. Every metaphysic postulates a connection between diverse fields of inquiry with their very diverse sets of results. The posited connection is an ideal which (if it is accurate) can regulate the pursuit of knowledge within the specific disciplines covered by the synthesis. Yet it guides scientific inquiry—when it does so at all!—not as an item of knowledge enjoying equal stature with the established theories within those disciplines, but rather as expressing the goal of the unity of knowledge that represents the final telos of any rational inquiry.

Some of the articles in the VO/CTNS volumes begin to reveal the outlines of this *tertium quid.* Take the discussion of the interpretation of quantum physics in *QM*, and consider in particular Jim Cushing's notion of "a partially interpreted formalism."[35] The phrase "*partially* interpreted formalism"

---

[33] See Franz G. Riffert and John Cobb, Jr., "Reconnecting Science and Metaphysics," in *Searching for New Contrasts: Whiteheadian Contributions to Contemporary Challenges*, Franz G. Riffert and Michel Weber, eds. (Vienna and New York: Peter Lang, forthcoming).

[34] Alfred North Whitehead, *The Function of Reason* (Boston: Beacon, 1974), 76, quoted by Owen Thomas, "Metaphysics and Natural Science," 11.

[35] Cushing, "Determinism versus Indeterminism in Quantum Mechanics," 99.

suggests a continuum from completely uninterpreted to fully interpreted. Doing quantum physics obviously requires enough interpretation of the mathematical formalism that it can be applied to observational data regarding micro-physical systems. Yet, as Cushing has convincingly shown, doing physics does not require the degree of interpretation that allows one to choose between deterministic and indeterministic theories of the world.[36] And it *certainly* does not require such a full interpretation that the metaphysical status of the observer is fully specified. When physicists debate the questions, "Can Schrödinger's cat collapse the wave function, or does it take a human being (say, Wigner's friend) to constitute an adequately robust observation of the system?" or "Would God's observation of a quantum mechanical system force an end to a quantum superposition?,"[37] they have moved into the domain of clearly metaphysical debate.

Because the various interpretive questions raised by quantum physics form a continuum, one can conceive of a point—or, better, region—between the minimal interpretation required to apply the formalism within science and the high level of interpretation involved in debates about a divine observer of the quantum world. Scientific metaphysics as I conceive it lies in this intermediate region. Only from within this region, I suggest, can one engage in rational discussion of the title topic for the VO/CTNS series, "scientific perspectives on divine action." At levels of abstraction below this region of the continuum, one can make scientific statements, but they are not really perspectives on divine action. (Many of the scientific essays in the series actually belong here.) And above this region, the perspectives on divine action under discussion are not really *scientific* perspectives at all.

It is from the perspective of scientific metaphysics, I believe, that we can best evaluate the VO/CTNS series and the work that remains to be done. The reader will recognize that some contributions to the volumes range downward to the point of intricate scientific detail, while others range upward to the most lofty claims of theology. A much smaller number of the contributions remain in the difficult region of intersection. As authors in the series, we too rarely held our noses to the grindstone at the points where the really hard work needed to be done. (It is all too easy to revert to what one knows best in his specific scientific discipline or field of theology.) It could also be that, in at least some of the VO/CTNS fields of inquiry, no real intersection point exists; given the present state of science any metaphysical work is insufficiently constrained and thus painfully speculative. Of the five volumes, this danger arises most visibly in the case of the quantum mechanics volume. Next in line, perhaps, is the neurosciences volume—not because the study of the human person leaves no place for an intersection of science and metaphysics, but because neuroscience in its present state is so poorly integrated with the human sciences that disciplined connections are difficult to draw. Cosmology and biology offer perhaps the greatest amount of intersection.

---

[36] See James Cushing, *Quantum Mechanics: Historical Contingency and the Copenhagen Hegemony* (Chicago: University of Chicago Press, 1994).

[37] See Raymond Chiao, "Quantum Nonlocalities: Experimental Evidence," in *QM*, 17-39.

In this proposal I have assumed that the line between scientific and metaphysical is not absolute. Or, put differently, I assume that the transition between them actually consists of a *number* of transitions. The discussion of quantum mechanics again provides a good example. Think of the multiple steps that are involved in the following progression: One writes down a set of quantum mechanical equations and does calculations. One interprets a specific experimental apparatus or result in light of the equations. One interprets these results as consistent or inconsistent with a broader interpretation of quantum mechanics (say, Bohm or Copenhagen). One defends Bohm or Copenhagen as "the best" interpretation of quantum mechanics. One comments on the significance of one of these interpretations for an overall understanding of quantum mechanical reality (à la Bernard d'Espagnat). One interprets these conclusions in turn for their contribution to our understanding of God and divine action.

When in this progression was a metaphysical dimension introduced? Was it only with the final step? Or did it begin already with the movement from the quantum mechanical formalism to the first interpretive statement about experimental results? An "either-or" interpretation of science and metaphysics leads to a black-and-white analysis that does not do justice to the complexity of the actual discussion. Conversely, to equate science and metaphysics obscures the gradual progression in which scientific factors come to play a smaller role and metaphysical factors a larger one. For this reason it is best to interpret the discussion as representing a continual progression along a spectrum from minimally to maximally metaphysical.

5 What Would a Successful Theory of Divine Action Have to Achieve?

Ideally, of course, one would specify criteria for distinguishing successful and unsuccessful theories in advance. Sometimes, however, one cannot formulate the standards for judging proposals in a field until she is sufficiently familiar with the actual proposals being advanced there. Consequently, an urgent task for the divine action debate is to review the major proposals that have surfaced and been debated in the five volumes. This retrospective survey should enable one to formulate criteria for judging between the various accounts of divine action. I limit the overview to eight major options, although other proposals also appear in the volumes.

5.1 Major Theories of Divine Action in the VO/CTNS Series

1. God can create changes in the physical world by resolving quantum indeterminacies according to God's will. God does this frequently but not in the case of every quantum state.

2. God is able to amplify God's quantum-level actions to the macro-physical level. This means that God is able to influence brain events, and thus the thoughts that humans form; God is able to influence classic physical systems, at least to the extent that they are the product of wave function collapse; and God is able to influence the process of evolution, which is a product of the dynamics of numerous (highly complex) physical systems. The

result of (1) and (2) is a theory of divine action that allows for God to influence outcomes in the physical world, yet without breaking physical law ("intervening") or abrogating human freedom.

3. Chaotic systems provide sufficient room for divine action. In such systems the role of God would be indetectable. Since "epistemology models ontology," and the ontology of chaotic systems does not allow for the prediction of future states by any finite agent, no future advances in science could ever "close the gap" for divine influence. In one version of this theory, chaotic systems are actually the locus for divine action in the world; in another, God timelessly causes human history as a whole, and chaotic systems are merely one of the means by which God timelessly achieves the outcomes God intends.

4. The process view of the world and of God is correct. Reality is fundamentally experiential. God is able to influence the decisions of the actual occasions which make up all reality, yet without constraining the outcome. Since what science thinks of as physical objects are actually aggregates of many individual moments of experience, and nothing in the nature of moments of experience excludes continual divine influence (unlike, say, classical physical systems that are determined by natural law and initial conditions), God is able to influence (though not determine) the development of the cosmos through time, indeed, in a manner fully consistent with natural science. Hence no science-theology tension exists.

5. I am not sure that one can find in these volumes any defense of what in section (1) above I called evangelical theism or an orthodox-theistic theory of divine action. This omission is unfortunate, since this has been the dominant view in the history of Christian theism and is the numerically dominant view among Christians outside of Europe today (including not only the United States but also most of the non-Western world).

6. Whatever may be the resolution of the science-divine action dilemma, God is at least responsible for the initial creation of the universe. This is especially clear in classical big bang theory, since physical law by definition breaks down at an initial singularity. But even if there is no initial point of time, no "t=0," the creative role of God can be preserved. The "fine-tuning" of the universe provides clear evidence of God's providential nature and creative intent. Admittedly, "multiverse" theories in cosmology seem to obviate the need for an initial divine creation. But such views are either scientifically less adequate than single universe theories (because no conceivable evidence could support them), or they are no better than creation theories, or the broader body of evidence (including the development of human

rationality, science, and the moral sense), taken as a whole, is enough to tip the scales in favor of creation.

7. Even if all natural events are explained by antecedent natural causes in an unbroken network of "secondary causes," a place for God remains, since God can still be the primary cause contributing to the outcome of each event. Thus, even if the reconciliation proposed in the previous cases is not successful, a sufficient place would still be preserved for divine action.
8. Emergentist panentheism. We return to this view below.

5.2 Introducing the Counterfactual Principle

If God is a contributing cause, or *the* cause, of some event in the world, must the divine role make a difference? To the non-initiate, this might seem like a silly question: of course it must. Assume, for the sake of argument, that at least deism is correct, and God created the world. (Clearly *this* action makes a difference: had God not carried out this creative action, there would be no world.) One can even assume general providence, that is, that God sustains the world in existence at every moment. We then ask, has God helped to cause any particular event within this world? If so, it would seem that the event in question must be in some way different than it would have been had divine action *not* been involved.

I find something compelling in this intuition. But it turns out that many of the positions on divine action in these five volumes *do not* presuppose it. That is, many authors seem to hold that the event in question would have been precisely the same in every detectable respect even if divine action had not contributed to its occurrence.

For those of us who recall the famous "university debate" in the late 1950's, there is a cost involved in the claim of divine action without detectability. Recall the famous closing words from Anthony Flew's challenge, which refer to John Wisdom's metaphor of the elusive gardener: "At last the Sceptic despairs, 'But what remains of your original assertion? Just how does what you call an invisible, intangible, eternally elusive gardener differ from an imaginary gardener or even from no gardener at all?'"[38] As the falsification debate showed, there is no outright contradiction in this position and it is not cognitively meaningless to assert it. Nonetheless, I think participants in the debate today would acknowledge that the view that divine action makes some difference, at least in principle, is more accessible to intersubjective discussion and critique than is the opposing position.

There is thus reason to prefer accounts of divine action that accept what we might call *the counterfactual principle*. Such theories accept the premise that, *had God not acted* in helping to produce some effect, or in directly bringing it about, the effect *would not have been identical* to the state of affairs that we in fact observe. The counterfactual principle may seem an obvious and

[38] See Anthony Flew, "Theology and Falsification," reprinted in *Readings in the Philosophy of Religion: An Analytic Approach*, Baruch Brody, ed. (Englewood Cliffs, N.J.: Prentice-Hall, 1974), 308.

noncontroversial principle to accept, but it does bring with it certain consequences. A difference in a physical system would be detectable, for example, if at least one particle in the system was discovered to have a different position or momentum than it would otherwise have had. In most cases, of course, the effects of divine action would be much more massive. If God contributes causally to curing Bill from AIDS, then an extremely large number of AIDS viruses in Bill's body will either have mutated so that they are no longer able to bind with sugars in his blood stream, or they will have disappeared altogether.

(Detection is rather more difficult at the level of human thoughts and actions. An action that is "out of character" for an individual need not have been divinely caused; characters are extremely complex things, and it belongs to human nature both to establish patterns and to break them. One may *sense* that a person could only have said or done something with divine assistance; indeed, persons sometimes sense this about themselves. But corroborating such divine action claims presents serious difficulties. In fact, there seems to be a continuum running from physics to the human sciences: the further one moves along the continuum, the more difficult it would become to verify that a divine action had in fact occurred.)

Assume for a moment that a divine action has occurred, such that some physical system is now different from how it would otherwise have been. This difference, it seems, is only possible through an infusion of energy into the system in question. If God has actually introduced energy in this fashion, the law of the conservation of energy will have been broken; the result will have negated the assumption of the causal closure of the physical world.

To assert that this has occurred is costly. In the VO/CTNS volumes, authors have tended to equate the abandonment of these two assumptions with approaches that they label "interventionist divine action." Since most authors have tended to eschew interventionist divine action, they have tended to insist that causal closure and the conservation of energy both be retained. I suspect, however, that most views on divine action that meet the counterfactual condition, such that divine action would make a detectable difference in the world, will entail that the world is not causally closed.

5.3 Can Information **Replace** *Energy?*

What then of the appeal to information instead of energy as the *modus* of divine action? A number of authors in this series, Polkinghorne and Peacocke prominently among them, have espoused the position that divine action involves imparting new information to a system, *rather than* energy. Since information is conceptually a different category than energy, in principle God might impart new information without breaking the causal closure of the world. In the natural world, and for that matter in culture as well, information is expressed through patterns, and patterns are detectible as structures or forms (e.g., the pattern of nucleotides in a DNA molecule or the pattern of events in history). Thus, the argument continues, the notion of information apparently offers the prospect of being able to bridge between the concrete empirical structures that science studies and God as the source and interpreter of all information.

I admit to a certain degree of skepticism about these approaches. Ian Barbour notes, for example, that "in radio transmissions, computers, and biological systems, the *communication of information* between two points requires a physical input and an expenditure of energy."[39] He then adds, "If God is omnipresent (including presence everywhere at the micro-level), no energy would be required for the communication of information."[40] But it is not clear why this would be the case. What Barbour's second sentence presupposes is an interaction between two levels: the divine level and the level of physical systems.[41] All agree that physical energy does not apply at the level of God and does apply at the level of physical systems. So which "logic" should apply in a two-level assertion of this sort? Does the divine level trump the physical level, such that divine laws only apply? Or, conversely, if effects are to be caused within the physical system, do *its* laws provide the controlling parameter? Finally, if information involves "pattern," in what does the pattern inhere if not in some form of matter and/or energy? I conclude that some additional arguments will have to be provided if the information model is to serve as the crux of one's integration of science and divine action.

6 Emergentist Panentheism and Divine Action

I close by briefly presenting the approach to the problem of divine action that I believe currently holds the greatest potential.

6.1 Where To Look for Divine Action

Emergentist panentheism begins with the assumption that language about God and divine action stand in some tension, at least initially, with the mindset and the conclusions of scientific research. In the VO/CTNS series authors have often spoken of this tension using the concept of *methodological naturalism*. This concept entails, to put it bluntly, that the last thing a scientist would or should appeal to in explaining some state of affairs in the world is the hand of God. Hence, if in the end one accepts a metaphysic other than metaphysical naturalism (say, orthodox theism), one's metaphysic will stand in some tension with the practice of science. Authors divide on whether this tension is merely superficial or whether it is fundamental, but most acknowledge that it exists.

The authors in the series who come closest to the latter position—that is, holding that the tensions between methodological naturalism and classical theism are fundamental—are probably Willem Drees and Wesley Wildman; their position is shared to a greater or lesser extent by Arthur Peacocke, John Haught, Paul Davies, and numerous other authors in the science-religion

---

[39] Ian G. Barbour, "Five Models of God and Evolution," in *EMB*, 434.

[40] Ibid.

[41] Of course, if Barbour accepts an idealist version of process metaphysics, whereby there only *seem* to be physical systems but in fact they are really only aggregates of experiencing entities ("actual occasions"), then there is no physical-mental gap. To take this position, however, is already to have rejected the conservation of energy and the closure of the physical world. (There are also standard objections to idealism that would have to be addressed.)

debate not included in these volumes, such as Terrence Deacon and Ursula Goodenough. For these authors there is an apparent incompatibility between most of the accounts of divine action summarized above and the actual practice (and results) of science. As a result, they advocate accepting religious naturalism in one or another form. For some naturalism represents the only adequate metaphysic in an age of science; for others, it is a stage or component in a metaphysic that is in the end theistic or panentheistic. Following the lead of the latter group, I wish to explore how the question of divine action is transformed if one begins with the assumptions of religious naturalism but supplements it with an emergentist theory of evolution.

Here is the argument in outline: the disciplines that are best suited to explain the evolution of systems at each particular level of evolution are the particular sciences devoted to the study of those systems. The physical world is best explained by physics—microphysically by quantum mechanics, and macrophysically by classical physics supplemented by relativity theory and the theories of the four forces; the chemical world by chemistry, the biological world by the neo-Darwinian synthesis in evolutionary biology, the brain by the neurosciences, and systems of human behavior by psychology, sociology and anthropology. Questions that do not fall within the proper domain of these disciplines, such as what caused the singularity that we call the big bang (if indeed it was a singularity), are not properly physical questions and thus are fair game for philosophers and theologians. But questions about the dynamics and evolution of physical, chemical and biological systems *do* fall within the proper domain of the special sciences. Hence the hypothesis of divine action within these spheres is misplaced.

This conclusion may well appear to the advocates of some of the positions covered above as a serious setback. Yet, I suggest, it does not have to be bad news for a theology of divine action, at least not to one that is already committed to developing its proposals in light of contemporary science. In fact, by making the assumption of naturalism at the outset, rather than looking for theological explanations that apply to the dynamics of physical systems, one may be able to develop a more compelling account of divine action in the end. (A panentheistic account of the natural world represents a crucial step in the overall argument. If the world is located within the divine, the lawlike functioning of the natural world would also be an expression of divine agency. I believe the dialectical or "both-and" logic of panentheism—events that are simultaneously expressions of natural law and divine agency—will turn out to be a necessary feature of any successful theory of science and divine action.[42]

Consider the contrast with the theory that divine action takes place at the quantum level (the Russell-Murphy-Tracy proposal). I am not convinced that Nicholas Saunders and other critics have shown the impossibility of this program;[43] it still seems conceptually possible that God would exercise this sort of influence. But it is an influence that could never be demonstrated either

[42] For more on this topic see Arthur Peacocke and Philip Clayton, eds., *In Whom We Live and Move and Have Our Being: Panentheistic Reflections on God's Presence in a Scientific World* (Grand Rapids, Mich.: Eerdmans, 2004).

[43] See Nicholas Saunders, *Divine Action and Modern Science* (New York: Cambridge University Press, 2002).

physically or theologically; hence it stands in tension with the counterfactual principle. Moreover, claims for this sort of divine influence are not likely to open the doors to warm dialogue with natural scientists. A gap for divine action that cannot be closed may represent a breakthrough for the theologian, but inserting God into this space, even if it is a permanent opening rather than a gap, may well seem like a sleight of hand to one's scientific discussion partners. In fact, there is some risk of creating the perception that one is treading with non-empirical theological boots on terrain that properly belongs to this or that specific science.

On the emergentist proposal, then, *divine action is to be located not within specific scientific disciplines, but in the interrelationships between them*. More particularly, divine action is best expressed as that meta-scientific or meta-physical position that results when one seeks to combine the various scientific disciplines together into a single, overarching picture or world view. The view that represents the best overlap set between the set of scientific conclusions and the set of theological conclusions, I suggest, is the emergentist view of the natural world. In the past we have often asked this question: What holes or gaps are left within the various natural sciences that science itself tells us could never be closed? On the present view, by contrast, one asks the question: What is the pattern that emerges when one looks at the progression from physics to chemistry to biology to zoology to neuroscience to the human sciences to...?

According to emergence theorists such as Peacocke, Ellis and the present author, what one detects at this level is a pattern of emerging complexity. This is not the place to develop the theory in detail,[44] although its basic features are probably already familiar to readers of the VO/CTNS series. The emergent features of the natural world become less and less like the energies and particles that we know through fundamental physics and that lie at the base of it all. Studying natural systems already provides good indications of the immense variety of these emerging structures and causes: epigenetic factors in cell functioning; fundamental drives in organisms; complex relations that emerge between animals and species, such as reciprocal altruism; *qualia* and self-consciousness in humans and perhaps other higher primates; and finally, that complicated set of motivations, thoughts and experiences that we refer to as the "spiritual" dimension of human existence. If there *is* some hint of divine action in cosmic evolution, if there are "intimations of transcendence" (as Peter Berger calls them) that are offered to us by the scientific study of the world, surely it is in an analysis at *this* level of generality that we will be able to discern them.

Not surprisingly, what it *means* to look for hints of divine action is altered by this move to a much higher level of generality. Evangelical theists look to particular physical systems for signs of the hand of God. If God has indeed directly healed a disease or a broken limb, for example, signs of this miracle should be directly detectable by physicians. By contrast, advocates of quantum-level divine action, who for physical reasons are unable to detect the individual divine acts directly, look to broader patterns in history for signs of divine direction and then extrapolate backwards to the quantum level as the

---

[44] I attempt to do so in *Mind and Emergence: From Quantum to Consciousness* (Oxford: Oxford University Press, 2004).

probable locus of influence. Emergence theorists also look to the broader patterns—the development of rationality, morality, purposiveness, love—for signs or intimations of divine involvement in cosmic history, but with an important difference. Imagine for a moment that we were to agree that such signs exist. In that case, I suggest, the emergentists have greater resources available to account for the results. The quantum mechanical theory requires God to bring about effects through microphysical influences, which are then "amplified" upward until they make a biological or psychological difference. By contrast, the top-down causal influence of the emergentists allows for influences that are already psychological in character (and, for some emergentists, biological as well), so that no "translation into another genus" (Aristotle) is required. From the beginning emergence theory emphasizes the patterns *between* the various disciplines, connecting them using the narrative of natural (and later, cultural) history. In this sense, the patterns that it focuses on—its hermeneutics, if you will—are meta-theoretical; they involve relationships between discrete sets of data and theory regarding cosmic evolution. This sort of analysis offers a closer match between the postulated divine actions and the manner or level at which they can be detected. Quantum theories, by contrast, postulate the divine causality at the lowest level and then have to trace its effects all the way up the ladder, so to speak, until they reach a level at which any differences can actually be discerned.

6.2 Emergence and Scientific Metaphysics

Emergence theory lies in the intersection region that I discussed earlier under the heading of scientific metaphysics (section 4.3). Specific appeals to emergence within particular sciences, such as Robert Laughlin's use of emergence in physics,[45] are not in and of themselves metaphysical claims (nor of course are they meant to be!). At the other end of the spectrum, the doctrine of "emergent evolution"—say, as advanced by C. Lloyd Morgan in the 1920's, or in the "new age" form in which it is advanced by Brian Swimme[46]—is insufficiently scientific to win a serious hearing from scientists today. Between the two, however, lies the crucial region. Here it is possible to debate a theory of emergence that proceeds under the control of data derived from specific sciences, while attempting to unify these data into a "key-category" (Owen Thomas), a root metaphor for making sense of reality as a whole. If discussions in this intervening field are to be made as rigorous as possible, one must mine downward into the scientific details, even to the point where the word emergence is no longer used. At the same time, one must speculate upward beyond the point where empirical details constrain the outcome of one's discussions. For scientific metaphysics depends on a double recognition. It recognizes that strictly scientific work ceases actually to address

---

[45] See Robert Laughlin, "Nobel Lecture: Fractional Quantization," *Reviews of Modern Physics* 71 (July 1999): 863-74; idem, "Emergent Relativity," to be published in the C. N. Yang Festschrift (Singapore: World Science, forthcoming).

[46] See Brian Swimme's cosmic evolution essay in *The Reenchantment of Science: Postmodern Proposals*, David Ray Griffin, ed. (Albany, N.Y.: SUNY Press, 1998), 47-56.

metaphysical questions directly, whereas the higher "flights of fancy" (as Kant called them) are no longer directly constrained by scientific results but are related to them only through a long (sometimes too long!) series of "inferences to the best explanation."[47]

The hope for this approach lies in the fact that the scientists can agree upon the phenomena in their various specific disciplines that I am here grouping under the label "emergence," *and* on the specific features that emergent systems have in specific domains, while disagreeing on the theological or metaphysical significance of these results. And this is as it should be. I cannot imagine that the correct scientific description of the world will ever compel a person to faith—or, for that matter, make faith impossible. The best that the theist can hope for is to develop a philosophy of science, and then a metaphysics of science, that meets two goals: first, that the description of the various scientific domains is accepted by the specialists in each field; and second, that her results are suggestive of a theological interpretation *that does not do damage to the facts in question*. The idea of God collapsing a wave function is not a part of normal physics. By contrast, the idea that it might be necessary to describe more complex systems in terms of more complex forms of causality and more complex properties clashes with no scientific discipline and no scientific results. And yet this core emergentist idea offers at least some support to the contention that the *telos* of the process of cosmic evolution is the emergence of God-like properties in the universe.[48] It is then open to the theologian to argue that the process of downward causation originates in a being who both encompasses the world as a whole and transcends it at the same time. Clearly, arguing in this fashion goes beyond anything one could say as a scientist. And yet the argument is potentially as fully grounded in the results of science (taken as a whole) as any metaphysical claim could be.[49]

---

[47] See Philip Clayton, "Inference to the Best Explanation," *Zygon* 32 (1997): 377-91.

[48] Cf. Samuel Alexander's notion of the "deification" of the world in *Space, Time, and Deity: The Gifford Lectures at Glasgow, 1916-1918*, 2 vols. (London: Macmillan, 1920).

[49] I am grateful to Steven Knapp for comments and criticisms of an earlier draft, and to the members of the "capstone conference" for invaluable discussions and criticisms that contributed substantially to the paper in its present form.

EMERGENCE, DOWNWARD CAUSATION, AND DIVINE ACTION

Nancey Murphy

1 Introduction

I shall begin this essay by mentioning what seem to me to be the most significant philosophical issues running through the Vatican Observatory/Center for Theology and the Natural Sciences (VO/CTNS) series of conferences, but then I shall narrow my focus to one issue, the conflict between reductionism, on the one hand, and anti-reductionism or emergentism on the other. The two points where this issue arises most directly are with regard to Arthur Peacocke's account of divine action in terms of top-down or whole-part causation and with regard to human nature—especially human freedom and moral responsibility. In short, this issue relates to both human and divine action.

My purpose in this paper will be to survey developments in understanding reductionism, downward causation, and emergentism in the course of the VO/CTNS project, and to set this discussion within a broader intellectual context—more precisely, to present some recent philosophical developments that I believe move beyond the discussion of these issues in the VO/CTNS series.

I shall then draw out the consequences of what I take to be the best account of downward causation for approaching the problem of divine action. I shall, regretfully, conclude that if my understanding of downward causation is correct, then Peacocke's account of divine action is not, in itself, successful, but needs supplementation with something like an account of divine action at the quantum level.

2 What Are the Philosophical Issues?

In reviewing the five volumes from this series of conferences, as well as *PPT*, I have drawn up a list of philosophical issues that have run through the series. I first mention in passing three obvious issues: the general relation(s) between theology and science; action theory; and the relation between God and the world. I also mention in passing issues germane to only one conference: the conference on biology raised the issue of teleology; the conference on the neurosciences raised the mind-body problem.

The more focused issues that run through the series pertain to the philosophy of language, epistemology, and metaphysics:

1. The nature of theological and scientific language.
2. The relations between theological and scientific methodologies.
3. The relation between science and metaphysics.
4. The intelligibility of the universe; the nature of mathematics.
5. The nature of time; temporality and eternity; infinity.
6. Chance and determinism; necessity and contingency.
7. The nature of causation; the nature of the laws of nature.
8. Reductionism versus emergentism.

The first four issues are interrelated. For example, the thesis that science and theology both aim for a critically realist account of their subject matter is both a linguistic thesis (#1) and methodological thesis (#2). The possible relations between science and metaphysics (#3) depend on how one understands the nature and epistemological status of science (#2) on the one hand and of metaphysics on the other. For example, Imre Lakatos's concept of a scientific research program understands the program as a whole to be the empirical development of a core thesis that is often metaphysical in nature. The nature of mathematics and its relation to the world (#4) is also an epistemological issue.

The next four issues are all within the province of metaphysics: time, causation, determinism. The topic of the nature of the laws of nature is central to this constellation of issues. It also serves as a link to the epistemological question of the intelligibility of the universe. I suggest that the laws of nature question is a topic that would merit further and more focused investigation, especially given its centrality in the origination of the problem of divine action. I am not in a position to pursue this issue here. Instead I shall focus on the role of reductionism and its counter-theses in these volumes.

3 Reductionism versus Emergentism

An important element of the worldview of this particular intellectual community—namely, scholars from the liberal side of Christian theology engaged in dialogue with science—is a conception of the created order as comprising a hierarchy of increasingly complex systems, more or less represented by the model of the hierarchy of the sciences. This hierarchical view of reality and of knowledge has been a prominent feature of the modern world, going all the way back to Thomas Hobbes at the beginning of the seventeenth century.

However, while most others have either assumed or pursued the reduction of higher levels to lower, the theology and science community is one of a few movements advocating a nonreductive view. My own introduction to the concept of the hierarchy of the sciences came in my undergraduate days, but it was not until I read the works of Ian Barbour and, especially, Arthur Peacocke in the 1980s that I encountered a nonreductive understanding. This hierarchical understanding of reality, along with the rejection of reductionism, is an important conceptual innovation for a variety of purposes. For example, philosophers of mind who attempt to understand the differences between humans and inanimate objects without appreciating intervening levels of complexity are severely handicapped.

In this section I shall first provide a brief history of the development of reductive and nonreductive views of the hierarchy of the sciences, and then a brief overview of contributions in this area by participants in the VO/CTNS project. Then I shall situate the VO/CTNS conversation within other current developments in this area. In my assessment, two of the most valuable contributions are Robert Van Gulick's account of top-down causation and Terrence Deacon's account of emergence.

3.1 History

When James McClendon and I were working together on criteria for distinguishing modern from postmodern thinking in terms of the philosophical assumptions employed we argued that individualism was characteristically modern.[1] McClendon called individualism not only a theory of the relation of the individual to society but also a *metaphysical* thesis. I did not at that time understand what he meant, but have gradually come to see individualism in the human sphere as but an instance of the pervasive metaphysical thesis of atomism.[2]

The Copernican revolution is seen as a major revision of Western thought because of the displacement of humans from the center of the universe. However, its truly radical element was the resulting change in conceptions of matter, from Aristotle's hylomorphism to something like Democritus's atomism. On this account, commonsense entities came to be seen as ontologically secondary when contrasted with the primary ontological status of the atoms.[3] If we combine the assumptions of atomism with the hierarchical conception of reality and Pierre Simon de Laplace's determinism we seem committed to the following argument:

1. All entities are (nothing but) arrangements of atoms;
2. Atoms have ontological priority over the entities they compose; and
3. The laws of nature are deterministic.

Therefore:

4. The behavior of complex entities is determined by the behavior of their parts; or perhaps equivalently,
5. The laws of physics determine the behavior of all complex entities.

The first significant counter-move to the reductionist ideology of modern science and philosophy was the emergentist movement in the early twentieth century.[4] This was an attempt in philosophy of biology to provide an alternative to both mechanism and vitalism. The best-known emergentists were British (Samuel Alexander, C. Lloyd Morgan, and C. D. Broad) but there were also substantial contributions from American philosophers such as Roy Wood Sellars, A. O. Lovejoy, and Stephen C. Pepper. Jaegwon Kim attributes the lack of influence of this school to the anti-metaphysical movements of mid-

---

[1] Nancey Murphy and James W. McClendon, Jr., "Distinguishing Modern and Postmodern Theologies," *Modern Theology* 5 (April 1989): 191-214.

[2] See Nancey Murphy, *Anglo-American Postmodernity: Philosophical Perspectives on Science, Religion, and Ethics* (Boulder, Colo.: Westview Press, 1997), chap. 1.

[3] Edward Pols, *Mind Regained* (Ithaca, N.Y.: Cornell University Press, 1998), 64.

[4] That is, the first attempt to defeat reductionism per se; there were, of course, much broader anti-mechanist movements such as Romanticism.

twentieth-century philosophy.[5] I believe that a more common account involves the lack of clarity in definitions of "emergence" and the sense that something "spooky" and anti-scientific was being proposed.

3.2 Developments in the Theology-and-Science Discussion

It was from the writings of Peacocke and Barbour that I became aware of the anti-reductionist movements in recent history, and I am guessing that their writings have served the same purpose for most of the rest of the theology and science community. Peacocke has made the most extensive use of the concept of the hierarchy of the sciences and was the first (to my knowledge) to propose that theology be considered the top-most science in the hierarchy,[6] although there is some precedent for this idea in Barbour's *Issues in Science and Religion*. Barbour claimed that "an interpretation of levels can contribute to a *view of man* which takes both the scientific and the biblical understanding into account." He thus implies that the religious perspective is an indispensable level of description of human life.[7]

It is interesting that one of the early emergentists, Roy Wood Sellars, had included the religious or spiritual as one of the levels of emergent reality—above the inorganic, the organic, the conscious or mental, and the moral. Sellars's account is not truly a precedent for Peacocke's for two reasons. First, in spite of his championing a nonreductive account of the levels studied by the natural sciences, he ultimately reduced religion to human values.[8] Second, of course, Sellars is talking about levels of emergent reality rather than the hierarchy of the *sciences* and has nothing to say (so far as I know) about the discipline of theology.

In addition to a more general account of anti-reductionist thinking, Peacocke is also to be thanked for introducing many of us to the concept of downward causation, as found in the writings of Roger Sperry and Donald Campbell. I believe that it was clear from the beginning of the VO/CTNS project that Peacocke's account of divine action via downward causation from the whole (God in relation to the world) to the part (the world) was one of only a few major theories of divine action worked out from within the sphere of science. Thus, Robert Russell could write in his "Introduction" to *CC* that within the theology and science community there were four approaches to divine action: those employing downward causation, bottom-up causation via the quantum realm, chaos theory, or the traditional distinction between primary and secondary causation. Given that entire conferences focused on chaos theory and quantum theory, it is odd, looking back, that there was no

---

[5] Jaegwon Kim, "Being Realistic About Emergence," in *The Re-Emergence of Emergence*, Philip Clayton and Paul Davies, eds. (Oxford: Oxford University Press, 2006), 190.

[6] See Arthur Peacocke, *Creation and the World of Science* (Oxford: Clarendon Press, 1979), 369.

[7] Ian Barbour, *Issues in Science and Religion* (San Francisco: Harper and Row, 1966), 360.

[8] Roy Wood Sellars, *Principles of Emergent Realism: The Philosophical Essays of Roy Wood Sellars*, W. Preston Warren, ed. (St. Louis, Mo.: Warren H. Green, 1970).

conference devoted specifically to Peacocke's proposal. Nonetheless, several authors have attempted to contribute to an understanding of anti-reductionism, emergentism, and downward causation.

In the remainder of this section I shall briefly survey Peacocke's work on nonreduction and downward causation in the papers he contributed to this project. Next I shall consider Philip Clayton's contribution on the topics of nonreduction and emergence, and then offer an assessment of both authors.

Peacocke's account of divine action via downward causation is nicely summarized in this volume. His earlier work on this topic is well represented by essays in two previous volumes: "God's Interaction with the World: The Implications of Deterministic 'Chaos' and Interconnected and Interdependent Complexity" in *CC*, and "The Sound of Sheer Silence: How Does God Communicate with Humanity?" in *NP*. In "God's Interaction" he refers to Campbell's and Sperry's concepts of top-down or downward causation, but notes that he prefers the terminology of whole-part constraint. His central example is that of the Bénard phenomenon. Peacocke says:

> In such instances, [reference omitted] the changes at the micro-level, that of the constituent units, are what they are because of their incorporation into the system as a whole, which is exerting specific constraints on its units, making them behave otherwise than they would in isolation. Using "boundary conditions" language, [reference omitted] one could say that the sort of relations between the constituent units in the complex whole is a *new* set of boundary conditions for those units.[9]

In this case, it appears that the burden of explaining what is happening in instances of whole-part constraint, as well as explaining why it does not abrogate lower-level laws,[10] falls on the concept of *boundary conditions*.

Peacocke's argument for the nonreducibility of relevant features of the whole depends on his critical realist thesis. If higher-level laws and theories are needed to account for the interaction of the system as a whole with its parts, then the higher-level entities to which the terms of those theories and laws refer ought to "be deemed putatively to exist at the various levels being studied."[11]

In "The Sound of Sheer Silence," Peacocke addresses reductionism and whole-part constraint at greater length. He makes an argument for anti-reductionism comparable to that in "God's Interaction," but one that is more nuanced.

> When the nonreducibility of properties, concepts, and explanation applicable to higher levels of complexity is well established, their employment in scientific discourse can often, *but not in all cases,* lead to a putative and then to an increasingly confident attribution of a *causal efficacy* to the complex wholes which does not apply to the separated, constituent parts...[12]

---

[9] Arthur Peacocke, "God's Interaction with the World: The Implications of Deterministic 'Chaos' and of Interconnected and Interdependent Complexity," in *CC*, 273.

[10] Ibid., cf. 283.

[11] Ibid., 273.

[12] Arthur Peacocke, "The Sound of Sheer Silence: How Does God Communicate with Humanity?," in *NP*, 219.

Peacocke refers again to Sperry's and Campbell's accounts of downward causation and notes that "there are imprecisions and a lack of generalizability in Campbell's example" of the evolutionary development of the termite jaw structure.[13] He refers again to the Bénard phenomenon as a better example, and affirms the value of the concept of whole-part causation as more general than Campbell's downward causation.

Peacocke points to the need for a concept of causality richer than that of a simple chain of events.[14] In addition to his own concept of whole-part causation he lists a variety of partially overlapping analyses of causation: One is Fred Dretske's contrast between structuring and triggering causes. He describes Karl Popper's concept of "propensities" as being the effects of Dretske's structuring causes "in the case that triggering causes are random in their operation."[15] Under the heading of "boundary conditions" he refers to Bernd-Olaf Küppers's essay in *CC*, to Michael Polanyi's use of the term to describe the design of machines, and to Wim Drees's emphasis on actual physical boundaries of systems. Two further resources he notes are the concepts of *supervenience* and *information.*

Clayton's contribution to the issue of reductionism is to be found in *NP*.[16] Here he develops a concept of *supervenience,* noting that its use in philosophy of mind is intended to grant the dependence of mental phenomena on physical phenomena while at the same time denying the reducibility of the mental to the physical.[17] He defines "weak supervenience" as postulating that physical structures and causes determine the emergence of the mental, but without fully determining the outcome of the mental life subsequent to its emergence.[18] Noting that early emergentists such as Lloyd Morgan sometimes used "supervenient" and "emergent" as stylistic variants, Clayton proceeds to attempt to clarify the concept of emergence.

One starting point for Clayton's account is Timothy O'Connor's definition of property emergence:

> Property P is an emergent property of a (mereologically-complex) object O if and only if:
>
> 1. P supervenes on the parts of O;
> 2. P is not had by any of the object's parts;
> 3. P is distinct from any structural property of O… [and]
> 4. P has direct ("downward") determinative influence on the pattern of behavior involving O's parts.[19]

---

[13] Ibid.

[14] Ibid., 220.

[15] Ibid., 223.

[16] Philip Clayton, "Neuroscience, the Person, and God: An Emergentist Account," in *NP*, 181-214.

[17] Ibid., 199.

[18] Ibid., 200.

[19] Ibid., 201; referring to Timothy O'Connor, "Emergent Properties," *American Philosophical Quarterly* (1994): 97f.

So here the crucial factor is the downward "determinative influence" (note the vagueness of this term) of the supervenient property. Clayton then moves immediately to Ansgar Beckermann's definition of "emergence" in terms of the non-deducibility of the emergent property:

> F is an emergent property of S if and only if (a) there is a law to the effect that all systems with this micro-structure have F; but (b) F cannot, even in theory, be deduced from the most complete knowledge of the basic properties of the components $C_1,\ldots C_n$ of the system.[20]

Clayton also quotes Godehard Brüntrup's account in terms of the unpredictability and the absence of "nomological necessity" of the emergent.[21]

Clayton's own conclusion, reflecting on these three sources, is that a property should be designated as emergent "only if laws cannot be formulated at the lower level that predict its occurrence *and* subsequent behavior."[22] However, at the end of the section, Clayton returns to the issue of causation. Speaking of mental phenomena, such as the experience of being in love, he says "such phenomena exercise a type of causal influence of their own."[23] He draws support for the supposition of the causal efficacy of the mental from several sources. One is by analogy: bottom-up explanation fails in the case of functional explanations in biology, and must be supplemented by means of top-down explanations involving the environment. In addition, it seems to be necessary to invoke intentions to explain human behavior.[24] So here we see an explicit relation between the concepts of emergence and downward causation.

3.3 Critiques

In *NP*, Clayton and I disagreed on the best terminology for describing an account of human nature that is neither dualist nor reductionist. I objected that his "emergentist monism" was vague because of the lack of consensus in the philosophical tradition on the meaning of "emergence." I also noted that use of the term still raised the hackles of some who were influenced by the vitalist-emergentist-mechanist debates in the philosophy of biology. I preferred "nonreductive physicalism" because "physicalism" specifies the nature of the one sort of substance, whereas "monism" does not. In addition, "nonreductive" is by itself ambiguous, in that there are a variety of reductionist theses. However, much has been done to disentangle and define these interrelated

---

[20] Ibid.; referring to Ansgar Beckermann, "Supervenience, Emergence, and Reduction," in *Emergence or Reduction? Essays on the Prospects of Nonreductive Physicalism*, A Beckermann, H. Flohr, and J. Kim, eds. (New York: W. de Gruyter, 1992), 94-118, 104.

[21] Ibid.; referring to Brüntrup, "The Causal Efficacy of Emergent Mental Properties," *Erkenntnis* 48 (1998): 133-145, 140.

[22] Ibid.

[23] Ibid., 202.

[24] Ibid., 205.

theses (largely within the theology and science community), and so it is possible to specify exactly what sort of reductionism is being denied.[25]

There are at least six related but distinguishable reductionist theses:

1. Methodological reductionism: a research strategy of analyzing the thing to be studied into its parts.
2. Epistemological reductionism: the view that laws or theories pertaining to the higher levels of the hierarchy of the sciences can (and should) be shown to follow from lower-level laws, and ultimately from the laws of physics.
3. Logical or definitional reductionism: the view that words and sentences referring to one type of entity can be translated without residue into language about another type of entity.[26]
4. Causal reductionism: the view that the behavior of the parts of a system (ultimately, the parts studied by subatomic physics) is determinative of the behavior of all higher-level entities. Thus, this is the thesis that all causation in the hierarchy is "bottom-up."
5. Ontological reductionism: the view that higher-level entities are nothing but the sum of their parts. However, this thesis is ambiguous; we need names here for two distinct positions:

   5a. One is the view that as one goes up the hierarchy of levels, no new kinds of metaphysical 'ingredients' need to be added to produce higher-level entities from lower. No "vital force" or "entelechy" must be added to get living beings from non-living materials; no immaterial mind or soul needed to get consciousness; no Zeitgeist to form individuals into a society.

   5b. A much stronger thesis is that only the entities at the lowest level are really real; higher-level entities—molecules, cells, organisms—are only composite structures made of atoms. This is the assumption, mentioned above, that the atoms have ontological priority over the things they constitute. I shall designate this position "atomist reductionism" to distinguish it from 5a, for which I shall retain the designation of "ontological reductionism." It is possible to hold a physicalist ontology without subscribing to atomist reductionism. Thus, one might want to say that higher-level entities are real—as real as the entities that compose them—and at the same time reject all sorts of vitalism and dualism.[27]

---

[25] Nancey Murphy, "Supervenience and the Downward Efficacy of the Mental: A Nonreductive Physicalist Account of Human Action," in *NP*, 147, no. 1.

[26] See John R. Searle, *The Rediscovery of the Mind* (Cambridge, Mass.: MIT Press, 1992), 112-14.

[27] Francisco J. Ayala distinguished among methodological, epistemological, and ontological reduction in his "Introduction," in *Studies in the Philosophy of Biology:*

Causal reductionism is the thesis that raises the most significant issues. While methodological reductionism is a useful strategy for scientific research, it is now widely recognized that in practice it needs to be complemented by systems approaches. Both logical and epistemological reduction are now seen to be impossible in many cases even if causal and atomist reductionism are true. Ontological reductionism is entirely unobjectionable. Atomist reductionism expresses more of an attitude than a philosophical thesis: it is difficult to state it without employing, as I have done, the nonsense phrase "really real," and so it is not clear what it could mean to refute it. In short, even if causal reductionism is true, there are a variety of reasons why methodological, epistemological, and logical reduction might still fail. Thus, showing that these other forms of reductionism fail tells us about the limitations of our knowledge, but not about what really makes things happen.

So I conclude that causal reductionism is the most significant issue. If causal reductionism is the thesis that all causation is ultimately bottom-up, then the appropriate counter-thesis must be that there is also something like downward or top-down causation. This brings us back to Peacocke's contributions.

3.3.1 Peacocke

Given the above, I conclude that Peacocke is exactly right to focus on the topic of downward causation. However, I have two objections to his work. The first is to his argument, via critical realism, for the reality of higher-level entities on the basis of our need for higher-level concepts in scientific explanations. The problem here is that we know of cases where the higher-level concepts are needed even when we assume complete bottom-up causal determinism. This is, in fact, what Peacocke is exploring in the appendix to "The Sound of Sheer Silence" in his discussion of theory- and process-autonomy,[28] and what leads him to qualify his argument from epistemological and definitional nonreductionism to causal nonreductionism. I would go further and reject such arguments wholesale: If a higher-level phenomenon is *not* causally reducible, it will not be epistemologically or definitionally reducible either. But if it is not epistemologically or definitionally reducible, this is not a reliable indicator of its casual nonreducibility. Often epistemological and definitional reducibility fail instead because higher-level concepts are wildly multiply realizable.[29]

My second criticism is more tentative. Peacocke has done an excellent job in surveying and analyzing earlier literature on downward causation and has offered a variety of examples in which downward causation must be occurring. My frustration has always been that I have not enough knowledge of the

---

Reduction and Related Problems, Ayala and Theodosius Dobzhansky, eds. (Berkeley and Los Angeles, Calif.: University of California Press, 1974), vii-xvi. Arthur Peacocke pointed out the ambiguity in the meaning of 'ontological reduction' in his *God and the New Biology* (Glouster, Mass.: Peter Smith, 1986).

[28] Peacocke, "The Sound of Sheer Silence," 245-7.

[29] Cf. Theo C. Meyering, "Mind Matters: Physicalism and the Autonomy of the Person," in *NP*, 168-9.

examples he provides (such as the Bénard phenomenon) to be able to understand *how* the larger system (the whole) affects the behavior of the parts. Thus, I have not been able, on the basis of his writings, to answer the question of how top-down and bottom-up causation interact without causal over-determination. (Sperry sometimes spoke—unhelpfully—of the properties of the higher-level entity *overpowering* the causal forces of the component entities.[30]) I appreciate the attempt to provide a more generalizable account than Campbell's but I have come to believe that the generality of Peacocke's whole-part constraint loses just the specificity needed to understand how to reconcile bottom-up and top-down causation. I shall argue below that what Peacocke loses in his generalizing is the element of *selection* found in Campbell's account. However, I shall also affirm several of the suggestions Peacocke has made for reconciling top-down and bottom-up causation: the role of boundary conditions, structuring causes, nonlinearity, and information.

3.3.2 Clayton

My first critique of Clayton is parallel to my criticism of Peacocke. Clayton moves through three definitions of emergence in a way that suggests their equivalence: O'Connor's downward determinative influence, Beckermann's non-deducibility of the emergent property from knowledge of lower-level properties, and Brüntrup's absence of nomological necessity linking the emergent property to lower-level properties.

I want to affirm what appears to be Clayton's final position, that a property can be defined as emergent if it has downward causal efficacy, but I would argue that the detour through definitions in terms of deducibility or predictability confuses the issue, just as does the similar detour through epistemic and definitional anti-reductionism to causal anti-reductionism. My second critique of Clayton's position is that I see no resolution of the problem of how downward causation can take place without "overpowering" the lower-level.

3.3.3 Murphy

I have so far been silent about my own writings on these issues. I addressed the concepts of supervenience and reduction in "Supervenience and the Nonreducibility of Ethics to Biology" in *EMB*, as well as in "Supervenience and the Downward Efficacy of the Mental" in *NP*. In both cases I argued for the value of the concept of supervenience in countering reductionist theses, but only insofar as "supervenience" is defined differently than in the standard account. The standard definition is nomological: S properties supervene on base properties (B) if there can be no difference in S without a difference in B. (The conditional is not reversible because S properties are often multiply realizable—there may be a difference in B without a difference in S.)

My definition of supervenience reflects a minority position in the literature: Property S supervenes on property B if and only if x's instantiating

[30] Roger W. Sperry, *Science and Moral Priority: Merging Mind, Brain, and Human Values* (New York: Columbia University Press, 1983), 117.

S is in virtue of (or constituted by) its instantiating B under circumstance c. The purpose of this definition is to state explicitly the *dependence* of supervenient properties on base properties, but to allow for the fact that the same base properties may constitute different supervenient properties *under different circumstances*. I still believe that this is a concept we need in order to escape causal reductionism, but it is not alone sufficient. It needs to be incorporated into an account of downward causation, and I shall make an attempt to do so below.

4 Current Developments

I turn now to developments outside of the theology and science community. Two authors, Robert Van Gulick and Terrence Deacon, I suggest, have made significant advances in understanding downward causation and emergence.

4.1 Van Gulick on Downward Causation

I suggested above that a problem with Peacocke's account of whole-part influence is the lack of a full explanation of how the influence from whole to part is compatible with—fails to conflict with or override—the part-whole causation. Van Gulick's account is based on *selection*.

From Peacocke's and Clayton's accounts we can note a recognition of the importance of *boundary conditions*, *structuring causes*, *functional roles*, *nonlinearity*, and *information* in understanding downward causation and emergence. For a complete account I would add the concept of *feedback*, and for one that applies to downward *mental* causation, I would add *representation* and *semiosis*. Van Gulick's emphasis on selection makes it possible to tie all of these factors together.

Van Gulick makes his points about top-down causation in the context of an argument for the nonreducibility of higher-level sciences—his argument is from causal nonreducibility to epistemological nonreducibility. The causal reductionist, he says, will claim that the causal roles associated with special-science classifications are entirely derivative from the causal roles of the underlying physical constituents of the objects or events picked out by the special sciences. Van Gulick replies that although the events and objects picked out by the special sciences are indeed composites of physical constituents, the causal powers of such an object are not determined solely by the physical properties of its constituents and the laws of physics, but also by the *organization* of those constituents within the composite. And it is just such patterns of organization that are picked out by the predicates of the special sciences. Thus, Van Gulick concludes, "we can say that the causal powers of a composite object or event are determined in part by its higher-order (special science) properties and not solely by the physical properties of its constituents and the laws of physics."[31]

---

[31]Robert Van Gulick, "Who's In Charge Here? And Who's Doing All the Work?," in *Mental Causation*, John Heil and Alfred Mele, eds. (Oxford: Clarendon Press, 1995), 233-56, 251. This essay is reprinted in Nancey Murphy and William R. Stoeger, S. J.,

So far we have nothing different from Peacocke's account: we could speak of Van Gulick's "patterns of organization" in terms of Dretske's distinction between triggering and structuring causes, or by saying that physical outcomes are determined by the laws of physics together with initial and boundary conditions.

Van Gulick's contribution is his emphasis on *selection* among lower-level *causal processes*. The patterns of boundary conditions picked out by the special sciences have downward causal efficacy in that they can affect which causal powers of their constituents are activated or likely to be activated.

> A given physical constituent may have many causal powers, but only some subsets of them will be active in a given situation. The larger context (i.e. the pattern) of which it is a part may affect which of its causal powers get activated...Thus the whole is not any simple function of its parts, since the whole at least partially determines what contributions are made by its parts.[32]

Such patterns or entities, he says, are stable features of the world, often despite variations or exchanges in their underlying physical constituents; the pattern is conserved even though its constituents are not (e.g. in a hurricane or a blade of grass). Many such patterns are self-sustaining or self-reproducing in the face of perturbing physical forces that might degrade or destroy them (e.g. DNA patterns). Finally, the selective activation of the causal powers of such a pattern's parts may in many cases contribute to the maintenance and preservation of the pattern itself. Taken together, these points illustrate that

> higher-order patterns can have a degree of independence from their underlying physical realizations and can exert what might be called downward causal influences without requiring *any objectionable form* [my italics] of emergentism by which higher-order properties would alter the underlying laws of physics. Higher-order properties act by the *selective activation* of physical powers and not by their *alteration*.[33]

Campbell's example of the production of the termite's jaw structure fits Van Gulick's account of downward causation and also illustrates the roles of function, information, and feedback in selective processes in the biological realm. It is, of course, *feedback* from the environment via differential rates of reproduction that does the selecting of the optimal jaw structure. The selection of one genome over another can only take place because the genes embody *information* about how to form the jaw and because those jaws are either good or bad at fulfilling their *function* in the termite's world.

4.2 Murphy on Supervenience and Selection

Warren Brown and I have developed Van Gulick's account of downward causation by cataloguing a broader range of low-level causal ingredients and

---

eds., *Evolution and Emergence: Systems, Organisms, Persons* (Oxford: Oxford University Press, 2007), 74-87.

[32] Ibid.

[33] Ibid, 252.

by attempting to further specify the bases upon which selection of those lower-level factors typically takes place.[34] First selection *of*: Van Gulick says that it is the causal powers of lower-level constituents that are selectively activated. Brown and I include selection of the lower-level entities themselves. For example, in Campbell's account, it is the termites themselves, with their varied genetic materials, that are selected. In addition, we can further specify possible sorts of causal powers. Any causal account requires specification not only of the laws in operation but also of the initial or boundary conditions of the system. So, for example, building a machine involves designing (i.e., selecting a design for) a physical apparatus in which the relevant natural laws produce the desired outcome. One can also speak in Dretske's terms of triggering and structuring causes. For example, a gas leak is the structuring cause that makes it the case that striking a match will trigger an explosion. The point of these examples is to show that downward causation, understood as selection, is entirely compatible with the undisturbed working of the lower-level laws because the laws themselves never provide a complete causal account. We always also need to know the context within which they are operating. Note that the means by which the selection takes place varies from case to case.

Now, selection *on the basis of* what? I shall be intentionally vague here: downward causation involves selection on the basis of how those lower-level entities, processes, structures, *fit into* a higher-level system. A higher-level system is a broader, more complex system that incorporates the lower. In this sense, Peacocke is correct in saying that we are looking at *whole-part* influences. The bases upon which selection is made depend on the sort of higher-level system involved. Examples of features selected *for* include *function, information content,* and *meaning.* For example, the immune system works by producing a wide variety of antibodies. When one of the antibodies is successful in attacking a microbe, a feedback mechanism triggers the mass-production of this particular cell. The system as a whole selects variants that fulfill a specific function. I shall come back to selection on the basis of information and meaning below.

I now hope to show the relevance of the account of supervenience I developed in two of my essays for the VO/CTNS project to the foregoing account of downward causation. An account of the causal efficacy of the mental will involve selection: selection among boundary conditions and structures, and selection on the basis of function. In addition, I propose that downward causation operates in the mental sphere by selecting among brain states and processes (and their associated casual roles and powers) on the basis of the supervening properties of information content, representational power, and meaning. All such properties supervene on brain states, but involve as well *particular relations to context or circumstances.*

The simplest example I can conceive of is the development of a neural connection between two cell assemblies through the process of conditioning. If, say, a flash of light is repeatedly paired with the sound of a bell, a neural connection grows or is reinforced between the cell assemblies that are

---

[34]Nancey Murphy and Warren S. Brown, *Did My Neurons Make Me Do It?: Philosophical and Neurobiological Perspectives on Moral Responsibility and Free Will* (Oxford: Oxford University Press, 2007), chap. 3.

activated by the light and by the bell, respectively. This neural connection has the property of being a *representation* (of the relation between the sight and the sound) under the circumstances of its having been formed in response to the fact in the environment of the co-occurrence of these two stimuli. Thus, I suggest that representational or informational (and, by analogy, meaningful) properties supervene on neural properties or states or events, under appropriate circumstances regarding their relationship to the environment. Thus, representational and mental properties are *contextualized* brain states or events. Mental properties are higher-level properties because they involve the brain state in relation to a broader, more complex system: the brain in the body in its environment and usually with a longer causal history. Mental properties are supervenient in my sense of "supervenience" but not in terms of the standard account because they can vary without a change in the base property (that is, due to changed circumstances). For example, a true belief may become a false belief if the world changes. The causal relevance of supervenient mental properties is a function of the way they serve to relate the neural base property to a broader causal system. It is in virtue of their informational, representational, or semantic content that the base properties are able to play the causal role in the world that they do. And it is in virtue of the supervenient informational, representational, and semantic content that the environment is able to exert a selective (downward) effect on neural connections. The supervenient property *is* the base property's relation to the broader system, which entails its enmeshment in a more complex set of causal processes. If downward causation is, in general, a matter of selection among lower-level causal processes, then the supervenient mental property is what constitutes the criterion for that selective process.

4.3 Deacon on Emergence

I believe that Terrence Deacon's work on emergence is the most significant, and elaborates in important ways on the topic of downward causation.[35] His account, as does my Van-Gulickian account in section 4.1, serves to tie together the concepts of initial conditions and structures, function, selection, feedback, information, and nonlinearity. One aspect he adds is "amplification."

Deacon sees the existence and relative autonomy of holistic properties, and their top-down influence over the properties and dynamics of system constituents, to be both the key defining character and the most criticized claim of arguments for emergence. He introduces his own discussion of emergence by noting the role selection plays in circumventing the law of entropy. When new and more complicated structures appear in the universe it is the result of extensive sampling of a vast array of alternatives and then keeping and improving on the alternative that best serves some higher-order need. This is what allows for the spontaneous increase of fitted order and functional complexity in both human designs and in nature.

---

[35] Deacon has now written a number of articles on emergence; I am following his "Three Levels of Emergent Phenomena," in *Evolution and Emergence: Systems, Organisms, Persons*, Nancey Murphy and William R. Stoeger, S. J., eds. (Oxford: Oxford University Press, 2007), 88-110.

Deacon extends existing accounts of emergence by tracing ways in which "nature can tangle causal chains into complex knots. Emergence is about the topology of causality."[36]

> What needs explaining is how some systems come to be dominated by higher-order causal properties such that they appear to "drag along" component constituent dynamics, even though these higher-order regularities are constituted by lower-order interactions. The secret to explaining the apparently contrary causal relationships is to recognize the central role played by *amplification processes* in the pattern formation occurring in these kinds of phenomena. Wherever it occurs, amplification is accomplished by a kind of repetitive superimposition of similar forms. It can be achieved by mathematical recursion in a computation, by recycling of a signal that reinforces itself and cancels the background in signal processing circuits, or by repetitively sampling the same biased set of phenomena in statistical analyses. In each case, it is the formal or configurational regularities that serve as the basis for amplification, not merely the "stuff" that is the medium in which it is exhibited. Amplification can be a merely physical process or an informational process (the latter usually depends on the former). Its role in the analysis of emergence is in explaining how certain minor or even incidental aspects of complex phenomena can come to be the source of its dominant features.[37]

Deacon uses "amplification logic" as a guide to distinguish three levels of emergent systems. The most useful architectural feature is whether this causal architecture is recurrent or circular across levels of scale. He analyzes relationships of recursive causality in which the feedback is from features of a whole system to the very architecture of its components and how these levels interact. He asks: "What happens when the global configurational regularities of a locally bounded open physical system are in some way fed back into that system, via effects propagated through its 'environment?'"[38] The three categories of emergence Deacon describes exhibit nonrecurrent-, simple-recurrent-, and recurrent-recurrent-trans-scale architectures.

Deacon applies the terms "first-order emergence" or "supervenient emergence" to systems in which lower-order *relational* properties are the constitutive factor determining some higher-order property. An example is the turbulence of large bodies of liquid. Such properties are emergent only in the sense that they exhibit a certain autonomy from the particulars at the micro-level.

Second-order emergence occurs when there is temporal development, or symmetry breaking, in a system.

> There is a simple self-similarity to liquid properties across time and position that is further 'smoothed' by entropic processes. In contrast, there is a self-differentiating feature to living and mental processes, which both retains and undermines aspects of self-similarity. This characteristic breakdown of self-similarity or symmetry-breaking is now recognized in numerous kinds of complex phenomena, including systems far simpler than living systems. These complex emergent phenomena

---

[36] Ibid., 94.

[37] Ibid., 95.

[38] Ibid., 96.

> share this characteristic of change of ensemble properties *across time,* and are often computationally unpredictable as well. So it would be useful to distinguish first order emergence from these more complex forms of emergent phenomena in which the cumulative stochastic canceling of configurational interactions exhibited by simple entropic systems is undermined, and where this contributes to development and change of both micro- and macro-properties across time.[39]

Chaotic and self-organized systems fall into this category. In chaotic systems certain higher-order regularities become unstable and

> an unpredictability of higher-order dynamics results...[T]his unpredictability derives from the fact that the regularities at lower levels have become strongly affected by regularities emerging at higher levels of organization. This can happen when configurational features at the ensemble level drastically change the probabilities of certain whole classes of component interactions.[40]

Whereas first-order emergent systems can be adequately described without taking their history into account, second-order systems cannot be because of their sensitive dependence on initial conditions and because perturbations are likely to be amplified rather than smoothed out as the system evolves.

An example of a second-order emergent system is the formation of a snowflake. One factor is the result of micro-properties of water; the shape of water molecules makes for hexagonal micro-structural biases in the growth of ice-crystal lattices. However, different temperatures and humidities interact with water-molecule-binding regularities to generate a handful of distinct patterns of ice lattice formation. So the final shape of the snowflake could not be explained without a history of the atmospheric conditions it has encountered as it fell. Here we have a clear instance of the interaction between bottom-up and top-down causation.

Another factor is that the shape of the crystal at any point in time places limits on the range of possibilities for future growth. Thus, "[t]here is a "compound interest' effect...The global configuration of this tiny developing system plays a critical causal role in its microscopic dynamics by excluding the vast majority of possible molecular accretions and growth points..." In this sense, the "configurational properties of the system self-amplify."[41] So here we see a parallel to Van Gulick's emphasis on selection, but with Deacon's additional insight regarding amplification, which makes the system not only self-sustaining but also self-modifying.

Third-order emergent systems involve, in addition, some form of information or memory. "The result is that specific historical moments of higher-order regularity or of unique micro-causal configurations can additionally exert a cumulative influence over the entire causal future of the systems."[42] That is, "constraints derived from specific past higher-order states

[39] Ibid., 99.

[40] Ibid., 101.

[41] Ibid., 103.

[42] Ibid., 105-6.

can be repeatedly re-entered into the lower-order dynamics leading to future states, in addition to their effects mediated by second-order processes."[43] Third-order emergence exhibits a developmental or evolutionary character, and occurs when there is "both amplification of global influences and reintroduction of them redundantly across time and into different realizations of the system."[44]

Deacon states,

> The representation relationship implicit in third order emergent phenomena demands a combination of multi-scale, historical, and semiotic analyses for adequate description. This is why living and cognitive processes require us to introduce concepts such as representation, adaptation, information, and function in order to capture the logic of the most salient of emergent phenomena. It is what makes the study of living forms qualitatively different from other physical sciences. It makes no sense to ask about the function of granite, or the purpose of a galaxy. Though the atoms composing a heart muscle fiber or a neurotransmitter molecule have no function in themselves, the particular configurations of the heart and its cell types of the neurotransmitter molecule do additionally beg for some sort of teleological assessment, some function. They do something *for* something. Organisms evolve and regulate the production of multiple second order emergent phenomena with respect to some third order phenomenon. Only a third order emergent process has such an intrinsic identity.
>
> So life, even in its simplest forms, is third-order and its products can't be fully understood apart from either history or functionality concerns.[45]

Third-order emergence constitutes the origination of information, semiosis, and teleology in the world. It is the "point where physical causality acquires (or rather constitutes) significance."[46] In sum, "third-order (evolutionary) emergence contains second-order (self-organizing) emergence as a limiting case, which in turn contains first-order (supervenient) emergence as a limiting case."[47]

I conclude that Deacon's account of emergence adds a great deal of precision and specificity to previous accounts of both emergence and downward causation. He confirms earlier accounts based on the role of structures and other holistic properties, as well as Van Gulick's emphasis on selection as the means by which higher-level configurational properties manage to influence lower-level processes without interfering with them. In addition, he shows the role of amplification in promoting self-modification.[48]

---

[43] Ibid., 106.

[44] Ibid.

[45] Ibid., 106-7.

[46] Ibid., 107.

[47] Ibid.

[48] Since completing this chapter I have found an even more useful resource: Alicia Juarrero's *Dynamics in Action: Intentional Behavior as a Complex System* (Cambridge, Mass.: MIT Press, 1999). She describes downward causation in terms of the imposition of context-sensitive constraints in complex dynamical systems.

5 Downward Causation and Divine Action

I mentioned above that there have been three basic proposals for understanding divine action in terms drawn from contemporary science—those involving quantum mechanics, chaos theory, and downward causation. It would have been appropriate to devote an entire conference to the issues Peacocke has raised, not only because of the prominence of his work but also because the topic of downward causation is so important for other purposes of theological interest, such as explaining human freedom and moral responsibility. Consequently I have chosen to focus my evaluation of the divine action project on Peacocke's work.

My question here will be to ask whether my account of downward causation in terms of *selection* among lower-level causal processes makes any difference in understanding divine action. I shall conclude, regretfully, that if Van Gulick's and my account of downward causation is correct, then, given the traditional Christian distinction between God and creation (in particular the rejection of pantheism), downward causation does *not* in itself provide a solution to the problem of divine action. I say "regretfully" because I agree with almost everything else Peacocke has written and because his overall position is so pleasingly coherent.

In brief, I shall argue, the problem lies in the fact that selection among lower-level causal processes in accordance with higher-level specifications requires some sort of causal effect on the lower-level entities. If God is the higher-level agent and the events in the world are the lower-level entities, then the original problem of divine action simply re-appears at this point.

5.1 On the Insufficiency of Downward Causation

In "God's Interaction" Peacocke writes that the world-as-a-whole can be considered as a total system, so that its general state can be a holistic constraint on what goes on at the various levels that comprise it. If that total system is "in God" (as panentheists argue) then God, on analogy to whole-part constraints within the world, can affect the state of the world-as-a-whole. Such effects could then have further downward effects on particular events within the world without abrogating the laws and regularities that pertain to them.[49]

Peacocke's preferred model (in both essays) for God's action is the action of the whole person, brain-in-body, on its constituent parts. Also in both essays he proposes that God's "input" might better be conceived as a flow of information rather than a flow of energy.[50]

Peacocke emphasizes that because of the ontological difference between God and the world, his account of divine action can only be *analogous* to an account of whole-part causation (or influence) *in* the world.[51] The question, then, is whether this ontological gap is such as to *entirely vitiate* the analogy.

---

[49] Peacocke, "God's Interaction with the World," 282-3.

[50] Ibid., 285; Peacocke, "The Sound of Sheer Silence," 236.

[51] See, for example, Peacocke's, "God's Interaction with the World," 285 and "The Sound of Sheer Silence," 236.

My suggestion (in section 3.3.1) was that Peacocke's shift in focus from earlier accounts of downward causation (Campbell's in particular) toward the more general category of whole-part constraint or influence represented a loss in specificity about *how* the whole affects the parts. I believe that we need the notion of *selection* to fill this gap. But, of course, "selection' is a broad and vague term as well. So we need to consider examples in order to understand how it happens in particular cases.

Let us consider Deacon's example of the formation of a snowflake. There are two sorts of downward causation involved. One is from the environment to the snowflake; the other is from a holistic property of the snowflake—its current shape—to its subsequent lattice formation. In the first case we have straightforward physical causation. The growth of the crystal depends on presence of additional water molecules and absence of heat. The shape of the flake at a given time influences future growth, as well, through the ordinary processes of crystal formation. It is simply that the next molecule will attach *here* rather than *there* due to the micro-features that together constitute its shape.

If God's action on the world-as-a-whole is analogous to either of these instances of downward causation, then it would seem that one of the following must be true: (1) God as "environment" affects the available supply of matter and/or energy; or (2) God must *be* the whole (as the snowflake is to its parts).

To put my critique in general terms: downward causation does not provide an adequate model for divine action because downward causation in the world is always mediated by ordinary physical forces. My one example does not bear the weight of "always," yet this *must* always be the case unless some other, higher-level forces are postulated instead. Given Christians' (and Jews' and Muslims') traditional understanding of the "ontological gap" between God and the world, the problem of divine action simply arises again when we ask *how* God manipulates the *causal* factors that result in the selection of specific lower-level causal processes or powers.

Does the concept of a flow of information help here? I believe it does not, and the reason is noted by Peacocke himself: while information flow is not the same as flow of matter or energy, "in actual systems no information flows without some exchange of energy and/or matter."[52] So the proposal that God arranges patterns in the world that convey information to humankind[53] is open to the question of how God *causes* one pattern of events to obtain rather than another. I believe Peacocke's attempt to use downward causation and whole-part influences to model God's action on the universe-as-a-whole fails for a second reason, one first pointed out by Robert John Russell.[54] The reason is that, unlike water in a pot, where the pot as the boundary of the water

[52] Peacocke, "The Sound of Sheer Silence," 225; see also idem, "God's Interaction with the World," 286-7.

[53] Peacocke, "The Sound of Sheer Silence," 236.

[54] Robert John Russell, "Arthur Peacocke on Method in Theology and Science and His Model of the Divine/World Interaction: An Appreciative Assessment," in *All That Is: A Naturalistic Faith for the Twenty-First Century*, ed. Arthur Peacocke and Philip Clayton (Minneapolis: Fortress Press, 2007), 140-51, esp. 148-50.

influences the motion of the water, the universe as understood in Einstein's General Theory of Relativity has no boundary for God to act upon.

5.2 Anticipated Objections

I anticipate two sorts of objections to my critique of Peacocke. One is that I have failed to appreciate the nuances of his argument—in particular his claim that the concepts of whole-part constraint and information provide *analogies* or *metaphors* for understanding divine action, and that Peacocke is just as aware as I am of the disanalogies. My claim, though, is that the one great disanalogy—God's ontological distinctness from the world—actually cuts to the heart of Peacocke's arguments.

The second objection is that by emphasizing the role of lower-level causal processes in downward causation I have effectively debunked the idea and have fallen back into causal reductionism. An instance of this sort of objection comes from William Hasker. In responding to an account of downward causation from the mental to the neural level in which I pointed to the mental as the means by which the neural system is able to interact with a broader causal system,[55] Hasker wrote:

> I take this to mean that the interaction between the specific physical system and others leads to different results than would occur if the system in question were left undisturbed, or were interacting in some other 'broader causal system.' Presumably, however, the interactions between the system and the other, surrounding systems will still occur according to the standard physical laws. In principle, then, all the events that occur are explainable in bottom-up fashion—that is, they are explainable mechanistically. The notion that 'downward causation' interpreted in this way represents an *alternative* to mechanism is sheer confusion.[56]

Hasker is right that the interactions between the system and the other surrounding systems will still occur according to standard physical laws, and this in two senses: the physical laws are not *violated* and there are no causal forces other than the physical. However, the literature on downward causation and emergence, invoking the roles of boundary conditions, structure, function, information, and so forth has been developed precisely to show why Hasker's next claim, that all events are explainable in bottom-up fashion, does not follow. Physical laws are necessary but not sufficient to explain emergent phenomena because they provide only half of the causal story.[57]

---

[55] See Nancey Murphy, "Giving the Nonreductive Physicalist Her Due: A Response to Hasker's *The Emergent Self*," *Philosophia Christi,* Series 2, Vol. 2, No. 2 (2000): 167-73.

[56] William Hasker, "Reply to My Friendly Critics," *Philosophia Christi,* Series 2, Vol. 2, No. 2 (2000): 197-207, 198-9.

[57] See Murphy and Brown, *Did My Neurons Make Me Do It?*, chap. 2 for our use of Juarrero's work to provide a more powerful response to this critique.

5.3 On the Essential Contribution of Downward Causation

Although I judge the concept of downward causation to be insufficient to solve the problem of divine action I want to note that it is nonetheless an essential ingredient in any complete account. Much of God's action in the world is mediated by human action, and downward causation appears to be necessary for understanding human agency.[58] God's influence on individuals comes most often through religious communities; this too is presumably an instance of downward causation. Another instance is, as Peacocke points out, God's work in nature and history constituting meaningful patterns that inform us of God's intentions.

I conclude, then, that downward causation provides a necessary but insufficient account of divine action. Furthermore, I believe that Peacocke's account is "hungry" for something like divine action at the quantum level to complete it. Peacocke emphasizes that God's influence is potentially located *everywhere* in space and time.[59] His panentheist understanding of God's relation to the world is intended to place equal stress on God's transcendence *and immanence*. If God acts immanently and everywhere in quantum events, selecting one of the possible outcomes for each, is that not also an instance of downward causation? The discussion of quantum divine action has moved far beyond my competence, so I trust others will evaluate these proposals.

6 Overview

The first goal of this paper was to assess the VO/CTNS research group's contributions to one particular philosophical topic, the cluster of issues relating to reductionism, downward causation, and emergence. I attended only to the writings of Peacocke and Clayton, and offer apologies to the others whom I have neglected.[60] While the VO/CTNS contributions have been significant, I claimed that important further resources are available in the works of Van Gulick and Deacon, which clarify and extend the concepts of downward causation and emergence.

My second goal was to contribute one small piece to the evaluation of the research group's work on the problem of divine action and, not surprisingly, I focused on Peacocke's proposal for divine action as whole-part constraint. I concluded that, while important, it does not get to the root of the problem; Peacocke's use of the terms "constraint" and "influence" does not get him off the hook of having to locate God's causal role in specific events. Thus, I ended with the hope that top-down and bottom-up accounts of divine action can be shown to be complementary.

---

[58] See Murphy and Brown, *Did My Neurons Make Me Do It?*

[59] See Peacocke, "The Sound of Sheer Silence," 236.

[60] For example, see George F. R. Ellis, "Quantum Theory and the Macroscopic World," in *QM*, 259-92.

THE DIVINE ACTION PROJECT, 1988-2003

Wesley J. Wildman

Does God act to achieve special providential aims in the world? Sacred texts of many traditions speak of intentional divine action. Some people pray expecting God to respond and answer their prayers. Religious liturgies express confidence in God's action in the past as well as God's ability and willingness to intervene in response to present concerns. Over the centuries, religious scholars have advanced theories of divine action in order to give intellectual support to traditional claims about God's special, intentional, purposeful action. What is the state of the art in contemporary theories of special divine action?[1]

On the pessimistic side, the recent book *Divine Action and Modern Science* claims that contemporary theology is in a serious crisis.[2] Its author, Nicholas Saunders, believes that Christianity desperately needs a sound theoretical account of God's action in the world if it is not to be swept away by demythologized secular worldviews with no need of the hypothesis of God's special providential activity. Unfortunately, according to Saunders, no adequate theory of divine action yet exists, and this state of affairs threatens to be devastating for contemporary Christian theology, as well as for the faith and practice of Christian believers.

By contrast, far more optimistic conclusions are flowing from what I shall call the Divine Action Project (DAP). Divine action was the major theme of a wide-ranging series of conferences and publications on theology and science jointly sponsored by the Vatican Observatory (VO) and the Center for Theology and the Natural Sciences in Berkeley (CTNS).[3] This research group culminated its work in a September 2003 "capstone conference" in Castel Gandolfo, near Rome, reflecting on its published output, identifying areas of agreement and disagreement. What I call the DAP refers to the work of this

---

[1] An earlier version of this essay published under the same title appears in *Theology and Science*, Vol. 2, No.1 (2004): 31-76.

[2] Nicholas Saunders, *Divine Action and Modern Science* (New York and Cambridge: Cambridge University Press, 2002). Saunders concludes, "Would it be correct to argue on the basis of the foregoing critique that the prospects for supporting anything like the 'traditional understanding' of God's activity in the world are extremely bleak? To a large extent the answer to this question must be yes. *In fact it is no real exaggeration to state that contemporary theology is in crisis*" (215; italics in original).

[3] In addition to the VO/CTNS volumes, a related book concerns the views of Pope John Paul II, together with expert commentary: *John Paul II on Science and Religion: Reflections on the New View from Rome*, Robert J. Russell, William R. Stoeger, S. J., and George V. Coyne, S. J., eds. (Notre Dame, Ind.: Notre Dame University Press, 1990). The *EMB* volume includes an address given by Pope John Paul II to the Vatican Observatory Conference at the 1996 meeting. Summaries of all 5 introductions and 91 articles, the work of 50 authors, are available on the CTNS website, http://www.ctns.org/books.html.

group as it bears on divine action. DAP participants have been trying to meet precisely what Saunders rates as Christian theology's single most pressing need. The major points of consensus among DAP participants were that there are significant theoretical grounds for confidence in the intelligibility of the concept of providential divine action, and that there are several technically and theologically feasible theories of it.

What could account for these spectacularly divergent conclusions? Is this more evidence that theology is a "think anything, say anything" discipline, with no prospects for broad consensus or rational understanding? Or can we triangulate the disagreement and explain it? The basic aim of this paper is to introduce the DAP and to summarize its results, especially as these were compiled in the capstone meeting, and drawing on the published output of the project where possible. This task consumes the bulk of the paper. The subsidiary aim is to evaluate criticisms of the DAP. I shall not devote space to my own criticisms beyond what I say in the next paragraph. I shall argue that Saunders does not provide a compelling argument for his pessimistic conclusions and that his otherwise insightful criticisms of some DAP theories silently depend on an interpretation of laws of nature that those he criticizes reject. Furthermore, referring to the argument of German philosopher Immanuel Kant that it is impossible to demonstrate freedom in terms of categories of causation, I shall try to diagnose the peculiar challenges facing the DAP, to analyze how participants handled those pressures, and to show how Saunders's criticism fits into this wider pattern. This attempt to explain the divergence of perspectives hopefully will extend the circle of consensus as far as possible and clarify the irreducible disagreements that remain.

I need to lay my cards on the table, particularly since credibility can be an issue both in summarizing complex projects and in addressing their critics. I have been involved in the DAP as one of the specialists in theology and religion, making contributions to *NP, EMB*, and *CC*. Unlike most other participants, I hold that the postulate of intentional divine action exacerbates the problem of theodicy to such a degree that we are justified in rejecting it for moral and theological reasons. This is the view of British theologian Maurice Wiles, also.[4] Although my view of God differs slightly from that of Wiles, I do concur with his judgment that the idea of intentional, discrete divine acts is incredible, no matter how much some strains of Christian piety and large tracts of the Vedas, the Bible, and the Qur'an take it for granted. I approach Saunders's book as a sympathetic and curious reader.[5] After all, if the argument of that book is sound, then (presumably against Saunders's intentions) I could venture to add incoherence of the concept of intentional divine action to the existing charge of moral repugnance. Likewise, though I participated in the DAP as one finally unconvinced about the reality of intentional divine action because of the problem of theodicy and the

---

[4] See Maurice Wiles, *God's Action in the World* (London: SCM, 1986).

[5] It may also be worth noting that, though I do not know Saunders personally, I have a great deal of respect for the Cambridge advisor of the dissertation from which the book flows, Sir John Polkinghorne, and for Saunders's sometime discussion partner, Arthur Peacocke, both of whom have been key participants in the DAP. Needless to say, I hope these warm thoughts and feelings do not interfere with my judgment.

incredibility of discrete divine acts, I am persuaded that the DAP has succeeded in demonstrating the coherence and technical feasibility of several theories of intentional divine action. It is fundamentally the view of God underlying the main proposals emerging from the DAP with which I disagree, therefore, rather than the proposed theories of divine action themselves.

1 Introduction to the Divine Action Project

In the late 1980's, Robert Russell of CTNS in Berkeley and George Coyne, S.J. of the VO, in discussions with Nancey Murphy and William Stoeger, S. J., discerned a need for research on the topic of God's action in the world. Rightly noticing that the question of divine action is one of the theological issues that most directly presupposes facts about the natural world and its governing laws, they reasoned that evaluating theories of divine action in relation to our best knowledge of the natural world was very likely to be fruitful. What would be the fruit of such a discussion? If it were done well, they supposed, it could produce confirmation or disconfirmation of certain theories, the development of new proposals for divine action, and at the very least a state-of-the-art survey of credible options. Such a survey might even make clear the theological presuppositions that would lead a thoughtful religious intellectual to affirm one theory of divine action over another. These were high hopes so their fulfillment would require a great deal of work and a well-designed process. Together, CTNS and the VO raised funds and devised a method for the discussions. In this way—and the details are more fascinating that this compact summary can hope to suggest—the DAP was born.

Most people involved in the DAP had been involved in science-religion dialogues enough to know how perilous any work plan would be. Here's a short list of challenges with associated implications for the project. First, theologians typically do not know enough science to do what was proposed so expert scientists would be needed. Second, scientists typically don't appreciate the nuances of theology well enough to have much patience for a theologically focused project so the design would have to involve education of the scientists as much as education of the theologians. Third, many scientific perspectives are relevant to divine action and scientists typically are expert in only one area of science so confusion might reign unless the conferences focus on one area of science at a time. Fourth, a relatively small core of people would have to be involved in all conferences to guarantee some continuity of thinking and this small group would have to be expanded with experts who could cover the specific theme of each conference adequately. Fifth, divine action is a broad topic and a focused discussion that can engage the sciences seriously would require the additional constraint that the target will be divine action in conformity with natural laws. Sixth, the world religions are so diverse in their views of Ultimate Reality that a focused discussion would be most likely if the religious thinkers were limited to experts in Christianity. Finally, the phenomenon of scientists and theologians talking past one another and utterly failing to engage the issue at hand is unpleasantly common in science and religion dialogue—it occurs in other interdisciplinary work, also—and so the method of procedure would have to involve intense interaction and a commitment on the part of project members to engage one another and to do the difficult work of learning new perspectives. It would also be vital to involve scholars such as Ian Barbour, Arthur Peacocke, John Polkinghorne, Bob Russell, and Bill Stoeger who had professional competence in both theology and science.

The outcome of these considerations was an intensive workgroup-style method with some unusual features. The series of conferences was united by the theme "Scientific Perspectives on Divine Action" and each conference in the series focused on a specific scientific topic: quantum cosmology and the laws of nature, chaos and complexity, biological evolution, cognitive neuroscience, and quantum mechanics. For each conference the organizing committee asked a new group of scientists and philosophers expert in the assigned topic to team up with the more or less unvarying group of philosophically minded theologians. Once the entire group was identified, the organizing committee distributed a packet of key readings that everyone in the group was expected to know, thereby establishing a kind of baseline level of shared knowledge. Once these were digested, a number of "pre-conferences" were held to discuss the topic in smaller groups and to formulate ideas for research papers. These pre-conferences were more or less local affairs, usually one somewhere in Europe and one somewhere in the United States.

Subsequently, those wanting to write a paper produced a draft and submitted it to the organizers. Here's where the genius of the method becomes evident. Organizers distributed these drafts to everyone involved in the working group and expected everyone to make written responses, which were in due course distributed to everyone. In this way, everyone together was involved in a rich discussion about the salient issues before the main conference even began. A second round of papers—some revisions, some new—were written and broadcast, followed by the second round of responses, once again with everyone reading all of the responses. Eventually those writing produced a definitive conference draft of their papers, suitably revised in light of the comments received, and these drafts were distributed and read in advance of the conference. This intense method of proceeding optimizes engagement with the issue under discussion and minimizes the problem of "parallel play" pseudo-dialogue.

At each main conference, where the entire group gathered for the first time, there was the refreshing policy of not reading any papers. Since all had been distributed and read prior to the conference, the author merely made a few introductory remarks, usually bearing on motivation or newly developed ideas, and then the group launched into vigorous discussion for ninety minutes or so. It was draining work but also work of the most intellectually rewarding sort. Some sessions in the conference were given over to relatively unstructured discussion of the main theme. After each conference, people rewrote their papers in light of the conference discussion and submitted the final draft to the editors of the conference volume. If the paper was accepted for publication in the volume, then the editorial suggestions were incorporated in yet one more round of revisions. The whole process, from beginning to end, took at least two years for each volume.

Nothing less than this intense working style can ever really come to grips with the complexity of a problem with such obvious interdisciplinary dimensions. This conference series was one of only two I have been involved in that produced a cautious yet real sense of making progress on a research topic.[6] There are many sorts of conference, of course, and not all have a research agenda. Time after time, however, conferences with a research agenda underestimate the difficulty of the task or attempt to spare participants from investing a lot of energy in the project, with the result that no progress is made. It is to the credit of the DAP organizers that they designed a procedure capable of supporting their research agenda.

Within the core group, the organizers included at various times process theologians (Ian Barbour, Charles Birch, Jack Haught), more-or-less Neoplatonists (Janet Soskice, Keith Ward, and myself), and more-or-less Thomists (George Coyne, Denis Edwards, Stephen Happel, Michael Heller, Bill Stoeger) of various stripes to challenge and complicate the dominant view of more-or-less personalist theism and panentheism (Bill Alston, Philip Clayton, George Ellis, Philip Hefner, Jürgen Moltmann, Nancey Murphy, Arthur Peacocke, Ted Peters, John Polkinghorne, Bob Russell, Tom Tracy). This complication arises mostly from within the sphere of advocacy of intentional divine action. But the organizers also included philosophers and theologians who do not advocate intentional divine action, such as Willem Drees and myself, to challenge the project's overall coherence. Of course, this list of names does not include the many historians of science, philosophers of science, and scientists who were involved at one time or another, nor the new theologians and philosophers present at the capstone conference.[7]

Other methodological decisions were equally important for success. The organizers framed the project within the science-religion dialogue to make the discussion more precise and to fulfill the conditions for detecting and understanding divergence among the views of participants. They conceived of a technical readership for the DAP volumes, thereby rejecting any tendencies toward oversimplification. They concentrated on tractable issues rather than spectacular ones, which temporarily deemphasizes traditional Christian themes such as the resurrection or miracles for the sake of allowing the project to move forward efficiently in other areas. They encouraged parallel research

---

[6] The other effective project was the Boston-based Crosscultural Comparative Religious Ideas Project, led by Robert Neville, Peter Berger, and John Berthrong. The output of this series of 25 day-long conferences was three volumes, all edited by Neville and published by SUNY Press in 2001: *The Human Condition*, *Ultimate Realities*, and *Religious Truth*.

[7] It is not easy to classify people, particularly in the case of participants such as Ernan McMullin and Fraser Watts. The list of specialist participants in the DAP not already mentioned and without categorization (and probably not error free), is as follows: Michael Arbib, Francisco Ayala, Michael Berry, Leslie Brothers, Jeremy Butterfield, Camilo Cela-Conde, Julian Chela-Flores, Michael Chiao, Chris Clarke, Anne Clifford, James Cushing, Paul Davies, Langdon Gilkey, Joel Green, Andrej Grib, Peter Hagoort, Chris Isham, Marc Jeannerod, Fergus Kerr, Bernd-Olaf Küppers, Joseph LeDoux, John Lucas, Gisele Marty, Theo Meyering, Michael Redhead, and Abner Shimony. One question the DAP did not answer is why this list of specialist participants is so improbably loaded toward the first half of the alphabet. Note that the capstone conference included participants on the theological side not previously involved in conferences: Paul Allen, Niels Gregersen, Owen Thomas, Kirk Wegter-McNelly (involved in editing several DAP volumes), and Mark Worthing.

agendas within the one overall project, as we shall see later. They steered away from practical ethics to avoid direct discussion of divisive moral questions that might diffuse the focus of research. Also, as noted above, they kept the theological focus on Christianity so as not to complicate the project with an interreligious dialogue agenda, and they focused mainly on personalist theism and panentheism, because it is with these ideas of God that traction with the natural sciences is strongest. Each of these organizing decisions limited the project and it is not difficult to imagine (indeed, I have heard) complaints about these limitations. But I think the decisions made were good ones because without the precious commodities of focus and efficiency, no advance would be possible.

The fruit of these labors was of several kinds. The DAP succeeded in stabilizing terminology that is key for understanding theories of divine action. It demonstrated the usefulness of certain strategies for inquiry. It generated a comprehensive classification of alternative theories of divine action. It diagnosed the way that theological and philosophical instincts and convictions affect debates over divine action. And it generated a number of credible theories of intentional divine action.

Stated quickly like this, this list of accomplishments may seem rather modest. As I hope the details will show, however, these are remarkable achievements in a theological climate marked by stormy disagreements over everything from legitimate strategies to acceptable conclusions. In what follows, I shall outline the most relevant and important details of these conclusions, drawing especially on discussions and papers from the capstone conference.

2 Results: Stabilizing Terminology for Discussion

Whenever a group of scholars can agree on terminological distinctions, they greatly improve the chances of advancing their research project and understanding disagreements among themselves. Sadly, this degree of clarity and terminological consensus is rare within theology and the humanities. I take these relatively stable distinctions also to be of great significance for the wider debate over divine action. I will refer to them freely in what follows, so I lay them out here. Grasping these distinctions is the most efficient way both to get into the mindset of DAP participants and to frame the state-of-the-art options within the contemporary discussion of divine action in the science-religion dialogue context.

One crucial issue affecting many terminological distinctions is the status of language about divine action. How literally can we treat the theological terminology we use? If we accept the traditional assumption that all language about God is analogical or metaphorical, then how can we render terms such as "divine act" or "divine intention" in the precise ways needed to make sense of the debates over feasibility of theoretical proposals for special divine action? It does seem clear that the DAP and theologians more generally treat language about God as having degrees of literalness, so that we can judge certain metaphorical or analogical statements to be closer to the truth of the matter, and others to be further from the mark. The issue of degrees of literalness of theological language, however, continues to be a difficult problem within theology. This is due fundamentally to the instability of any standard we might invoke to assess literalness or accuracy of conceptual formulation. If we wish to speak literally about God forming intentions to act, for example, the theological schema that makes sense of divine intentions will be subject to questions about literalness.[8] These questions in turn call for a metaphysical vision that is capable of stabilizing a theological schema of the divine nature. But now our metaphysical theory is setting conditions on what can count as a viable portrayal of God, whereas arguably it should be the other way around. The resulting choice seems to be between vicious regress and vicious circularity. The specter of theologians vainly fumbling after stable theological terminology has stirred mystical theologians (among whom I count myself) for centuries. It has also been one inspiration for the aggressive attacks on the so-called ontotheological tradition in recent decades. But if we turn to poetic indirection instead of metaphysics, we face other problems, including the fact that poetic play is itself parasitic upon already existing terminology and symbols, which liturgical practices and metaphysical theology stabilize. Several DAP participants urged greater consciousness of such problems upon the group,[9] though discussions seemed to proceed without satisfying resolutions. For now, we must simply assume that we can speak with "some degree" of literalness about features of the divine nature needed to make sense of the terms of the debate over divine action. With that in mind, we turn to the key terminological distinctions.

First, the DAP concluded that the distinction between general and special divine action is important but that theologians draw it differently depending on their particular interests. This occurs because there is considerable entanglement of the general and the special in such concepts as continuous creation, divine sustaining of reality, and ubiquitous intentional divine action. Thus we must stabilize these terms by stipulation, and this causes no confusion providing that we are clear about which way of distinguishing the terms we are using. One common path not taken here is to say that general divine action refers to the act of creation itself together with divine acts that occur at every place and time within created reality, whereas special divine action refers to acts that have effects in some places and times but not others. This way of drawing the distinction is useful for keeping the focus on one-time, one-place special divine acts but does not easily comprehend a number of views affirming universal special divine action (most clearly Murphy's). It also

---

[8] St. Thomas's analogy of being famously must confront difficulties about human ignorance of the divine mode of being. Karl Barth's analogy of faith begs questions of biblical authority and sources for theological knowledge. Communitarian theologies that try to stabilize theological terms with reference to practices and beliefs of a community pretend at epistemological self-containment and independence from other forms of knowledge, whereas this is demonstrably not the case.

[9] See Stephen Happel, "Metaphors and Time Asymmetry: Cosmologies in Physics and Christian Meanings," in *QC*, 105-35; idem, "Divine Providence and Instrumentality: Metaphors for Time in Self-Organizing Systems and Divine Action," in *CC*, 177-203; Michael Heller, "Generalizations from Quantum Mechanics to God," in *CC*, 107-22, esp. 191-3; and Fraser Watts, "Cognitive Neuroscience and Religious Consciousness," in *NP*, 327-46, esp. 340-1. Soskice was particularly conscious of issues about the status of theological language.

obscures relevant questions about the divine nature. For the purposes of this paper, therefore, I will draw the distinction differently, in terms of presuppositions about God's character, as follows:

- General divine action (GDA) is the creation and sustaining of all reality insofar as this does not necessarily presume any specific providential divine intentions or purposes.[10]
- Special divine action (SDA) is specific providential acts, envisaged, intended, and somehow brought about in this world by God, possibly at particular times and places but possibly also at all times and places.

This way of drawing the distinction captures the conclusions of the DAP more effectively than the alternative.

Second, over the years DAP members became increasingly aware of the variety of ways in which theological theories can relate to the natural sciences. Russell has presented a rich view of these relations in an elaborate diagram, which serves to underline how internally complex each of theology and science is, and thus how rich the mutual interaction between them can be in principle.[11] Within the DAP, two kinds of relation proved most important, both of which we called forms of "traction" between theology and science, by which we mean formal or informal logical connections that yield both intelligibility and potential for correction and improvement.

- Traction as consonance envisages theological assertions that are vague enough to be consistent with several competing scientific theories yet elaborated in such a way as to register and resonate with large areas of scientific theory.
- Traction as consistency envisages theological propositions so specific that they can conflict directly with scientific assertions.

These two kinds of traction correspond to two kinds of sub-projects within the DAP that coexisted more-or-less happily, though not always without confusion and misunderstanding, especially in the early years of the project. The DAP was strongly committed to maximizing traction in the different ways that each sub-project demands, on the grounds that traction increases the credibility of theological proposals.

---

[10] This survey does not cover the DAP's discussions of GDA, mostly because the focus was on SDA. The *QC* volume contains the essays that paid most attention to GDA, and the key paper on this topic is probably Robert John Russell, "Finite Creation without a Beginning: The Doctrine of Creation in Relation to Big Bang and Quantum Cosmologies," in *QC*, 291-326.

[11] See Robert John Russell, "Bodily Resurrection, Eschatology, and Scientific Cosmology," in *Resurrection: Theological and Scientific Assessments*, Ted Peters, Robert John Russell, and Michael Welker, eds. (Grand Rapids, Mich.: Eerdmans, 2002): 3-30, the diagram and associated discussion are on 10-17. Alternatively, see idem, "Eschatology and Physical Cosmology: A Preliminary Reflection," in *The Far Future Universe: Eschatology from a Cosmic Perspective*, George F. R. Ellis, ed. (Philadelphia: Templeton Foundation Press, 2002): 266-315, the diagram and discussion are on 275-9, 284-8.

Third, the DAP project tried to be sensitive to issues of theological consistency. For example, the idea of God sustaining nature and its law-like regularities with one hand while miraculously intervening, abrogating, or ignoring those regularities with the other hand struck most members as dangerously close to outright contradiction. Most participants certainly felt that God would not create an orderly world in which it was impossible for the creator to act without violating the created structures of order. This widely shared conviction led the main line of the DAP's research efforts to seek an account of SDA that was in accord with created structures of nature, which underlies the following key distinction:

- A noninternventionist special divine act is in accord with created structures of order and regularity within nature.
- An interventionist special divine act involves abrogating, suspending, or ignoring created structures of order and regularity within nature.

Some DAP participants were content to imagine that SDA is usually noninterventionist but on rare and providentially portentous occasions, such as the resurrection of Jesus Christ perhaps, or the consummation of the universe, God could act in interventionist fashion. Others entertained noninterventionist or "more than interventionist" readings of such events. For instance, Russell wondered whether we could interpret the incarnation and resurrection of Jesus Christ not so much as God violating existing regularities of nature but as God transforming those underlying regularities so that a new reality can emerge out of the existing one.[12] Still others such as myself, when entertaining this issue in the (to me) alien terms of personalist theism, thought the inconsistency of miraculous interventionism a small thing, and estimated the theological hubris of confining God to noninterventionist action a greater danger. Despite these differences, there was agreement that accounts of SDA as noninterventionist enjoy the greatest traction with the natural sciences, in the sense of consistency, because they allow scientific accounts of natural regularities to constrain what is possible in theological assertions about SDA.

Fourth, there was general consensus among DAP participants that miracles received insufficient attention in the course of the project. This was due to the theological interest in noninterventionist theories of SDA and the methodological interest in maximizing traction between theology and the natural sciences. Nevertheless, we did discuss the idea of miracle from time to

---

[12] Russell states that he does not wish to disparage interventionism but only resist its unnecessary use, acknowledging that interventionist readings of such events as incarnation and resurrection "might be justifiable and necessary after suitable nuancing, since when the domain of God's action is eschatological the 'laws of nature' (i.e. God's faithful action) themselves will be different and 'intervention' may cease to be a useful concept" ("Special Providence and Genetic Mutation: A New Defense of Theistic Evolution," in *EMB*, 200, fn. 21). We might construe this idea as interventionist, non-interventionist, or even more-than-interventionist, depending on our point of view. Thus, Russell's view of eschatologically momentous events shows that the categories of interventionism and noninterventionism are somewhat inflexible. Yet the distinction remains useful for many purposes.

time. Ward offered the richest set of distinctions among ideas of miracle, as follows:

- Miracles might be suspension or abrogation of nature's law-like regularities.
- Miracles might be ways of speaking of apparently providential events that strike us as important and surprising.
- Miracles might be the activation of latent features of natural objects that do not show up within the theoretical framework of our existing natural sciences. For example, theopoiesis (theosis or divinization) might be a natural but latent feature of human beings that God can activate despite the fact that our science knows nothing of this possibility, and perhaps never could recognize it.[13]

The first sense of miracle as suspension or abrogation of nature's law-like regularities is precisely as clear as the reigning view of those regularities themselves (the same is true of the concept of intervention), which in turn directs our attention to laws of nature, to which we will come in a moment.

Fifth, related to the various interpretations of miracles, the DAP experimented with several distinctions bearing on the epistemology of situations in which a divine act is noticed or not noticed as such. The published distinction between divine acts that are apparent with (or without) religious presuppositions[14] finally proved less compelling than Tracy's three-fold distinction among:

- special divine acts as objectively significant,
- special divine acts as causally (but not objectively) significant, and
- special divine acts as subjectively (but neither objectively nor causally) significant.

In the sense of this distinction, objective special divine actions have a counterfactual logical structure: if God had not acted then nothing significant would have occurred. Subjective divine actions have a similarly crisp meaning: they have to do with perceptions only and presume nothing at all about God. The middle position is trickier but still needed because of the logical possibility that some events may have special providential significance even when God does not act specially to cause them. This providential significance might even be anticipated by God (if a suitable interpretation of divine knowledge is in place) and yet not intentionally caused.

Sixth, the DAP adapted to its own purposes the distinction between compatibilism and incompatibilism that has proved useful in the context of philosophical debates over human freedom in the world. Wegter-McNelly

---

[13] Ward cites the view of Rom Harré that we should interpret the regularities of nature not as laws but as inherent capacities or dispositions or tendencies of objects. This view underlies Ward's conception of the potential for transformation within human beings.

[14] See the diagrams in the "Introductions" to *CC* and *QM*.

furnished the most comprehensive view of the potential for this distinction by noticing three variants:

- Anthropo-physical (in)compatibilism asserts that human freedom is (in)compatible with physical determinism.
- Anthropo-theological (in)compatibilism asserts that human freedom is (in)compatible with divine determinism.
- Theo-physical (in)compatibilism asserts that divine freedom is (in)compatible with physical determinism.

Determinism here means that, given that the world is a particular way at one moment, its unfolding thereafter is fixed and inflexible. In the case that this inflexibility is due to the laws of nature, we would be speaking of physical determinism. When it is due to God's action or God's will or perhaps even God's knowledge, then we would be speaking of divine determinism. Philosophical writings on determinism are filled with subtleties but this definition will serve my purposes here.

Each of the three ways of distinguishing between compatibilism and incompatibilism has its own characteristic debates and a wealth of literature. The third—theo-physical—was the main concern of the DAP. In this context, now dropping the qualification "theo-physical," we used the distinction in the following way:

- Incompatibilism assumes that physical determinism entails the impossibility of non-interventionist SDA (equivalent to NISDA $\rightarrow \sim$ PD).
- Compatibilism assumes that non-interventionist SDA is consistent with either physical determinism or physical indeterminism (equivalent to NISDA $\rightarrow$ PD or $\sim$ PD).

Incompatibilists adopt the strategy of showing that the physical world is indeterministic, because this is a necessary condition for non-interventionist SDA. This leads to strong interest in gaps, especially uncloseable gaps, in the world's causal nexus. By contrast, compatibilists have nothing to gain by demonstrating or assuming physical indeterminism. Because compatibilist theories of SDA remain untouched by debates over physical determinism, their proponents have no need to locate uncloseable gaps in the world's causal nexus. This strategic difference was evident throughout the DAP. In fact, there is a tight correlation between these strategies and the types of traction defined above. Incompatibilist proposals seek traction as consistency, which is to say they achieve intelligibility by exposing theological propositions about SDA to direct potential contradiction by physical propositions about the world's causal nexus of events. By contrast, compatibilist proposals seek traction as consonance, which is to say that their theological propositions about SDA are immune from direct conflict with physical propositions about the world's causal nexus of events but can still achieve intelligibility by richly registering the scientific portrayal of physical reality.[15]

---

[15] Among DAP participants, Tracy was particularly concerned to defend the possibility that one might be a compatibilist in respect of one sort of divine action and an incompatibilist with respect to another. If Tracy is correct, then of course theo-

Seventh, the DAP hosted a variety of approaches to developing incompatibilist theories of SDA (these are the theories that have something to gain by locating causal openness in physical reality). One strategic disagreement expresses different appraisals of the near certainty that our current scientific descriptions of the structures and regularities of nature are only approximations.

- An adequate incompatibilist theory of SDA should seek causal openness in nature not only as described in existing physics but also as intimated by the incompleteness and provisional character of current science.
- An adequate incompatibilist theory of SDA should seek causal openness in nature only as described by existing physics.

Theories of SDA following the first approach (represented especially by Polkinghorne) accept a short-term reduction in concrete intelligibility by entertaining speculative future physics, in exchange for greater long-term robustness in the face of anticipated changes in the scientific portrayal of nature. Theories of SDA following the second approach maximize the concrete intelligibility of the theological theory in the short term by focusing solely on existing physics while leaving the theological theory more vulnerable to falsification by future changes in science. Thus, this strategic distinction expresses different ways of balancing two theoretical virtues that are desirable in a theological theory: short-term concrete intelligibility through dealing only with existing science and steering away from speculative physics, versus long-term robustness in the face of ever-changing scientific portrayals of the world's causal nexus.

Eighth, the DAP generated a strong consensus around distinctions pertaining to interpretation of the laws of nature. The basic distinction is between the following:

- The laws of nature have descriptive status only; they refer to regularities and patterns that we discern in natural objects, relationships, events, and processes.
- The laws of nature have ontological status; they refer to ontologically independent principles to which natural objects, relationships, events, and processes are subject.

There are a number of ways to picture the mode of existence of principles to which natural objects, relationships, events, and processes are subject on an ontological interpretation of the laws of nature. Such principles may exist independently of nature yet be known through the study of nature, they may exist independently of nature and be rationally intuited without empirical observation, or they may be vested in nature as deep formal or structural

---

physical (in)determinism would be importantly different than anthropo-physical (in)determinism. Specifically, unlike divine action as Tracy imagines it, we must conceive of human action either in compatibilist or in incompatibilist terms, but not both, because human beings are always subject to the regularities of nature. This is one example of the care the DAP took to notice failures of the analogy between divine and human action, as well as similarities between the two.

principles. The distinction here is so framed that we can safely bracket these questions of ontology. The key point is that exclusively descriptive approaches to the laws of nature deny that these laws refer to existent principles.[16] In the case of stochastic or probabilistic laws, this second possibility is ambiguous. This leads to a finer, three-fold distinction for stochastic laws, as follows:

- Stochastic laws of nature have descriptive status only (as above).
- Stochastic laws of nature have ontological status, in the sense of referring to principles or deep structures of nature that statistically govern large ensembles of events but not each individual event within an ensemble of events.
- Stochastic laws of nature have strong-ontological status, in the sense of referring to principles or deep structures of nature that statistically govern each individual event within an ensemble of events.

While there are many perspectives on laws of nature in the philosophy of science literature, the DAP consensus was that the distinctions above capture the features of those wider debates that are relevant for evaluating theories of SDA. Some participants argued that the strong-ontological interpretation of stochastic laws was incoherent, and it is certainly true that no theory of SDA proposed within the DAP makes use of it. Yet it is important to acknowledge that this view of stochastic laws exists. In fact, it is one reason[17] for the widespread intuition that quantum-level SDA must be interventionist, violating the probabilities that constrain quantum measurement events. This view appears in criticisms of proposals for quantum level SDA both within the DAP and beyond.

Finally, with regard to the divine nature, the DAP predictably found itself grappling with the perennial question of the distinctions among personalist theism, classical theism, and panentheism. These words are used in so many ways that they can only be stabilized by stipulation, and even then any stipulation begs a horde of detailed questions. Nevertheless, for the sake of maximizing clarity in an inevitably murky situation, I present here the working set of distinctions that guided DAP terminology. First, assume that we can stabilize the distinction between God as complete without the creation and God as incomplete without the creation. Further, assume that we can make out a distinction between God being changed by the world and God remaining

---

[16] The ontological options expressed here are translations into this context of the major options in the medieval debates over universals: the realists spoke of universals as having independent reality (this corresponds to the ontological approach, which has empiricist and rationalist versions), whereas their nominalist opponents denied any reality to universals (this corresponds to the descriptive approach). Meanwhile, the Scotists affirmed the independent reality of universals but insisted that they were only ever present in "contracted" fashion as the form of the actual concrete being of particulars (this view is also included here under the ontological approach).

[17] Another reason (Stoeger's and the Thomists') for this intuition is that quantum-level SDA makes God an ordinary (secondary) cause like other causes. Of course, this concern potentially applies to SDA in any mode.

unchanged. Then, while these are by no means simple assumptions, they do allow for a grid of positions, as follows.

	God can be changed by the world	**God remains unchanged by the world**
God is complete without the creation	Personalist Theism	Classical Theism
God is incomplete without the creation	Panentheism	

Figure 1: Types of Theism

Note that the empty position on the grid has representatives, though not within the DAP. For example, Robert Neville's view of *creatio ex nihilo*, an intensification of John Duns Scotus's stress on the primacy of the divine will over the divine nature, has "God" determining both the world's nature and the divine nature in the primordial creative act. Thus, God has no nature apart from creation and so, in that sense, is incomplete (in fact, is nothing) without it. Neville belongs in the empty square, accordingly.[18] Note, also, that these distinctions assign a tightly constrained definition to panentheism. In its more literal sense, it affirms simply that all of reality is in God, by contrast with pantheism, which literally means that everything is God. Yet the narrower construal of panentheism introduced here is not an unusual usage in our time.

3 Results: Classification of Options

The positions advocated within the DAP can be summarized at two levels. First, taking the widest possible view, we can picture each position within the project, as well as positions not represented within the project, as making a series of key decisions that characterize the final outcome. We can represent the outcome of analysis at this level as a decision tree, one version of which is supplied in Appendix A. This view of matters is helpful for placing the main focus of the project—on personalist theism and panentheism—in a wider context that is sensitive to issues of broad concern in comparative theology.

Second, we can tighten the focus onto the major proposals of the DAP and characterize them using the distinctions introduced above. The result is a different kind of diagram containing the views that I will discuss in more detail below. This diagram also includes several actual or hypothetical critics of the major views defended within the project—Saunders, Wiles, and Neville—with their names in parentheses.

The major result of the DAP is simply that all of the views of objective SDA defended by participants are feasible and coherent in most respects. This is not the happy agreement it might seem on the surface. A number of participants, including me, reject the entire concept of SDA as morally intolerable, though not thereby unintelligible. There are also fierce disagreements about whether compatibilist or incompatibilist approaches are

[18] See Robert Cummings Neville, *God the Creator* (Chicago: University of Chicago Press, 1968). Note that John Scottus Eriugena also speaks about God determining the divine nature in the act of creation.

more promising, whether individual views are framed in the optimal way, and whether a strong-ontological interpretation of stochastic laws of nature is possible and desirable. But through all of these and other disagreements, the

	Views of Special Divine Action (SDA)						
Characteristics of Views	Non-objective	Objective					
		Compatibilist	Incompatibilist				
			Laws of Nature as:			Laws of Nature as:	
			Approximate	Descriptive	Ontological	Strong-Ontological	
Holders of Views	Davies Drees Wildman (Neville) (Wiles)	Clayton Peacocke Soskice Stoeger Ward	Polkinghorne	Murphy Russell	Ellis Tracy	(No one within DAP)	(Saunders)
Noninterventionist	N/A	Yes	Yes	Yes	Yes	No	Yes
Notes	These views seek traction with the natural sciences through consonance and richly registering scientific details rather than through direct consistency constraints		This view is effectively compatibilist but can invite causal joint proposals	Universal SDA; divergence on whether God plays a constitutive role in events	SDA occurs only in some events and God plays a constitutive role in none	Tracy does explore this view as a thought experiment (miracles)	This is the view that Saunders seems to want to achieve

Figure 2: Views of Special Divine Action

consensus is that we can give several theologically feasible and scientifically coherent accounts of God's action in the world. The details have proved fascinating, however, so I now turn to a discussion of general trends and particular positions.

Views on the left side of the diagram enjoy the least traction (in the sense of direct consistency constraints) with the natural sciences, whereas traction of this sort increases as we move rightwards through the columns. Indeed, the non-objective (i.e. merely causally or subjectively significant) views and the compatibilist views seek traction of the other sort, as consonance with the natural sciences. There is an issue of taste or rational style here and it is fair to ask about why people diverge on such issues. I think two competing theological instincts are at play and understanding them can illumine the divergent judgments about the best type of traction to seek.

- One instinct is to seek *concrete intelligibility of theological assertions*, which drives theories of divine action toward incompatibilist, non-interventionist approaches, discourages speculation regarding future science, and encourages theological speculation to maximize intelligibility-yielding traction between theology and science.

- Another instinct is to demonstrate *credibility of basic faith claims*, which drives theories of divine action to seek long-term stability by avoiding over-commitment to existing science, and to limit theological speculation to the minimum needed to establish the rational feasibility of faith claims, resisting further speculation so as to avoid tying Christian claims too closely to unduly specific, credibility-stretching hypothetical proposals.

If concrete intelligibility were the only worthy goal of a theory of divine action, then everyone would crowd towards the right side of the diagram. With the concern for credibility and the associated worries about the destabilizing effects of undue speculation thrown into the mix, however, there is an impulse to move toward the left side of the diagram. Balancing these two theological virtues is a matter of art and each DAP participant found his or her own distinctive way to do that.

3.1 Compatibilist Theories of SDA

Compatibilist views within the DAP were of several kinds. Clayton and Peacocke articulate versions of panentheism and seek to articulate special divine action in such a way as to register contemporary scientific discussions of the evolutionary emergence of complex systems, including especially the realms of life and mind, from earlier and simpler forms of organization in our universe. The idea of emergence provokes complex debates about ontological levels, causal powers and forces, and supervenience relations, as well as causation of the whole-part and top-down kinds that we might make plausible in the natural sphere to as to stabilize analogies for accounts of SDA. Soskice, Stoeger, and Ward defend variations of classical theism, in rather neo-Platonist, rather neo-Thomist, and rather theopoietic forms, respectively. Thus, they all presume the distinction between the secondary causes of natural objects or processes and God as the primary cause of all events in nature, lying behind and supporting the efficacy of secondary causes. These views are compatibilist because there is no difficulty imagining God acting freely and noninterventionistically in nature regardless of how causally closed the scientific portrayal of nature is. Yet all think, on other grounds, that the world is ontologically indeterministic in important ways that make for meaningful human freedom and responsibility.

I will not venture to discuss the compatibilist proposals in any detail here, in part because their consonance with the natural sciences' portrayal of the complex process of biological evolution is difficult to evaluate conveniently, and in part because they are still very much under development.[19] But it is important to note that the DAP became a forum for complex and penetrating debates over the ideas of top-down causation and whole-part constraint, and that there remains fairly significant disagreement over precisely how best to understand these terms. This lack of consensus directly correlates with the proliferation of proposals for understanding complex biological systems in the wider literature. Other conferences and future research doubtless will lead to gains in clarity around these issues.

3.2 Incompatibilist Theories of SDA and Causation

The remainder of this review and analysis will focus on incompatibilist theories of SDA. At the outset, however, it is important to note that there is

---

[19] Peacocke's view is laid down in *Theology for a Scientific Age*, enlarged ed. (London: SCM Press, 1993). Clayton intimates his view in a few existing publications but the main work has not yet appeared.

widespread debate (if not outright confusion) in the philosophy of science and in metaphysics over the nature of causation. This presents a serious difficulty to any incompatibilist theory of SDA because the very distinction between compatibilist and incompatibilist strategies depends on a concept of causation. In practice, this difficulty is less severe than it might seem at first, so long as we can grant that there is, *in some sense*, a workable concept of "ordinary causation" to which we can appeal in devising an incompatibilist theory of SDA. Bracketing philosophical debates over the precise meaning of ordinary causation in this way allows theological proposals to get started. But they cannot travel very far before having to face two major disputes in the theory of causation that have direct consequences for a theory of SDA.

The first dispute concerns Alfred North Whitehead's proposal of the process model of causation.[20] By making God's action one element in every event (in the technical sense of an actual occasion), Whitehead appears to solve the theological problem of SDA and the philosophical problem of causation simultaneously. So why, we must ask, has not everyone flocked to affirm Whitehead's view? While some theologians have done so, many others have not, primarily because Whitehead's view requires abandoning the conception of God as creator, proposing instead that God and the world are mutually codetermining, coeternal realities. Philosophers of science are unenthusiastic about Whitehead's proposal for causation because it appears unrelated to the natural sciences. Of course, Whitehead intended this, deeming it desirable for a metaphysical theory to be consistent with any possible discovery of the sciences. For their part, however, most philosophers of science are more interested in detailed consonance and, where possible, direct consistency constraints between scientific theories and their proposals about causation. While some DAP participants accepted Whitehead's theory of causation (Birch, Barbour, and Haught), most did not. But everyone advancing an incompatibilist theory of SDA had to make this decision one way or another so as to select a specific causal context within which to frame the distinction between compatibilism and incompatibilism.

The second dispute concerns the feasibility of types of causation other than ordinary, low-level causation. If theoretical portrayals of top-down, whole-part, or mind-to-mind causation correspond to genuine natural processes, then the meaning of an incompatibilist theory of SDA might be quite different than would be the case if ordinary, low-level causation were the only type operative. To illustrate this, suppose we conceive of top-down causes as ontologically distinct from ordinary causes organized in complex systems. Then we would be able to explain (in principle!) the top-down and bottom-up effects that we observe around us not in terms of the way complex systems constrain and marshal ordinary causes, but in terms of direct action from a higher level of nature to a lower level, without any need for ordinary causation at all. This, in turn, would open the speculative possibility of God being able to act at any level of nature using an analogous causal joint, one appropriate to God's top-down relation to the world. There would be no intervening in the laws of nature on this view because the divine mode of top-down causation

[20] See Alfred North Whitehead, *Process and Reality*, corrected ed. (New York: Macmillan, 1978).

would be as ontologically independent of our familiar laws of nature as are other forms of top-down causation (such as strong views of mental causation).

This might seem like a promising option for a theory of SDA but no DAP participant unambiguously embraced it. I think there are two main reasons for this. On the one hand, from the point of view of theoretical criteria for the adequacy of SDA proposals, the "concrete intelligibility" virtue mentioned above is not satisfied when we invoke a special causal joint that is not related to the ordinary laws of nature. Of course, this special God-world, top-down causal joint may have its own laws but they would lie outside the realm of the natural sciences. On the other hand, there may be no such thing in nature as top-down causation in this strong sense, as ontologically independent of ordinary causes marshaled in complex systems. If there is no such thing in nature, then there is no good reason to propose a special God-world, top-down causal joint, either. But how would we settle the question of the reality of top-down natural causes in the strong sense? This takes us back to the same debates over supervenience, emergence, and reductionism that show up in compatibilist theories of SDA, and I have already given my reasons for not surveying these here. It is enough to note that there was disagreement among DAP participants over the best way to explain the top-down, bottom-up, whole-part, and part-whole features of the behavior of complex systems and that no DAP participant advanced an incompatibilist view of SDA affirming top-down divine causation in this strong sense.[21]

By contrast with the widespread disagreement over the status of top-down causation, the DAP enjoyed consensus on one issue regarding whole-part constraint, which I shall explain in what follows. Whole-part constraint describes an obvious feature of many complex systems whereby larger structures constrain the behavior of constituent parts. Of course, explanations for whole-part (as well as part-whole, top-down, and bottom-up) behaviors vary, as just described. But it is clear that merely describing the behavior using the phrase "whole-part constraint" is not in itself an explanation of it. For a satisfying explanation, we have to turn to ordinary causation marshaled by complex systems, to top-down causes ontologically distinct from ordinary causes, or to something else. Similarly, in the context of incompatibilist

[21] Of course, the phrase "top-down" is used in many ways, which makes a precise discussion difficult to achieve. For example, my working hypothesis is that ordinary causes marshaled by complex systems are sufficient to explain all of the behavior we see around us, including mental causation, and I also hold that the dignity and value of minds and ideas is secured by their ontological status as emergent realities, even though this weak sort of emergence does not entail any strong form of mental causation as new causal powers ontologically irreducible to ordinary causation operating within the complex world of brains-in-contexts (and I care more about securing the value of minds and ideas than I do about their ontological status). Clayton disagrees, taking his stand on the view that certain emergent levels of reality do confer new and irreducible causal powers, pointing especially to life and mind, but taking the final decision on precisely where these new levels of causal power emerge to be largely an empirical question. While these are very different views in respect of their presuppositions about the ontology of causation, both are characterized as affirming "top-down" causation at one point or another in the confusing literature on the subject. From this it follows that, in this area as in so many others, names for positions are invidious and that only patient conversation can discriminate genuinely different views in satisfying ways.

theories of SDA, we must specify a causal joint for God's action if we imagine it operating in the mode of whole-part constraint. Having agreed on this much, the DAP consensus ended. Perhaps God influences wholes (boundary conditions) through causal joints described in other SDA proposals, such as those to be discussed below. Perhaps it is through top-down causation in the strong sense discussed above. Perhaps it is through the primordial creative act itself. Or perhaps the incompatibilist approach of specifying causal joints is simply not the best way to go here, so that we can consider ourselves free to assert that God influences boundary conditions somehow, even if we cannot say how. Peacocke and Clayton appear to take this last, compatibilist path, which leads them to seek traction between their SDA proposals and the natural sciences not through direct constraints but through large-scale consonance.

The final dispute concerns direct mind-to-mind causation, which presumes a dualist account of human nature, whereby human beings have minds and souls somehow related yet not causally reducible to one another. Most DAP participants deny this dualist account of human nature. Ward has great sympathy for this view of human nature, however, and so is willing to entertain direct mind-to-mind causation as one possible mode of divine action.

3.3 Incompatibilist Theories of SDA and Chaos Theory: Polkinghorne

As noted above, one of the major differences among the incompatibilist views represented in the DAP concerns how to deal with the inevitability that our current understanding of the laws and processes of nature will change in the future. Polkinghorne makes a virtue of this, treating current physical laws as approximations to a suppler, subtler underlying reality, within which God can act freely and noninterventionistically. He applies this approach especially to chaos theory, thought of as a physical as well as a mathematical theory.[22] Polkinghorne further proposes that God can insert pure active information with no energy cost into physical systems thanks to the features of chaotic systems, especially the infinite closeness of trajectories within chaotic attractors. Then the sensitive dependence of chaos allows macroscopic effects to emerge from these low-level changes. Thus, God's action is not about moving mass-energy so much as somehow prompting the emergence of new forms of organization.[23]

This first part of this proposal—laws of nature as approximations—is perfectly intelligible, but there has been controversy about the kinds of arguments that might legitimately count for and against it.[24] In particular,

---

[22] See John Polkinghorne, "The Metaphysics of Divine Action," in *CC*, 147-56. He writes, "The deterministic equations from which classical chaos theory developed are then to be interpreted as downward emergent approximations to a more subtle and supple physical reality. They are valid only in the limiting and special cases where bits and pieces are effectively insulated from the effects of their environment. In the general case, the effect of total context on the behavior of parts cannot be neglected" (153).

[23] Ibid., 154.

[24] Saunders has suggested that Polkinghorne's position has been widely misinterpreted but most of his citations are to positions that I think interpret the [Polkinghorne] view correctly and merely dispute the arguments Polkinghorne uses to

critics have targeted the "realist" strategy by which Polkinghorne argues that epistemological limitations in chaos theory entail ontological openness in nature.[25] The epistemological limitations are well understood: they result from the eventual unpredictability of chaotic systems, which in turn is due to the fact that the way they repeatedly stretch-and-fold their input domains produces extreme sensitivity to initial conditions.[26] Polkinghorne is committed to the program of critical realism, a commitment he says most scientists share, consciously or unconsciously.[27] This program seeks the "maximum correlation between epistemology and ontology...Its motto is 'epistemology models ontology'; the totality of what we can know is a reliable guide to what is the case."[28] This instinct lies at the root of the ontological interpretation that most scientists attach to the epistemological limitation expressed in Heisenberg's Uncertainty Principle and Polkinghorne argues that "it is a rational and attractive option to pursue the same strategy in relation to other intrinsic unpredictabilities which we discover in nature."[29] The epistemological limitations of chaos theory, therefore, "signal that ontologically much of the physical world is open and integrated in character."[30]

Polkinghorne's critics typically are critical realists themselves in one way or another, so the fulcrum of this debate is not the epistemology-models-ontology maxim itself but how it is applied. The eventual unpredictability of chaotic dynamical systems absolutely requires a deterministic environment (in the way that the mathematical environment is). It is impossible to take the epistemic limitations of chaos seriously without taking the deterministic framework equally seriously. Yet Polkinghorne detaches the underlying deterministic elements of chaos that his theory needs to explain eventual unpredictability from the ignorance forced on us by the eventual unpredictability of chaos itself. Then he applies his "epistemology follows ontology" strategy only to the ignorance part, despite the dependence of ignorance on determinism (again, in the sense that a mathematical chaotic system is deterministic).[31] Polkinghorne's move seems inappropriately

---

support it; see Saunders, *Divine Action and Modern Science*, 186-96, esp. 186 and 196. In fact, Saunders makes the same critique of Polkinghorne's epistemology-models-ontology argument that Drees, Murphy, Tracy, Russell, and others have made. I do not see evidence in Saunders's presentation to support his claim of widespread misinterpretation.

[25] Polkinghorne, "The Metaphysics of Divine Action," 147-9 and 153.

[26] See the description in Wesley Wildman and Robert John Russell, "Chaos: A Mathematical Introduction with Philosophical Reflections," in *CC*, 71-4.

[27] Polkinghorne, "The Metaphysics of Divine Action," 147-8.

[28] Ibid., 148.

[29] Ibid., 148-9.

[30] Ibid., 153.

[31] Polkinghorne sometimes speaks as if the determinism insisted on by his critics is the physical determinism expressed in Newton's mechanics, as when he writes of "the claim that chaos theory is intrinsically deterministic" both that "[t]his claim was made by several contributors to *Chaos and Complexity*" and that this claim "depends upon taking the Newtonian equations as given and unquestionable—a decision that simply preempts the metaphysical issues from the start" ("Physical Process, Quantum Events,

arbitrary to his critics within and beyond the DAP project, even while most agree that nothing blocks the central metaphysical hypothesis of Polkinghorne's view that the laws of nature are approximations.[32]

The second part of Polkinghorne's proposal concerns the causal joint of divine action. Polkinghorne's use of the concept of active information presupposes the detailed continuance of mathematical chaos theory into the supple, subtle physical reality that he postulates beneath our current laws of nature. For example, both chaotic attractors and sensitive dependence need to exist in nature for Polkinghorne's view to make sense. But these are delicate features of purely deterministic mathematical systems that we have little reason to think could survive, and every reason to believe would vanish, in an indeterministic setting. Something extremely complex is going on in nature, of course, but it is unlikely to be chaos in any of its mathematical senses, all of which require very precise conditions that only the controlled environment of mathematical dynamical systems seem suited to furnish. To put the point ironically, nature seems far too messy for chaos.

So far, then, I have defended the feasibility of Polkinghorne's main hypothesis about the laws of nature as approximations, and I have raised questions both about the arguments he advances to support this hypothesis, and about the proposed causal joint for divine action. But these difficulties may not be decisive and we do well to consider the matter more carefully.

For the sake of argument, therefore, let us grant Polkinghorne's hypothesis that laws of nature are downward emergent approximations to an indeterministic underlying reality. Furthermore, let us determine to use mathematical chaos as a conceptual model for describing this indeterministic ontology, including the causal joint of divine action. To pull this off despite the difficulties just described, we would have to commit ourselves to a dramatic revision of the basic ideas of chaos. The concepts of strange attractors, infinitely fine fractals, and sensitive dependence as we get them from mathematics would have to be rendered usable in an indeterministic physical environment that seems grainy at the level of the very small, and lacking the infinite fineness of mathematics. The most promising place to look for such a

---

and Divine Agency," in *QM*, 189, including no. 8). While I am entirely sympathetic to resisting any preemptive settling of metaphysical questions of determinism, Polkinghorne's characterization of the views expressed in *CC* blends physical and mathematical determinism in a deeply misleading way. Chaos theory only requires mathematical determinism. The question is how this affects our *subsequent* attempts to decide whether physical reality is deterministic or indeterministic.

[32] One critique has gone further. Murphy argued in *CC* not only that the reasoning of Polkinghorne's argument is logically flawed but also that the position itself cannot be saved: "Is this move in Polkinghorne's thought simply an instance of using a bad argument for a position that may well be defensible on other grounds? I think not" ("Divine Action in the Natural Order: Buridan's Ass and Schrödinger's Cat," in *CC*, 327). This more aggressive, second phase of Murphy's critique raises an excellent point, namely, that Polkinghorne's thesis of divine input of active information is difficult to translate into a concrete setting (her amusing example is of Father Murphy trying to save his school in a high-stakes game of pool). This critique deserves an answer but does not clinch Murphy's argument. As far as I can see, nothing can block the hypothesis that our existing laws are approximations to the deep structures of nature.

reformulation is in quantum chaos, which is why Polkinghorne rightly stresses the importance of quantum chaology for future thinking on this subject.[33] Unfortunately, these basic concepts of chaos theory are so completely dependent on the determinism of mathematics that quantum chaos seems a vain hope to me; I think chaos in the strict sense is a mathematical abstraction not directly relevant to the physical world.[34] Nevertheless, there are enough open questions in quantum mechanics itself that hope remains for a new kind of chaos theory.[35]

Alternatively, if we do not want to wait to see if quantum chaology proves successful, perhaps we could try to find ways to loosen the basic dependence of our proposal on mathematical determinism. For instance, because it is impossible in principle to show that chaotic attractors do not occur in nature, we might just continue to assert the thesis that chaos does occurs in nature despite the fact that it seems so improbable and we don't yet have the sort of chaos theory that helps us make sense of such a claim. Or we might find a way to accommodate the likelihood that chaos does not occur in nature, perhaps by abandoning the concept of active information in chaotic attractors as an energy-free way for God to intervene, or perhaps by applying the concept of active information more vaguely to the world of nature underlying our existing laws.[36]

The weight of consensus within the DAP is that chaos theory is poor evidence for ontological openness in reality.[37] In fact, the impact of chaos

---

[33] For example, see Polkinghorne, "Physical Process, Quantum Events, and Divine Agency," 189.

[34] If entering a contest to predict the future of quantum chaos, I would place my bet on there being no future for it at all. I think it is most likely that chaos occurs nowhere in nature, that it is a misleading artifact of non-linear dynamical systems in mathematics that we find exciting in the same way that the abstract ideas of points and lines and planes are intriguing but occur nowhere in nature. It is a beautiful artifact, to be sure, but chaotic orbits and attractors of chaotic systems have never been used successfully in physical modeling and never can be because of intrinsic limits on testing. We can use non-linear dynamical systems for models but we can never take the chaotic part of those models seriously. Moreover, the lumpiness of nature at the atomic level, and beneath that the quantum level, suggests that we are going to have to tell the incredibly rich and intricate story of complex systems in nature without recourse to chaos in nature (Saunders makes the latter point about the impossibility of realizing in nature the delicate fractal geometry of chaotic attractors; see Saunders, *Divine Action and Modern Science*, 186-96, esp. 194-5).

[35] It may be, for example, that deterministic "hidden-variables" interpretations of quantum mechanics, such as Bohm's, or modal interpretations, have potential for elaborating a quantum theory of chaos that the standard interpretation does not, which may in turn count as evidence in favor of such interpretations.

[36] Polkinghorne himself notes the problem with his view of the causal joint, thanking Saunders for the insight. While Polkinghorne provides no solution, he does indicate that he prefers keeping the concept of active information but reformulating it to accommodate the breakdown of the fractal geometry of chaotic attractors, "at least at Heisenberg energy uncertainties"; see "Physical Process, Quantum Events, and Divine Agency," 189.

[37] If chaos theory were the deepest inspiration for a reenergized contemporary defense of ontological indeterminism, then the *practical* epistemological limitations

theory on debates over determinism in nature is two sided. On the one hand, chaos theory enhances any deterministic metaphysical agenda by promising to extend deterministic explanations to complex phenomena that previously were difficult to handle. This serves to remove evidence against metaphysical determinism while adding no further evidence for indeterminism. On the other hand, chaos theory also ensures that the case for metaphysical determinism can never be completely decisive, because chaotic systems can never be tested with the perfect precision needed.[38] Construing chaos theory as evidence for metaphysical indeterminism may be futile, but the traditional motivations to affirm indeterminism are by no means any weaker. Human freedom and moral responsibility, as well some interpretations of quantum measurement, strongly suggest ontological indeterminism. And chaos theory does not block these familiar considerations. Inspiration for defense of ontological indeterminism arises far more naturally from these springs than from the deterministic well of chaos theory. Because the inspiration is there, however, the central hypothesis of Polkinghorne's view, that the laws of nature are approximations, remains perfectly feasible. Though the concept of active information in chaotic attractors seems too dependent on determinism and on the reality of chaos in nature to work in its current form, I have argued that these considerations serve merely to constrain the ongoing development of Polkinghorne's view.

3.4 Incompatibilist Theories of SDA and Quantum Mechanics: The Possibilities

The remaining views on the diagram all develop their positions with reference to quantum mechanics. This only makes sense, because traction as consistency increases as we move rightwards through the diagram, and the existence of stochastic laws governing measurement events allows quantum mechanics to constrain theological claims about divine action more directly than in other spheres of science.

The views of SDA at the quantum level (hereafter, QSDA) differ most basically on the question of the status of stochastic laws of nature. The three-fold distinction discussed above captures the relevant features of this disagreement: stochastic laws may be descriptive, ontological (in the sense of statistically constraining only ensembles of quantum measurement events), or strong-ontological (in the sense of constraining even individual quantum measurement events). The disagreements among DAP participants led to a number of different but viable proposals for objective, incompatibilist, noninterventionist theories of SDA, and I will endeavor to make clear the reasons DAP proposals diverged.

Despite the friendly disagreements, there was complete agreement on one point, namely, that the strong-ontological interpretation of stochastic laws of nature was not the right environment for developing an objective, incompatibilist, noninterventionist theory of QSDA. Tracy thought that the

---

suffusing the deterministic mechanics of Newton already would have inspired enthusiasm for ontological indeterminism. It did not work that way with Newton and it should not work that way with chaos theory either.

[38] See Wildman and Russell, "Chaos," esp. 75-83.

strong-ontological interpretation of stochastic laws of nature in quantum mechanics is probably incoherent. Most others thought the idea probably is coherent but rejected the strong-ontological interpretation as hostile to the goal of a noninterventionist account of SDA. Of course, some people have claimed both that the strong-ontological interpretation of stochastic laws of nature in quantum mechanics is coherent and that theories of QSDA should adopt it, or are somehow committed to it regardless of their protestations to the contrary.[39] Yet none of the defenders of QSDA affirms the strong-ontological interpretation of stochastic laws.

The DAP did reach consensus on the various possibilities for QSDA. I will present these options in what follows. As we get started on this, note that a basic grasp of the mathematical formalism of quantum mechanics, and especially quantum measurement, is quite important to avoid errors, so I pause here to give a quick, non-technical summary.

Our current knowledge of the quantum world seems to allow for several places at which God might act, arguably without violating any laws of nature. Most revolve around the great mystery of quantum mechanics: measurement events. The Schrödinger equation and its relativistic equivalent the Dirac equation describe with great precision the apparently deterministic evolution of quantum systems between measurement events. The model for a quantum system in these equations is a continuous function, a so-called wave function, which assigns a complex number to every spacetime point in such a way that, roughly speaking, the square of the wave function is a probability distribution expressing the likelihood that the system will be in a particular state if it were measured.

The mystery of quantum measurement is expressed in Max Born's "projection postulate." Quantum systems when measured are always found in certain basic states, depending on the quantity being measured, whereas the quantum formalism represents the state of an evolving quantum system just prior to measurement as a complicated superposition (linear combination) of these basic states. Every meaningful sort of measurement is associated with a distinctive set of basic states. The projection postulate supposes that, when a measurement occurs, the quantum system irreversibly collapses onto one and only one of the basic quantum states associated with that sort of measurement. If the same type of measurement is performed repeatedly on similarly prepared quantum systems, the resulting statistics show that the probability of obtaining a particular post-measurement state corresponds precisely to the weighting of the post-measurement state in the superposition of basic states that is the pre-measurement wave function.

The quantum formalism is usable and clear on its own terms yet the ontology of the situation is utterly obscure. How does a superposition of basic states suddenly lurch into one and only one of those basic states, conforming all the while to the probabilities specified by quantum theory? What kind of theory predicts measurement outcomes with perfect accuracy using a model of quantum systems that is ontologically indecipherable? As unsatisfactory as this

---

[39] Within the DAP, this criticism was advanced most notably by Peacocke. Some others shared his intuition on this point. Outside of the DAP, Saunders appears to hold this view; see below for the details.

situation is at the level of philosophical interpretation of quantum mechanics, the mathematical formalism itself continues to be experimentally robust and extremely accurate.

DAP participants have pointed out that we must assess the prospects for ontological openness capable of supporting a causal joint for non-interventionist QSDA relative to the various philosophical interpretations of the quantum formalism. There are a couple of dozen such interpretations. Many of these are relevant to the question of QSDA but I do not have space for a comprehensive treatment. Without any discussion, therefore, I make the following simplifying assertions. Recall that DAP proposals for QSDA sought incompatibilist and noninterventionist theories, which require indeterminism in nature. We can exclude nonlocality and the Heisenberg Uncertainty Principle[40] as suitable places to look for ontological openness, because both are properties of the deterministic Schrödinger equation, which makes the prospects for ontological openness dim.[41] Likewise, we can exclude all theories that extend the determinism of the Schrödinger equation to the theory of quantum measurement using nonlocal hidden variables (such as David Bohm's) because of their rejection of indeterminism, and all views that postulate direct divine manipulation of wave functions because of the interventionism involved. We have to exclude interpretations that face serious theoretical difficulties at the level of the physics, such as continuous spontaneous localization or decoherence theories. Finally, I contend that all of the relevant philosophical issues surface in two of the remaining interpretations, so that we can neglect the others. Specifically, the space for non-interventionist divine action arises in two basic ways: either when the deterministic evolution of a wave function is disrupted at a non-reversible measurement event, as this is described within the standard interpretation of the quantum formalism; or in the strange spaces between quantum worlds that result from measurement events on various quantum-branch scenarios.

We can picture the options for indeterminism in quantum mechanics by means of the following diagram, which portrays measurement events as if they were several frames in a movie: just before the measurement event, at the onset of the measurement event, the selection of outcome(s) within the measurement event, and the post-measurement state(s).

[40] Insofar as the Uncertainty Principle describes measurement outcomes, it is included in the measurement problem, which is the focus of attention in the analysis below.

[41] Note, however, that Wegter-McNelly's recent dissertation, *Created Wholeness*, argues that quantum entanglement and nonlocality furnish analogies that are useful for articulating a *compatibilist* theory of SDA.

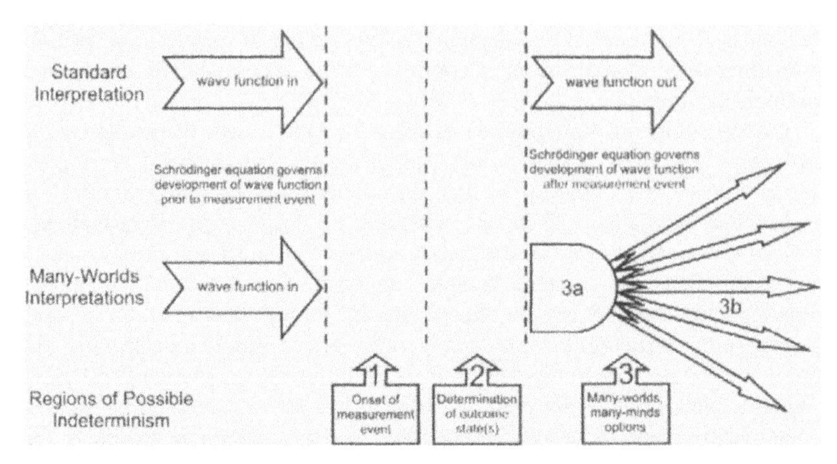

Figure 3: Regions of Indeterminism in Quantum Measurement Events

Region 1, the onset of measurement events, is not well understood in quantum mechanics. Does it require a quantum system to interact with a macro system? With another quantum system? With consciousness? Most importantly, does it have a stochastic element? Only in the case that there is some stochastic element involved in the onset of quantum measurement can we understand region 1 as involving indeterminism. Decoherence theories explicitly propose this but it is not difficult to imagine that a stochastic element could be involved on other interpretations, also. The indeterminism of region 2 appears in the stochastic process governing the selection of outcome events (the projection postulate). Indeterminism in region 3 is more difficult to make out. There are no prospects for indeterminism in region 3 on the standard interpretation but the many-worlds family of interpretations has more flexibility. In the diagram, the rounded shape (marked 3a) is supposed to suggest a stage between determination of outcome states (region 2) and the reality of split states (region 3b). We can imagine God acting there to select among potential worlds before they become actual. Similarly, after quantum splitting has produced actual worlds (region 3b), it is possible to imagine God acting between these worlds, providing we can construct a view of indeterminism that applies to the strange ontological spaces between and among these split worlds.

The three possible locations of indeterminism in this portrayal of a quantum measurement event lead to six possibilities for objective, incompatibilist theories of QSDA that we might claim are non-interventionist. I list these six options here and defer discussion of the question of their status as noninterventionist because this is their most controversial aspect.

Within the region marked (1), there is one possibility for QSDA in the indeterminism that might exist if there is a stochastic element in the onset of quantum measurement events.

- [OPTION #1] God could initiate measurement events.

Within the region marked (2), there are two possibilities for QSDA grounded in the indeterminism of measurement events:

- [OPTION #2] God could adjust probabilities to make an outcome state more likely.
- [OPTION #3] God could select an outcome state.[42]

Within the region marked (3), there are three possibilities for QSDA grounded in the possibilities for indeterminism associated with many-worlds interpretations:

- [OPTION #4], within region 3a: God could behold the array of worlds produced by a measurement event "just before" they become real, evaluate them, and then select one world to become actualized while letting the others never come into being.
- [OPTION #5], within region 3b: God could attend to some worlds and neglect others.[43]
- [OPTION #6], within region 3b: On the many-minds version of the many-worlds interpretation,[44] God could change consciousness so that we are able to construct a different consistent history of reality than otherwise would be possible.

---

[42] OPTION #3, that God selects outcome states, seems closely related to OPTION #2, that God adjusts the probabilities of measurement outcomes, with the difference being merely that God assigns a probability of 100% to one possible outcome and 0% to all others. We cannot simply collapse the distinction between the two views, however, because OPTION #2 seems to assume a strong-ontological interpretation of stochastic laws of quantum mechanics as constraining individual quantum measurement events, whereas OPTION #3 is more neutral on the question of how to interpret the stochastic laws of quantum mechanics.

[43] While this option is not a causal joint proposal like the first four options, and while it exacerbates the theodicy problem rather bluntly, it is worth pondering. OPTION #5 imagines that God allows all universes to exist but eventually loses interest in the "failures" and pays attention to and answers prayers only in the worlds that prove interesting to God. This is a kind of natural selection of worlds where the outcome is not survival of the fittest but maintaining the divine focus of attention and action. This view focuses on the divine intention rather than causal joints, much as the primary-secondary causation model discussed above focuses on the divine intention with regard to events within one world. While some might well complain about the severity of the theodicy problem inherent in a view that paints God as negligent of parts of creation, I think that this hypothetical theodicy problem is precisely as difficult as the one we actually have in this world, and on the same terms, the familiar charge being that God is negligent of certain parts and people of our world, not acting when acting would seem to ease pain, prevent cruelty, and increase justice and love, and educate human beings in much needed ways.

[44] David Z. Albert gives a clear description of the many-minds view in *Quantum Mechanics and Experience* (Cambridge, Mass.: Harvard University Press, 1992). This view offers significant advantages over the ordinary many-worlds interpretation and the price paid is modest (in the context of the alternatives): accepting an unanticipated role for consciousness. A more general framework for understanding this version of the many-minds approach is the consistent histories approach, on which see Chris Clarke, "The Histories Interpretation of Quantum Theory and the Problem of Human/Divine Action," in *QM*, 159-78.

> This view locates indeterminism in the communication between divine and human consciousness rather than in quantum measurement, but does not explain how this is possible.[45]

The fifth and sixth options yield meaningful perspectives on QSDA. Whether we can give OPTION #5 a non-interventionist rendering would depend on what else we wanted to say about the way God "pays attention." As far as a causal joint is concerned, this view reduces to other views, which in the quantum context means OPTIONS #1-#3. We could argue that QSDA in the sense of OPTION #6 is non-interventionist on two grounds. On the one hand, we could suppose that no laws of nature (not even the Schrödinger equation) govern this sort of between-the-worlds action on the many-minds interpretation. On the other hand, we could postulate that the interaction between divine and human consciousness has indeterministic elements. In any event, neither OPTION #5 nor OPTION #6 was formally defended by DAP participants. All of the debates over QSDA swirled around OPTIONS #1-#4, accordingly. Among these four, OPTIONS #1-#3 apply to most interpretations of quantum measurement whereas OPTION #4 only arises in the context of the many-worlds family of interpretations.

The trick in rendering each of the first four options non-interventionistically is to ensure that the ruling interpretation of stochastic laws of nature in quantum mechanics is consistent with noninterventionist divine action within the spheres of indeterminism that each proposes. We have discussed three interpretations of stochastic laws of nature and each has some consequence for theories of SDA, as follows.

First, suppose we interpret laws of nature (including stochastic laws) descriptively, with Murphy and Russell. Then, assuming God's action conforms to these laws (noninterventionism), they describe not only nature's operations but also God's actions within nature, whenever, wherever, however often, and in whatever mode those actions occur. This requires the theologian to view SDA not only as accomplishing God's providential purposes but also

[45] OPTION #6 deserves serious consideration in the context of the many-minds interpretation in which measurement events do not describe ontological splits but rather the synchronizing of conscious observers and events within an unimaginably complex superposition of states. All possible quantum "worlds" thus coexist within a single superposition and consciousness just happens to be the sort of thing that selects out an intelligible world for observation. On this interpretation, special divine action might be a kind of mind-to-mind influence in which God triggers subtle shifts in our consciousness so that we see a slightly different kaleidoscopic cross-section of the vast superposition that is reality. It may even be possible for observers within the world to modify their own consciousness in such a way as to skip to other world-synchronizations. This view also allows for the possibility that people capable of observing one another could yet see other features of the world differently, which possibility could serve as a speculative framework for articulating the distinction between conventional and ultimate reality that certain Buddhists and Hindus deploy to describe enlightenment and stages along the way to that ultimate state of liberation. If forced to choose among the various options for QSDA, I would choose this one, not because of its interpretation of the quantum formalism but because of the elegant way its rendering of SDA corresponds to long-held beliefs in the world's great spiritual wisdom traditions.

as sufficiently regular and mathematically intelligible that scientists can frame the laws of nature. It also strongly invites the theologian to treat the universal aspects of God's action as the ultimate explanation for all of the regularities of nature, an invitation that both Russell and Murphy accept. Once this invitation is accepted, however, given that the laws of nature describe the overall pattern of divine action regardless of level or mode, a meaningful connection between SDA and the quantum world may be in danger. In fact, the only reason to suppose that the quantum level is a distinctive locus for divine action would be that we specify a causal joint for divine action specifically at the quantum level. It is important, therefore, to engage the details of quantum mechanics and make a causal joint proposal if we want both a descriptive account of laws of nature and a concretely intelligible theory of QSDA.[46]

Second, if we interpret stochastic laws of nature ontologically, in the sense of constraining only ensembles of events (with Ellis and Tracy), then these laws do constrain God's action in a minor way, if it is to be noninterventionist. God must be sure to make experiments in which scientists gather quantum statistics come out right. This is definitely an awkward constraint in the sense that human beings can more or less force God to act in a particular way, constraining the divine freedom. Yet in practice it is not so severe, particularly if God only acts in some but not all events, because providentially relevant events are unlikely to include anything about which scientists can gather quantum statistics.[47]

Finally, if we interpret stochastic laws of nature in the strong-ontological sense, which constrains each individual quantum measurement event, then any action of God will violate those laws and noninterventionist versions of QSDA will be impossible. No DAP participant defending a noninterventionist theory of QSDA holds this view of the laws of nature, obviously, but some DAP participants arguing against QSDA (such as Peacocke) appear to believe that this view of stochastic laws of quantum mechanics is the correct one.

3.5 Incompatibilist Theories of SDA and Quantum Mechanics: Two Critiques

Since Saunders has gone to great lengths to critique existing proposals for QSDA,[48] it is worthwhile pausing to assess his arguments. We can also generate a hypothetical but illuminating critique of efforts to develop theories

---

[46] I conjecture that it is partly because Murphy does not like these consequences of a descriptive approach to the laws of nature that she speculates about OPTION #4, which ties her theory of SDA decisively to the quantum realm in a way that a descriptive approach to the laws of nature requires, thereby gaining valuable theoretical intelligibility.

[47] I think it is partly because of this difficulty that both Ellis and Tracy entertain divine action only in some providentially relevant events, rather than in all events. This gives the strongest and least contrived answer to the theological complaint that we can't have human beings constraining divine freedom to act based solely on whether scientists happen to be gathering quantum statistics.

[48] See Saunders, *Divine Action and Modern Science*, chaps. 5-6. Also see his "Does God Cheat at Dice?: Divine Action and Quantum Possibilities," *Zygon* 35:3 (2000): 517-44.

of QSDA based on Kant's analysis of the powers of human reason. I will discuss both of these critiques in what follows.

First, we turn to Saunders. *Divine Action and Modern Science* makes a significant contribution to debates over SDA especially because Saunders gives detailed attention to so many of the relevant philosophical, theological, and scientific issues. In particular, he gives a clear and reasonably accessible presentation of most of the features of quantum mechanics relevant to deciding whether a proposal for QSDA is consistent with what scientists believe is the case in quantum mechanics.[49] His presentation is not quite comprehensive in its coverage of indeterministic interpretations of the quantum formalism,[50] which matters because his argument that contemporary theology is in crisis depends on not overlooking any promising alternatives. Nevertheless, his presentation of quantum mechanics should prove useful to theologians who want an accessible introduction to quantum mechanics and its interpretation.

In relation specifically to QSDA, we might ask how Saunders constructs his argument. Though he does discuss the proposals of Ellis, Murphy, Russell, and Tracy,[51] his primary argument unfolds independently of the consideration of particular positions.[52] This is helpful because it minimizes debates over fine points of interpretation and keeps the focus on the conceptual structure of proposals for QSDA. Saunders follows the incompatibilist strategy of seeking indeterminism in the quantum realm (which requires selecting an interpretation of the quantum formalism) and then trying to develop a noninterventionist form of QSDA within that indeterministic space. For our limited purposes, it is sufficient to consider his discussion of indeterminism and QSDA relative to the standard interpretation of the quantum formalism, because all of his disagreements with QSDA proposals within the DAP emerge in this context.

Rightly rejecting divine manipulation of wave functions between measurements as bluntly interventionist, Saunders identifies three ways in which the standard interpretation of quantum mechanics may be indeterministic and thus three ways to conceive of QSDA. These correspond to OPTIONS #1-#3 discussed above (OPTIONS #4-#6 do not arise for the

[49] See Saunders, *Divine Action and Modern Science*, 127-48, which is the first part of chap. 6; see also matching parts in idem, "Does God Cheat at Dice?"

[50] Specifically, Saunders's treatment of divine initiation of quantum measurement events does not discuss the possibility that there may be some stochastic element involved, even outside the continuous spontaneous localization or decoherence theories, which is the key to making OPTION #1 feasible. He alludes to OPTION #4 in passing but does not analyze it fully and so misses its potential for an incompatibilist account of QSDA; see Saunders, *Divine Action and Modern Science*, 162. Also, OPTION #5 does not appear in his discussion of the many-worlds approach (159-62), though he does note correctly that compatibilist approaches (of which OPTION #5 could be one if suitably interpreted) have some room to breathe in the many-worlds view. Moreover, he gives no consideration to the strange and possibly indeterministic spaces *between* worlds that result from quantum splitting in the many-minds approach, or the potential for a mind-to-mind causation approach, both of which come into play in OPTION #6.

[51] See Saunders, *Divine Action and Modern Science*, 110-26, which is the last part of chap. 5.

[52] Ibid., esp. 149-56, but continuing through to the end of chap. 6; see also matching parts in idem, "Does God Cheat at Dice?"

standard interpretation of the quantum formalism). So far, then, there is agreement with the DAP on the options for QSDA. What does he say about each one?

Regarding OPTION #1, which proposes that God may make measurements on a quantum system, Saunders points out three difficulties. First, the empirical adequacy of quantum mechanics depends on there being no unexpected collapses intervening between measurements, as there might be if God initiated measurement events. Second, because measurements involve the interaction of parts of nature, God can't contrive to cause an interaction without intervening. Third, God cannot control the outcomes of a measurement event using this approach, but only trigger a measurement event and leave the outcome to chance, which is theologically awkward.

The first of these objections is unimportant in practice, as there is no reason to think that God would be providentially initiating measurement events in physics experiments. Anyway, the vagaries of any experimental apparatus entail that assessing anomalous data is a statistical process, which leaves plenty of room for masking divine initiation of measurement events in the statistical noise of experimentation. The third objection is spurious because nothing prevents combining OPTIONS #1 and #3, thereby allowing God both to trigger measurement events and to select their outcomes. The second objection is more serious because, on that view of the onset of measurement events, there is no indeterminism whatsoever. But recall that the DAP conclusion about OPTION #1 was that noninterventionist divine initiation of quantum measurement events is possible only if there is some stochastic element in the onset of measurement and if the relevant laws of nature are not given a strong-ontological interpretation. There is no reason at this stage in the development of quantum theory to rule out stochastic elements within the onset of quantum measurement events. Unfortunately, Saunders does not address this possibility. But it matters little because no DAP participant consistently affirmed OPTION #1.[53]

Regarding OPTION #2, that God may alter the probability of obtaining a particular result in a quantum measurement event, Saunders's main objection coincides with that entered here, namely, and in my terminology, that this option presumes a strong-ontological interpretation of stochastic laws of quantum mechanics and thus is interventionist. Of course, the fact that Saunders does not use the same terminology as the DAP at this point complicates judgment slightly, but the arguments seem to match.

Regarding OPTION #3, that God may select the result of a measurement event, Saunders is similarly pessimistic. Unlike the unimportant disagreement over OPTION #1 and the agreement over OPTION #2, however, the disagreement with DAP conclusions in the case of OPTION #3 is serious indeed and we very much need to understand it. Saunders's lack of clarity about statistical constraint of ensembles of events versus constraint of individual events in the interpretation of stochastic laws of quantum mechanics lies at the root of the disagreement. The categories he uses in his analysis of laws of nature concern whether probabilities are ontologically prior to or

[53] Russell alludes to it in some places while rejecting it more clearly in others; see below for the specific details of DAP proposals for QSDA.

derivative from measurement events. But these terms are too coarse in respect of expressing only two options rather than the three that are relevant (descriptive, ontological, strong-ontological). They are also too vague in respect of being untranslatable into the language of statistical constraints on ensembles or theoretical constraints on individual events. Nevertheless, in an attempt to understand what can be understood, consider Saunders's key argument about OPTION #3 [my comments are in square brackets]:

> The technical substance of this approach is to deny that Born's probability interpretation of the wave function [i.e. the projection postulate] has any ontological priority and assert that it is simply an approximate relationship between ensembles of identical systems for a given measurement repeated a large number of times. The next move is to interpret quantum laws in a regularitarian methodology [i.e. as descriptive]—a move that is quite at odds with the position of every proponent of quantum SDA considered above [including Ellis, Murphy, Russell, and Tracy].[54]

There appear to be several problems here. First, an ontological interpretation of the projection postulate does not deny all ontological priority to the probabilities; it regards them as "ontologically prior" but as constraining ensembles rather than individual events. Saunders here suggests without argument that we have to embrace a strong-ontological interpretation of quantum laws of nature. But if we really do have to embrace a strong-ontological interpretation of the laws of nature, then we don't need these two chapters of his book or a bunch of conferences to conclude out that a non-interventionist account of QSDA is impossible. That is why DAP participants advancing theories of QSDA reject the strong-ontological interpretation of stochastic laws of nature. Second, Saunders mistakenly assimilates the ontological and descriptive interpretations of stochastic laws of nature, thereafter arguing against both by attacking only the descriptive interpretation. Third, in the final sentence, Saunders simply mischaracterizes the positions of proponents of QSDA. As a result of these considerations, Saunders's discussion of OPTION #3 never achieves optimal clarity, his invidious "either-or" choices overlook more feasible alternatives, and he produces no arguments capable of unsettling the conclusion of the DAP that OPTION #3 is feasible so long as we reject a strong-ontological interpretation of the relevant stochastic laws of nature.

This is a key lapse in a book that usually is argued closely and well. It serves as evidence for Saunders's own point, namely, that approaching analysis of options for SDA with inadequate distinctions (in this case, pertaining to stochastic laws of nature) can produce misleading results. How did this happen? I would conjecture that Saunders has an intuition that the strong-ontological interpretation of stochastic laws must be correct. Yet instead of simply arguing for this and then drawing the obvious conclusion that noninterventionist QSDA is impossible, he presents an unnecessarily complicated argument with the same conclusion, without materially new conceptual elements. Saunders does not adequately explain his attachment to the strong-ontological interpretation of stochastic laws of nature, though we

---

[54] See Saunders, *Divine Action and Modern Science*, 155.

should expect to find such an explanation given the way his argument against QSDA proposals depends on it.[55]

Moving from Saunders back in time to Kant, we find the inspiration for an elegant tetralemma argument that can serve as a lens for viewing the strategic moves made within the DAP as participants sought to develop theories of SDA. This argument is particularly illuminating in the case of theories of QSDA.

To frame the tetralemma argument, let us assume (contrary to fact, for most DAP participants) the most demanding criterion for an adequate theory of SDA, namely, the following conjunction of four propositions:

> {objectivity} and {incompatibilism} and {noninterventionism} and {strong-ontological view of laws of nature}.

Then the argument concludes that:

> All theories of SDA fail to meet this criterion.

If this argument against the possibility of SDA is valid—and no view of SDA that I am aware of challenges the entailment—then we can protect SDA only by weakening or rejecting one of the four propositions defining the criterion for success. Of course, no DAP participant accepts this four-fold criterion as the desirable goal for a theory of SDA. But the various moves within the DAP can still be analyzed along these lines, and there is a payoff in insight for making the effort.

With the tetralemma argument now in place, let us consider the secret of its validity (note: not its soundness!). It is an application to the specific context of QSDA of a more general point that Kant made long ago. Kant argued that we never can give a causal analysis of reality in such as way to justify our intuitions of human freedom and moral responsibility.[56] Whenever we try to use our ordinary reasoning powers (from causes and other such categories) to articulate and justify human freedom, we find ourselves illustrating his antinomy of pure reason: we simply can't get there from here. For Kant, we can only reconcile categories of causality and human freedom in compatibilist fashion by postulating human freedom as a condition for the possibility of our experience; we can never demonstrate its consistency with a causal, scientific account of nature.

---

[55] Ibid., chap. 3. Unfortunately, there is no clear discussion of the scope of stochastic laws of nature governing quantum measurement events (do they apply to ensembles only? to individual events?) in the otherwise helpful treatment of laws of nature in that chapter, yet this distinction seems to be operating silently in the critiques of chapter 6. It is similarly unfortunate that the *reasons* for rejecting the strong-ontological interpretation are not presented clearly in existing DAP publications. But there is no mystery about this: Ellis, Murphy, Russell, and Tracy are right to believe that the strong-ontological interpretation of stochastic laws instantly destroys their proposals and they are justified in spending their energy arguing for the feasibility of their own views of the laws of nature and SDA.

[56] See Immanuel Kant, *Critique of Pure Reason* (New York: Macmillan, 1929, 1933). The argument about freedom and determinism is in the Transcendental Dialectic, Book II, Chapter II, Section 9.

Kant's transcendental philosophy has not survived the transition away from deterministic Newtonian physics as well as Kant might have wanted, but the deep point he makes about the antimony of pure reason remains difficult to dismiss. In our era, we still encounter the phenomenon Kant described: the more precise our articulation of causal joints, the more elusive ontological openness becomes, and the more we confirm our suspicions that we need to switch away from an incompatibilist approach in order to make the postulate of human freedom credible. In one respect, Kant's insight is independent of the apparatus of the transcendental dialectic of the *Critique of Pure Reason*. Science is causal language from beginning to end and only fitted to describe the causal web of reality. Where the scientific project of detailing the causal web runs aground, as it does in the quantum measurement problem, science lapses into silence; there can be no scientific account of the workings of indeterminism. Within this silence, however, the speculative metaphysical urge to continue the scientific project of explaining how things work lifts its voice and keeps us busy, despite its intrinsic limitations. In our time, it is arguments over the philosophical interpretation of laws of nature that express this urge. When we accept the strong-ontological interpretation of laws of quantum mechanics, we drive the controlling powers of scientific laws of nature all the way into each individual quantum measurement event, thereby subjecting even ontologically indeterministic processes to rigid (probabilistic) laws that destroy the incompatibilist project of locating human freedom to act in ontological indeterminism. That's why Kant was a compatibilist. Moreover, while Kant's immediate concern was human freedom to act, his argument applies equally well to divine freedom to act.

With Kant in mind, we can portray the DAP in general terms as trying to satisfy the theological urge to locate human and divine freedom in the causal web of nature while struggling with Kant's insight about the antinomy of pure reason. The struggle is least compelling when we reject objective SDA or when Kant's own or another compatibilist approach relaxes the tension between categories of freedom and categories of causation. It is more compelling when we adopt an incompatibilist approach, and more difficult still when we force our inquiry to abide by conditions that maximize traction between metaphysics and science, such as ontological interpretations of the laws of nature and non-interventionism. As our causal account of the joint of divine action becomes most precise, as it does if we embrace the four-fold criterion of the tetralemma argument, we will discover exactly what Kant implied that we would, namely, that the vision of divine freedom to act vanishes the way a mirage in the desert vanishes when we approach it. To avoid this, we will have to pull out of the intellectually suicidal dive that speculative metaphysics takes us on: we will have to relax the constraints that maximize traction between metaphysics and science or else collide with the immovable fact that we can never argue from categories of causality to categories of freedom. Our choice is when to pull out of the dive. Which of the traction-maximizing constraints on our inquiry will we relax?

Within the DAP, the problem does not arise for Davies, Drees, and others who deny the objectivity of SDA. Clayton, Peacocke, Soskice, Stoeger, and Ward see the writing on the wall early on and submit more completely than others to Kant's strictures on speculative metaphysics.[57] The others stay on course for Kant's predicted collision longer by embracing objectivity, incompatibilism, and noninterventionism. Within this metaphysically more aggressive sphere of theological work, traction (as consistency) between theology and science is strongest, yet there are still ways to position theories of SDA that avoid Kant's antinomy. That is, we can still ask precisely where and how in our speculative metaphysical interpretation of quantum measurement events God is supposed to act. In this way, we will entertain a more detailed causal story (though of course elaborated not scientifically but metaphysically) and more heavily constrain our theological assertions about divine freedom. Polkinghorne weakens the incompatibilist commitment by stressing the provisional status of existing laws of nature. Murphy and Russell treat the laws of nature as describing God's universal (or in Russell's case almost universal) action in nature. Ellis and Tracy accept an ontological interpretation of laws of nature, which statistically constrains ensembles of quantum events but not individual measurement events, and thereby enjoy the greatest degree of traction, but they too avoid the Kantian specter of the collapse of speculative metaphysics. If we maximize traction still further by accepting (with Saunders?) a strong-ontological interpretation of the stochastic laws of quantum mechanics, then the inability of causal categories to comprehend categories of freedom will force us to conclude that prospects for theories of QSDA are grim, and Kant's prophecy of doom will be fulfilled.

If the argument here is correct, then the relaxation of constraints—constraints that theories of SDA use to generate traction (as consistency) between science and theology—is unavoidable, for Kant's reasons. But it is also unproblematic so long we regard the demanding four-fold criterion as dispensable. In terms of this Kantian perspective, I think Saunders's critique amounts to holding an inevitable outcome of a speculative metaphysical venture against the theorists, as if they could with more care or more imagination or more attention to detail somehow avoid it. With that in mind, we might consider steering away from representing the relaxation of ideal constraints on theories of SDA as failures to reach a goal, which after all is illusory anyway. Rather, having established the technical feasibility of SDA proposals, we should celebrate the artistry of these intellectuals, and understand this artistry in terms of the way traction is first maximized for the sake of concrete intelligibility of theological proposals, and then relaxed in a variety of ways for the sake of avoiding the collapse of theories of SDA under the implacable weight of the tetralemma argument and the limitations of human reason that it describes.

3.6 Incompatibilist Theories of SDA and Quantum Mechanics: Specific Proposals

Having articulated the options for QSDA and defended the feasibility of most options in general terms, I now briefly analyze similarities and differences

---

[57] On this, see especially Stoeger's elegant discussion in "Describing God's Action in the World in Light of Scientific Knowledge of Reality," in *CC*, 239-62.

among the four specific proposals made within the DAP: those of Ellis, Murphy, Russell, and Tracy.[58]

These four views have in common the view of the causal joint of divine action: it is OPTION #3, above. This option hypothesizes that God selects the outcomes of measurement events. Russell has experimented in his published writings with OPTION #1, that God initiates measurement events,[59] and Murphy has experimented with OPTION #4.[60] No DAP participant formally defended OPTION #5 or OPTION #6, though I explored the latter in the final

---

[58] We might think that Polkinghorne's view counts as a fifth proposal for QSDA. After all, consistency demands that Polkinghorne, rather than arguing against QSDA, should make the same hypothetical proposal in relation to epistemological limits in quantum mechanics that he makes in relation to the epistemological limits of chaos theory, namely, that the relevant laws of nature are downward emergent approximations to a suppler, subtler physical reality within which God acts freely. He approaches this toward the end of his essay, "Physical Process, Quantum Events, and Divine Agency," 188-90, but, even in the context of a magnanimous review of the debate over QSDA, he implies that beginning speculation about SDA from chaos theory and starting from quantum mechanics are competing approaches. I have not yet grasped how this can be so, even if we accept the problematic causal joint proposal of active information changing complex systems through zero-energy alterations of particle trajectories on chaotic attractors. In fact, I would think that Polkinghorne's approach would apply to every domain of science and every level of reality and every physical law.

[59] At one point Russell clearly rejects the idea "that God … makes measurements on a given system" (Russell, "Divine Action and Quantum Mechanics: A Fresh Assessment," in *QM*, 296). Yet at the beginning of the same essay, he writes, "[I]f quantum mechanics is interpreted philosophically in terms of ontological indeterminism…, one can construct a bottom-up, noninterventionist, objective approach to mediated direct divine action in which God's indirect acts of general and special providence at the macroscopic level arise in part, at least, from God's objective direct action at the quantum level both in sustaining the time-development of elementary processes as governed by the Schrödinger equation and in acting with nature to bring about irreversible interactions referred to as 'quantum events'" (Ibid., 293). Here and in earlier writings, Russell appears at least to entertain the possibility that God could initiate quantum measurement events. I argued above that this possibility is not as problematic as he seems to conclude.

[60] In informal discussions surrounding the DAP, but not I think in writing, Murphy has experimented with the idea that God may select outcomes of measurement events in the sense of OPTION #4. She intends this as a thought experiment intended to resolve a difficulty in the many-worlds interpretation of the quantum mechanics formalism, at the same time as finding in quantum mechanics an ingenious source of support for the idea of special divine action of the pervasive "all events" sort. She points out that a few other people have explored this idea in conversation. While most views asserting universal intentional divine action usually remain vague on the question of the causal joint of action, Murphy's thought experiment is quite specific and especially interesting because of that. In particular, specifying a causal joint using OPTION #4 offers a way for Murphy to limit God's action to the quantum level that is less arbitrary than OPTION #3 alone. After all, her descriptive approach to laws of nature is indifferent to the locus in nature of God's action, apart from independent specification of a quantum-level causal joint. Alternatively, of course, Murphy might prefer to capitalize on the compatibilist tendencies of a descriptive approach to laws of nature. These same considerations apply to Russell's approach.

conference meeting.[61] And OPTION #2 is not promising, as noted above. Of all DAP participants, Tracy has most clearly resisted the idea that one has to choose one approach rather than others; in fact, he has clearly stated that he does not hold that God acts only at the quantum level, or even exclusively in causal gaps in nature.[62]

The differences among the four DAP participants advancing specific quantum-level proposals for SDA pertains to the scope of divine action, in three senses. First, there is the question about whether God acts specially at the quantum level always and everywhere or only at particular places and times. This difference is most evident in the comparison of Murphy's affirmation of divine action at every time and place[63] with Tracy's view of intermittent quantum special divine action.[64] Ellis sides with Tracy on this issue. Interpreting Russell's view on this point is more complex but his developed view is quite clear: God acts in all quantum measurement events and only refrains from acting when a free, conscious agent acts instead.[65]

---

[61] Though this is not my theological territory, recalling the need for an interpreter to maintain credibility, I should state my preferences among these and related options for SDA. To that end, consider the following chain of counterfactuals. If hypothetically forced to accept that God is a personal being who can form intentions and act on them, then I strongly prefer a theory of SDA that affirms miraculous abrogation or suspension of natural laws, owing to its protection of the divine freedom against human pretensions to understand and control it. If further forced to accept the goal of noninterventionism, then I prefer a compatibilist approach for the same reasons. If still further compelled to select among the options available to me within the sphere of objective, incompatibilist, non-interventionist theories, then I would elect the many-minds interpretation of the quantum formalism and OPTION #6 as the mode of divine action because of the flexibility it offers in talking about divine action and because of its consonance both with south Asian and east Asian interpretations of the religious quest for enlightenment and liberation, and with west Asian interpretations of sanctification and divinization.

[62] In explaining his hypothesis that God acts in indeterminacies at the quantum level, Tracy goes so far as to declare that, "I am *not* saying that God acts *only* through the gaps in the causal order of nature" (Tracy, "Particular Providence and the God of the Gaps," in *CC*, 319, italics in original). In fact, DAP participants usually took an open-ended approach to theorizing about SDA, focusing more on establishing the feasibility of their favored proposal rather than arguing that their view is the only possible one. Tracy was particularly concerned to develop what he calls a "tool box" of options for understanding SDA, within which he thinks his theory of QSDA has a rightful place, along with compatibilist possibilities and even miracles.

[63] See Murphy, "Divine Action in the Natural Order," esp. 340-2, where she writes the line, "To put it crudely, God is the hidden variable" (342).

[64] See Tracy, "Particular Providence and the God of the Gaps." Tracy discusses the issue of the scope of God's action at the quantum level on 320-2.

[65] Though Russell's contribution to *EMB* focuses on quantum-level special divine action in genetic mutations, at some places he seems to affirm comprehensive divine action. The most extreme example is when Russell writes, "I think that indeterminacies in quantum behavior arise in a much more pervasive way than the term 'measurement' suggests. Instead, they arise *constantly*, everywhere and at all times, in every part of the universe. If so, this claim can increase the theological intelligibility of our faith in general providence of the Triune God who is *everywhere and at all times* at work in and through all of nature" ("Special Providence and Genetic Mutation," 214).

Second, in relation to those theories postulating that God acts in all quantum events (Murphy and Russell), there is the question of whether God's special action is necessary to those events, and in what sense.[66] Everyone in this debate wants to avoid the theologically and morally objectionable view of occasionalism, whereby God is the only actor in reality and the laws of nature merely describe divine action rather than structures of reality created by God to have some measure of independence. Occasionalism represents the strongest sense in which God's special action might be necessary for ordinary events (in any sense of "event"): God is the ontological condition for the possibility that ordinary events occur at all and determines every event. Near the opposing end of a spectrum of views of the necessity of God's action (see the diagram, below) lies Russell's view, namely, that God does not need to act in any event, beyond merely sustaining creation in existence, but acts intentionally in every event anyway in order to accomplish providential purposes. Of course, rather many quantum measurement events in cosmic history may seem providentially irrelevant, but part of divine providence on Russell's view is God's faithful maintenance of patterns of regularity in nature.

In between these two extremes are a number of views that reject occasionalism yet affirm that God's special action is ontologically necessary in some sense. For example, toward the Russell end of the spectrum, one interpretation of the process metaphysics account of causation has God playing an ontologically essential but not a constitutive role in every event, in the sense that God must exercise influence in some events or the world as we see it would not be possible (the ontologically essential part) yet, if God did not furnish an initial aim to a particular actual occasion, concrescence of that occasion would still occur (the non-constitutive part). On another view of process causality, God's action is more than ontologically essential; it is constitutive though still not determinative for every event because it conveys structured possibilities to each actual occasion, without which concrescence would not occur. Toward the occasionalism end of the spectrum lies the view of divine determinism, which posits some degree of independence of the created world from God (as in process metaphysics) and assigns God the role of selecting every outcome of every event (as in occasionalism). The precise sense in which God is necessary to every event differs among the several variants of this view. Karl Heim thought that God determined every aspect of

---

Unfortunately, Russell does not elaborate on this suggestion, which seems to expand the notion of quantum event considerably. In Russell's contribution to *QM*, he clearly sides with Murphy's proposal that God acts in all quantum events and somehow can convey special meaning through some of these events. See Russell, "Divine Action and Quantum Mechanics," esp. 316-7. Of course, Russell's proposal also involves God contracting the sphere of special action in the presence of conscious creatures, which we will discuss below.

[66] DAP publications typically show great care in the use of the word "event." In the following discussion, I use the term "event" in multiple ways. In some cases the reference is to quantum measurement events, in others to events in the sense of process metaphysics, which have no clear relation to quantum measurement events. In still other cases, I do not specify a metaphysical framework for stabilizing the concept. This usage is merely for the sake of convenience and the context makes clear which sense of "event" is meant.

the world through action at the quantum level.[67] William Pollard seems to argue much the same, despite his more comprehensive way of expressing himself.[68] In these cases of omnideterminism, the necessity of God's action seems to amount to the strong claim that God must play this role if nature is to function properly. Within the DAP, Murphy seems closest to Heim and Pollard (as the phrase "hidden variable" and other references suggest).[69] Murphy also refers to the principle of sufficient reason in justifying her construal of God's universal action at the quantum level, which is another type of necessity. Yet Murphy is clear that God voluntarily respects the "natural rights" of all created entities, which probably evades divine determinism and certainly affirms the beauty and order of God's creation.[70]

Third, there is the question about the situations in which God acts at the quantum level, and whether God constrains divine action in order to support the flourishing of free, conscious creatures such as us. Russell imagines a shift in divine strategy. Prior to the emergence and outside the realm of free, self-conscious creatures, Russell proposes that God works in all events to bring about the divine will for the natural world, from cosmos to ecosphere. Russell further hypothesizes a grace-filled, kenotic contraction of this divine activity in order to create the possibilities needed to make freedom meaningful for self-conscious moral creatures such as human beings. Divine providence withdraws to allow such creatures the freedom they need to explore their moral potential, and God no longer acts in all quantum events but only in some.[71] Both phases of divine activity involve the same mechanism of divine action and differ only

---

[67] See Karl Heim, *The Transformation of the Scientific World* (London: SCM Press, 1953).

[68] See William G. Pollard, *Chance and Providence: God's Action in a World Governed by Scientific Law* (London: Faber and Faber, 1958).

[69] See Murphy, "Divine Action in the Natural Order." There Murphy makes the necessary relation between God and quantum events clear when she writes, "My proposal is that God's governance at the quantum level consists in activating or actualizing one or another of the quantum entity's innate powers at particular instants, and that these events are not possible without God's action" (342). Russell clearly rejects the omnideterminism of Pollard; see Russell, "Special Providence and Genetic Mutation," where Russell writes, "Though I strongly support Pollard's advance over Heim's formulation of the thesis, I do *not* support their advocacy of divine determinism" (209).

[70] See Murphy, "Divine Action in the Natural Order," 342. Murphy's view is strongly reminiscent of Jonathan Edwards's view of God's providence—God determines the world but beautifully, respectfully, and gloriously so. See his *The Freedom of the Will* (Morgan, Penn.: Soli Deo Gloria Publications, 2003).

[71] In "Special Providence and Genetic Mutation," Russell usually refers to biological evolution rather than cosmic history. For example, when he expresses the contraction of divine activity in relation to biological evolution, he writes as follows, "We may think of God as acting in all quantum events in the course of biological evolution until the emergence of organisms capable of even primitive levels of consciousness. From then on,...God may abstain from acting in those quantum events underlying bodily dispositions, thereby allowing the developing levels of consciousness to act out their intentions somatically" (215). I think divine action on a cosmic, indeed universal, scale is implied, however, and not just in biological evolution. This point is clarified in "Divine Action and Quantum Mechanics."

in scope.[72] Moreover, every quantum event has either God or a created conscious agent acting in it, yet this action is not constitutive of these events, in the sense that events would still occur even if God did not act. Russell's motivation for asserting the universality of divine action while not following Murphy in making divine action constitutive of quantum measurement events is theological in character: he wants to secure the omnipresence and omni-activity of God. George Ellis proposes that the principal (perhaps exclusive) mode of God's action is at the quantum level. He construes God's purpose to be the communication of divine inspiration and guidance to human beings, along the lines of the Quaker belief in the experience of the light of God within, all the while respecting the freedom of human beings to make their own choices.[73]

The following diagram presents the dimensions of variation among views of QSDA. The diagram uses rows to express the relevant characteristics of proposals for QSDA and columns to line up combinations of characteristics with the eight particular views mentioned above.

<table>
<tr><td colspan="3">God can act freely</td><td colspan="2">God can only persuade</td><td colspan="3">God can act freely</td></tr>
<tr><td colspan="6">God acts specially in all events</td><td colspan="2">God acts specially in some events but not in others</td></tr>
<tr><td colspan="5">God's action is universal in scope</td><td>Everywhere except realm of sentience</td><td>Anywhere providentially relevant</td><td>Especially human minds</td></tr>
<tr><td colspan="4">God's special action is constitutive of events (events cannot occur unless God acts specially)</td><td colspan="4">God's special action is not constitutive of events (events occur whether or not God acts)</td></tr>
<tr><td colspan="2">God determines every event (divine omnideterminism)</td><td colspan="6">God respects the created rights and ontological independence of creatures and does not determine every event</td></tr>
<tr><td>God is the only actor</td><td colspan="7">God sustains the activity of other creatures and processes</td></tr>
<tr><td>Occasionalism</td><td>Heim, Pollard</td><td>Murphy</td><td>Process (strong)</td><td>Process (weak)</td><td>Russell</td><td>Tracy</td><td>Ellis</td></tr>
</table>

Figure 4: Characteristics of Views of Special Divine Action at the Quantum Level

Though the territory is complex, the DAP's conclusion is clear: there are several viable possibilities for theories of SDA at the quantum level.

[72] Russell is sharply aware of the questions his view invites about God's goodness and believes that a Trinitarian theology of creation, redemption, and consummation is necessary finally to address them.

[73] See George F. R. Ellis, "Ordinary and Extraordinary Divine Action: The Nexus of Interaction," in *CC*, 359-95, esp. 379-82. Ellis's later essays appear to broaden the scope of divine action, though the emphasis remains on intermittent rather than universal action with a focus on human beings; see Ellis, "Quantum Theory and the Macroscopic World," in *QM*, 259-92.

4 Conclusion

I have here presented my reading of the DAP's conclusions and analyzed the disagreements among participants, with special reference to the papers and discussions of the recent capstone meeting. I have also addressed a couple of criticisms of proposals made within the project. We can now return to the question with which we began, namely, whether contemporary theology is in crisis because of a failure to present a credible theory of SDA, as Saunders alleges, or whether in fact theology has made significant progress in this area.

I have tried to explain the root of the disagreement between many DAP participants and Saunders, tracing it back to the insensitivity of his analytical categories to the three-way distinction among interpretations of stochastic laws of nature used within the DAP. If Saunders were to allow that correction to his otherwise impressive argument, it seems that his conclusion about SDA at the quantum level could not help but be more positive. Moreover, Saunders seems optimistic about Polkinghorne's general strategy of treating the laws of nature as approximations to an underlying indeterminate reality within which God can act freely, even though he shares my concerns about Polkinghorne's way of using chaos theory to articulate a causal joint. Saunders also seems sympathetic toward compatibilist proposals, especially Peacocke's. On the terms of his own argument, therefore, and quite apart from the key dispute over interpretation of laws of nature, I cannot quite see why Saunders is so pessimistic. From my point of view, theological theories of SDA are as strong as they have been at any time since Hume and Kant, and this is largely because of the contributions of the DAP.

In closing, it is important to note that many issues discussed in the DAP do not register at all in my survey to this point. Perhaps most prominent among the issues so far unmentioned is the problem of good and evil, and particularly the related problem of theodicy. There was wide agreement among DAP participants that any postulate of SDA exacerbates the theodicy problem,[74] so a lot of energy was expended in trying to deal with this. In particular, the tendency to affirm universal divine action, whether in process metaphysics (Barbour, Birch, Haught) or in personalist theism (Murphy, Russell[75]) seemed motivated in part by the desire to minimize, without eliminating, the severity of the theodicy problem. Likewise, some (including Ellis[76]) invoked the concept of kenosis outside its original sphere of application in Christology to explain why God does not act more often to ease pain and to educate us wayward creatures who so obviously need more guidance than we get. In this way, kenosis was used to strengthen the best-world, free-will, and free-process defenses of God's goodness that various participants defended (especially

---

[74] Russell's frank statement of this point is admirable: "I believe the problem of theodicy is stunningly exacerbated by all the proposals, including my own" ("Special Providence and Genetic Mutation," 216).

[75] See Murphy, "Divine Action in the Natural Order"; and Russell, "Special Providence and Genetic Mutation."

[76] See Ellis, "Ordinary and Extraordinary Divine Action."

Tracy[77]). Some participants thought that no amount of rational reflection would yield a satisfying answer to the question of God's goodness on best-world or free-will grounds, and argued for a return to one traditional Christian approach (not strictly a solution) to the problem, which has God somehow sharing in the suffering of the world through the incarnation and crucifixion of Jesus (Edwards, Moltmann, Russell[78]). Still others (Drees, Wildman) regard the problem of theodicy as crippling to all proposals of SDA and so propose interpretations of ultimate reality that reject the idea that God can form intentions and act specially at all. It was obvious to all within the DAP that the problem of good and evil, and the related problem of theodicy, require more attention than we were able to give them in the context of our study of SDA. But that is no surprise. The road of theological inquiry goes ever onwards.[79]

5 Appendix A: Decision Tree Diagram of Options in Theories of Special Divine Action

The virtue of a decision tree is to draw attention to key choices that theorists of SDA make, explicitly or implicitly, on their way to settled views. The disadvantage is that all interesting views are always subtler than a diagram can represent, balancing many complex considerations. It follows that a diagram is no substitute for a detailed understanding of the textured views themselves. I defined the terminology in the decision tree diagram within the body of this essay. Some features of the diagram are difficult to grasp without commentary, however, so this Appendix includes a brief explanation of the diagram.

The decision tree links related positions using a simple line and bracket system. For example, the decision to represent Ultimate Reality theistically or non-theistically corresponds to the left-most bracket on the diagram. The next decision arises within the theistic context, and pertains to God's nature: either God can or cannot (literally) act intentionally. The diagram records the fact that I hold the latter view (labeled "God as Ground of Being or Being Itself" on the diagram) while most DAP participants hold the former. In general, the names of views are printed in small italic type, the holders of views in small bold type, and the propositions characterizing the content of views in larger normal type. The diagram's presentation of any particular position as the outcome of a sequence of decisions means that we can describe each view as a conjunction of the propositions characterizing the decisions made. For example, Ellis's view affirms {Ultimate Reality as God} and {God can (literally) act intentionally} and {God does act intentionally} and {God Can

---

[77] See Tracy, "Evolution, Divine Action, and the Problem of Evil," in *EMB*, 511-30.

[78] See Jürgen Moltmann, "Reflections on Chaos and God's Interaction with the World from a Trinitarian Perspective," in *CC*, 205-10; Russell, "Divine Action and Quantum Mechanics"; and Denis Edwards, "The Discovery of Chaos and the Retrieval of the Trinity," in *CC*, 157-76.

[79] I am deeply indebted to the many members of the DAP, and especially to the core group of philosophers and theologians, for my understanding of the issues surrounding the concept of SDA. Their influence suffuses this survey. Any mistakes of logic or interpretation are exclusively my responsibility.

Choose to Act Intentionally Only in Some Events (yet May Act in All Events)} and {God Causally Initiates Specific Events} and {God Acts in Conformity with Natural Laws} and {God's Mode of Action is Rationally Approachable} and {Causal Joint of Divine Action Can Be Discussed to Some Degree} and {God's Action Requires Ontological Openness for Causal Joint}. The diagram does not describe the particular causal joint that Ellis defends; see the body of the essay for that.

We can imagine a more complex diagram with further decisions drawn in under most views. Several other diagrams within this paper elaborate the more complicated parts of the tree so as to draw out subtler distinctions among DAP views lumped together in this diagram. The view that {God Causally Initiates Specific Events} receives the most attention because that is where most debate within the DAP took place. The diagram shows how decisions specifying this view are either interventionist or noninterventionist, and how the latter views are compatibilist or incompatibilist. It is important to note, however, that other parts of the diagram could be elaborated in the same way. For example, the view that {God Necessarily Acts Intentionally in Every Event} includes both compatibilist and incompatibilist options, but these are not distinguished in the diagram itself.

The diagram has a number of more serious limitations. For example, Tracy's attempt to defend the intelligibility of several modes of divine action means that his name appears several times on the diagram. Murphy's view is awkwardly distant from other quantum-level proposals on the diagram because it asserts that "God Necessarily Acts Intentionally in Every Event." The distinction between more personalist (Thomistic) and more mechanistic (Aristotelian) variations on the primary-secondary causation model is important, but these views are listed separately from the "God as Ground of Being or Being Itself" view, which is unfortunate. Moreover, the distinction between Non-Theistic and Ground-of-Being views is difficult to stabilize. Finally, the concept of Ultimate Reality is extremely contentious within religious studies.

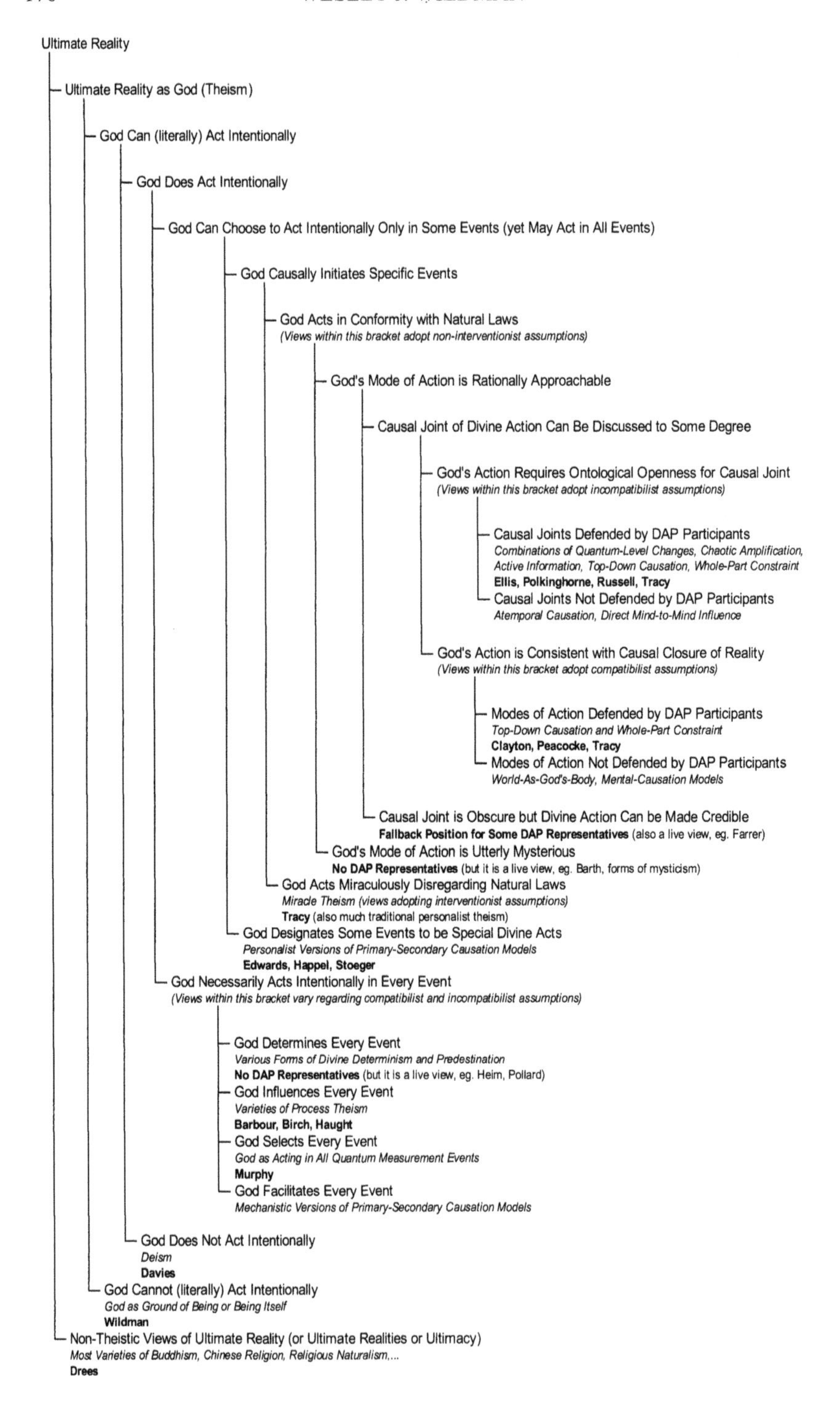

Figure 5: Decision Tree Diagram for Theories of Special Divine Action

III. THEOLOGICAL ANALYSIS OF SPECIFIC ISSUES IN THE SERIES

Niels Henrik Gregersen
Arthur Peacocke
William R. Stoeger, S.J.
Thomas F. Tracy
Keith Ward
Kirk Wegter-McNelly
Mark Worthing

SPECIAL DIVINE ACTION AND THE QUILT OF LAWS: WHY THE DISTINCTION BETWEEN SPECIAL AND GENERAL DIVINE ACTION CANNOT BE MAINTAINED

Niels Henrik Gregersen

1 Introduction

Between 1993 and 2001 the Vatican Observatory (VO) and the Center for Theology and the Natural Sciences (CTNS) produced an impressive series of volumes devoted to one pivotal issue: How can a scientifically informed Christian believer conceive of God's objective interaction with the world of nature *within* the constraints and opportunities offered by the natural sciences? I am convinced that when the history of the new dialogue between science and religion some day is going to be written, the five-volume VO/CTNS series on "Scientific Perspectives on Divine Action" will count as the first genuine example of a long-term collaborative research program within the field of science and religion.

In what follows I offer a critical assessment of this ambitious project. As an observer to the dialogue I have been one of many who have used the series as *the* important research volumes for reflecting upon the problem of divine action through the lenses of theology and science. In the following sections, I begin by pointing briefly to the theological context of the project and to the epistemological program that I see as constitutive for the core vision of the project (sections 2-4).[1] Against this background, I focus on what I see as the core critical issue from a theological perspective: Can general divine action (GDA) and special divine action (SDA) be separated from a theological point of view? I come to the conclusion that a too principled distinction between GDA and SDA may be part of the problem, and not of the solution. In section 5, I thus argue that SDA and GDA cannot be contrasted to one another, nor be treated as two distinct classes of divine action. Rather, any divine action must be treated as both special and yet as falling within the over-all pattern of divine self-consistency. What counts as 'special' and what counts as 'general' depends on our presupposed conceptual schemes, in particular our philosophical assumptions concerning the status of 'laws of nature' (section 5). But what happens if we address the theological divine causation within a richer concept of natural laws? In section 6, I aim to show that this is possible in close contact with current philosophy of science. Contrary to the early 20th century view of 'covering laws,' it may be possible to combine instances of nomological universality (at the level of fundamental physics) with a much richer tapestry of explanatory models that we find in disciplines such as biology, psychology and sociology (section 6).

---

[1] Hereby, of course, I will not be able to give the appropriate credit to all voices in the dialogue.

2 The VO/CTNS Series as a Coordinated Research Program

From the beginning common goals for a systematic inquiry have been set up. A first and preparatory goal has been to *explore the formally possible ways of relating divine action* to the world of nature as described by the sciences. Early on Robert John Russell developed a typology, showing a clear logical order of ontological commitments from the side of theology; this typology has served as a common framework for later discussions.[2] While Russell's typology gives priority to possible loci for divine action in areas of nature not fully determined by scientific laws, the philosophers William Alston and Thomas Tracy have later pointed out that options for affirming responsive actions of God are also available on the basis of deterministic laws.[3]

Within these formally possible options, *specific priorities of research* have been made. A second goal has thus been to develop theoretical proposals that seek to overcome two-language approaches, which on the one hand speak piously about divine actions in the biblical world, and yet on the other hand presuppose a scientific understanding of the world that eventually does not leave room for genuine divine action. Positively, the methodological goal has been to attain the highest possible level of *mutual interaction* or *"traction"* between science and theology as theoretical disciplines. In this venture a combination of analytical philosophy and constructive metaphysics has served as the philosophical meeting place for science and theology. After all, no scientific theory flows unsupported by metaphysical assumptions about the way the world is, and no convincing theological proposal can be formulated without conceptual clarity.

Third, and most importantly, a goal has been shared by a core group of researchers: how can one find space within our current scientific worldview(s) for conceiving of God's *objective and special* activity in the world as *mediated* by the laws of nature (as we know them through science), *without a divine intervention* that violates these selfsame laws?[4] Not all the participants in the series, however, have shared this third aim. Some have further re-emphasized the Thomistic distinction between first-order divine causality that creates and sustains the complex law-like order of nature, while the sciences investigate nature at the level of the second-order particular causes.[5] Others have found a

---

[2] See in particular the "Introduction" by Robert John Russell to *QC*, 4-10, and the typologies in the introduction to *CC*, 9-13, and *QM*, ii-viii.

[3] William Alston, "Divine Action, Human Freedom, and the Laws of Nature," in *QC*, 185-206; Thomas Tracy, "Creation, Providence, and Quantum Chance," in *QM*, 235-58.

[4] See Russell, "Introduction", in *QM*, i-xxvi and viii: "Such an option ["objectively special noninterventionist divine action"] promises to combine the best of both liberal (noninterventionist) and conservative (objectively special) options into a single, new approach to divine action." Russell here refers in particular to the 'bottom-up' quantum path of divine action, explored by Russell, George Ellis, Nancey Murphy, and Thomas Tracy in particular, but the commitment to this stance is also shared by proponents of various 'top-down' contextualist approaches within the series.

[5] This applies in particular to the contributions by William Stoeger, Ernan McMullin, and Stephen Happel. I shall later come back to Stoeger's position on laws of nature.

naturalistic worldview religiously satisfying in itself, though divine creation may still explain the very framework of the system of natural causes.[6] Surprising affinities have emerged between the Thomist position and a religious naturalism: both understand science as the search for understanding the causal network of natural events. Similarly I shall suggest below (in section 5) that there may also be substantial overlaps between classic Reformation theologies and a religious naturalism; in both cases natural events are seen as *identical with* divine actions; affirming the world of nature *as* a divine field of activity, the interest in distinguishing general and special divine action recedes into the background.

Whether one prefers to refer to the VO/CTNS project as expressing one unified research program, or rather as hosting and coordinating a variety of theoretical proposals is a matter of definition. Do we by 'research program' understand a tightly coordinated set of theories for solving a particular set of problems in competition with other proposals, or do we by 'research program' understand the enterprise of facilitating a systematic dialogue between distinct theories, some of which supplement one another, whereas others stand in a certain tension, if not conflict, with one another? It seems to me that the VO/CTNS project can best be said to constitute a research program in the latter (and more loose) sense.[7] The project has hosted a family of theories, which—much like brothers, sisters, and cousins within the same clan—are close in kin, but nonetheless (or perhaps therefore?) often end up in rivalries. As the series has proceeded, the reader cannot fail to notice a certain competition between at least three proposals, even though they intersect in many ways.

1. The 'quantum proposal' claims that special divine action may take place in a divine micro-management of otherwise unguided quantum processes (especially in determining the collapse of the wave function), a micro-management which is then amplified in a bottom-up manner within the macroscopic world. The boldest proposals have here been developed by Robert John Russell, George Ellis, Nancey Murphy, and (more tentatively) by Thomas Tracy.

2. A variety of 'complexity proposals' argue that God may be at work in a top-down or whole-part manner either (a) by influencing the boundary conditions of the natural world as a whole *or* (b) by acting through intersystemic interactions that constantly produce

---

[6] This view is represented by Paul Davies and Willem B. Drees; the latter also highlights the socio-ethical functions of religious life.

[7] The question is then, of course, whether the VO/CTNS series can be said to be a progressive research program in the sense of Imre Lakatos: Which kinds of "novel facts" have been predicted, in this case by redescribing scientific facts and theories in a way that is fully consistent with science, yet allows for higher-order theological explanations? And how have the different proposals fared in a ten-year perspective? I believe one can say that the proposals have progressed in terms of specificity and conceptual clarity. Since divine action is not empirically traceable, however, the proposals can, as far as I can see, only be tested indirectly, via (a) their conceivability, and (b) their coherence with scientific theories and philosophical assumptions.

unexpected higher-order entities during evolution, *or* (c) by acting at the level of conscious embodied persons. The main architect of this view has been Arthur Peacocke, whose pioneering work has been further pursued by Nancey Murphy, Philip Clayton, and others, including myself.[8]

3. The 'chaos proposal' argues that God may act in the subtleties of dynamical non-equilibrium chaotic systems, subtleties manifest in the sensitivity of chaotic systems to initial conditions. On this view, the iteration of the deterministic logistic equations only approximates a reality, which is much more flexible than indicated by the equations *per se*. A persistent proponent has here been John Polkinghorne.

These main proposals cannot be reviewed in detail here, but they will be addressed in the following, in so far as they bring in characteristic viewpoints on the issues of methodology, special divine action, and the concept of laws to be discussed below. Let me in advance point out that *the three proposals do not necessarily compete with one another, since they operate at different levels of nature.* They only compete, in so far as a proponent of a particular locus of divine action denies the possibility of other channels of divine influence. Comparing the three proposals, however, the quantum proposal has the strength of locating divine action within the perspective of a *fundamental theory* of modern physics; the complexity path proposes no such specific channels of divine influence since it argues that the divine influence takes place across the hierarchical order of the universe; finally, the chaos proposal (at least on its best interpretation) [9] is really not so much about the

---

[8] The technically most precise account of Peacocke's proposal is probably found in "The Sound of Sheer Silence: How Does God Communicate with Humanity?," in *NP*, 215-47. Peacocke here defends his notion of the world as "the Systems of systems," understood as a reality on its own, and not merely as an 'abstract description,' as I have earlier objected (see 227, no. 44 and 45); yet at the same time Peacocke now *also* refers to the constant production of novelty in the *inter-systemic* relation "between two interacting type-different systems" (228, cf. 245-7). This potential of this location of divine action does not reappear in Peacocke's own account of divine action (235-40), but it seems to me that new qualities often emerge in the interplay between distinct systems. Human consciousness, for instance, emerges out of the interface between a highly complex physical system (the brain and the central nervous system) and a cultural system of signs.

[9] I here concur with Nicholas Saunders's diagnosis in *Divine Action & Modern Science* (Cambridge: Cambridge University Press 2002): "The central claim Polkinghorne makes is to question whether the mathematical representation of chaos is an approximation to reality…*in essence a metaphysical postulate made by Polkinghorne*" (190). Earlier I have argued—in continuation of the line taken by Karl Popper, *An Open Universe: An Argument for Indeterminism* (London: Hutchinson, 1982)—that in order to infer a metaphysical determinism from the deterministic equations of chaos theory, one needs to be a naive realist; being a critical realist is not enough (see Niels Henrik Gregersen, "Three Types of Indeterminacy," in *The Concept of Nature in Science & Theology*, Part 1, Niels Henrik Gregersen, Michael W. S. Parsons, and Christof Wassermann, eds. (Geneva: Labor et Fides, 1997), 165-86). I

mathematical chaos theory *per se* as about metaphysics, more precisely about an appeal to a real-world complexity beyond the grasp of current science.

In this context it might be helpful to distinguish between *semantic* and *causal explanations* of divine action.[10] Causal explanations of divine action seek to answer the question how God, within the nexus of natural causes, can (or could) bring about the changes that Christians suppose God is doing; the quantum and the chaos proposal thus point to specific locations that could serve as manifestations of a divine influence. Semantic explanations, by contrast, aim to redescribe nature (as already described and partially explained by science) in terms of divine action, thereby arguing that the world of nature as a whole does not fully "make sense" without God being the ultimate source of reality; the 'complexity proposals' seem to make exactly this claim. The two sorts of explanation, however, are supplementary rather than exclusive. Also primarily causal explanations will need to explain how the divine causality "makes sense" in term of a larger picture of the universe. Likewise also semantic explanation brings in ontological commitments concerning the real efficacy of divine action in an ever complexifying world. However, a semantic explanation of divine action does not necessarily assume that divine action means that a separate "causal" divine action is added to the flow of natural causes. Rather, a semantic explanation could be saying that divine and natural causes coincide, so that God is at work in, through and under the guise of natural causes. This, in fact, seems to me to be the main thrust of Arthur Peacocke's argument.

3 The Critical Issue: Can Special and General Divine Action Be Separated?

Let me begin by pointing to a core theological issue that I see emerging out of the efforts. The project has been to find avenues for divine action through the strongest possible interaction with established scientific theories. Thereby the

---

therefore do not find the criticism of Polkinghorne by Willem B. Drees convincing, since Drees bases his critique on the assumptions of a critical realism at the level of theories, whereas Polkinghorne in fact makes an appeal to metaphysics (see Willem B. Drees, "Gaps for God?," in *CC*, 227). Neither do I share the conclusion of Wesley Wildman and Russell that "the hypothesis of metaphysical determinism is strengthened by chaos theory" (see Wesley J. Wildman and Robert John Russell, "Chaos: A Mathematical Introduction with Philosophical Reflections," in *CC*, 82). In fact, chaos theory tells us nothing about the efficient causes of non-equilibrium systems, since it only offers a descriptive mathematical model, say, of the correlation between prey and predators in vast ecosystems. Nor does 'chaos theory' offer any independent evidence of indeterminism (though it certainly points to the crucial role of initial conditions and ever changing boundary conditions, already well-known for other systems).

[10] See my essay, "Critical Realism and Other Realisms," in *Fifty Years in Science and Religion: Ian G. Barbour and His Legacy*, Robert John Russell, ed. (Aldershot: Ashgate, 2004), 77-96. An example of a semantic explanation (as here understood) in terms of philosophical theology is Keith Ward, "God as Principle of Cosmological Explanation," in *QC*, 247-62; while Philip Hefner, "Biocultural Evolution: A Clue to the Meaning of Nature," in *EMB*, 329-56, Denis Edwards, "The Discovery of Chaos and the Retrieval of the Trinity," in *CC*, 157-75, and idem, "Original Sin and Saving Grace in Evolutionary Context," in *EMB*, 377-92, esp. 386-92 may be construed as offering semantic explanations in terms of Christian theology.

project has naturally gravitated around *localist approaches* to divine action (in particular the quantum proposal). I share this core vision of the project in so far as the task is to overcome a reduction of SDA to GDA. I would add, however, that it is uniformitarianism that constitutes the over-all theological problem. Uniformitarianism is the conviction that God always and everywhere does the same job of creating-and-upholding an already established universe. This was the standard view of the classic liberal theology of Friedrich Schleiermacher, who reduced the divine guidance of the world to the divine conservation of the world, which again was reduced to God's general creativity.[11] This homogenizing position reappears in more recent representatives of liberal theology such as Maurice Wiles and Gordon Kaufman.[12]

A central motivation for developing a robust notion of SDA lies in the attempt to overcome a uniformitarianism, which presupposes a causally closed world and thus makes Christian practices such as petitionary prayer meaningless. The task has therefore been to conceptualize ways in which God may bring about significant changes of affairs in particular nexuses of nature. On reflection, however, I have come to the conclusion that *a too principled distinction between GDA and SDA may be part of the problem, and not of the solution.* In section 5, I shall thus argue that SDA and GDA cannot be contrasted to one another nor be treated as two distinct classes of divine action. Rather, any divine action must be treated as both special and yet as falling within the over-all pattern of divine self-consistency. What counts as 'special' and what counts as 'general' depends on our presupposed conceptual schemes, in particular our philosophical assumptions concerning the status of 'laws of nature.' But what happens if we address divine causation within a richer concept of natural laws? In section 6 I aim to show that this is possible in close contact with current philosophy of science. Contrary to early 20th century view of 'covering laws', it may be possible to combine instances of nomological universality with a much richer tapestry of explanatory models that we find in disciplines such as biology, psychology and sociology.

---

[11] See, for instance, Friedrich Schleiermacher, *The Christian Faith*, H. R. MacIntosh and J. S. Stewart, eds. (Edinburgh: T & T Clark, 1989) [German Original 1830, 2nd ed., #6.2], 173: "there will be a complete coincidence between the two ideas–namely, the unqualified conviction that everything is grounded and established in the universality of the nature-system, and the inner certainty of the absolute dependence of all finite being on God." This crucial expression of a compatibilist view of the divine-nature relation is in Schleiermacher combined with the uniformatarian view that God's guidance of the world equals God's conservation of the world system as a whole, see the quotations in my essay "Providence in an Indeterministic World, " *CTNS Bulletin* 14:1 (1994): 16-31.

[12] See the essays by Wiles and Kaufman in Owen C. Thomas, ed., *God's Activity in the World: The Contemporary Problems*, American Academy of Religion Studies in Religion 31 (Chico, Calif.: Scholars Press, 1983); on Gordon Kaufman, see Kenneth Nordgren, *God as Problem and Possibility: A Critical Study of Gordon Kaufman's Thought toward a Spacious Theology*, Uppsala Studies in Faiths and Ideologies 13 (Uppsala: Uppsala University, 2003).

4 The Theological Context of the VO/CTNS Series

The VO/CTNS project should be seen in the context of 20th century theological developments. More than once has Robert John Russell cited the poignant thesis of Langdon Gilkey's 1961 essay, "Cosmology, Ontology, and the Travail of Biblical Language," where Gilkey put his finger on the crisis of God-talk in the neo-orthodox era of dialectical and historical-critical theology: "Thus the Bible is a book descriptive not of the acts of God but of Hebrew religion...[It] is a book of the acts Hebrews believed God might have done and the words he might have said had he done and said them—but of course we recognize he did not."[13] Seen from a theological-contextual point of view the series can thus be seen as a constructive attempt to overcome the crisis of both liberal and neo-orthodox Protestant theology. Throughout most of the 20th century theological imagination had been captivated by approaches to Christianity, in which divine action was affirmed only in the form of personal or communal encounters with God: spiritual experiences triggered by biblical witness and informed by liturgical language. Existentialist, personalist, narrative or Wittgensteinian theologies have thus spoken a "paradoxical unity" between creaturely and divine action (Rudolf Bultmann), a "double agency" of divine and human action (Austin Farrer), or have referred to the idea of distinctive "language games" (Wittgensteinians) or "plain readings" of biblical stories (Hans Frei). Through these demarcation lines, potential conflicts between science and religion have been preventing thus securing a safe space of religious life, immune from the external criticisms of science and historical-critical research.

This VO/CTNS project may have a related, though also somewhat dissimilar *Sitz im Leben* in the context of modern Catholic theology. For sure, many insights of the Second Vatican Council (1962-65) would not have been achieved without the new anthropological foundation of Catholic theology in the wake of the phenomenological re-appropriations of Thomism in the writings of, say, Karl Rahner and Bernard Lonergan. However, Catholic theology has not been plagued by the dichotomization of human and non-human nature to the extent that we find in 20th century Protestant theology. Catholic theology, however, is still struggling to come to terms with mainline neo-Darwinian explanations of evolutionary history, when it comes to the spiritual capacities of the human person. Also scientific explanations in terms of chance and probability constitute new challenges to a classic Thomistic framework of natural entelechies as well as to Teilhardian concepts of orthogenetic development. Viewed from a Catholic context, the VO/CTNS series may thus be seen both comforting and as challenging. It is comforting in so far as the series has offered reasons for reclaiming the relevance of the classic Thomistic distinction between divine first-order causality (Thomas: *causa universalis*) and the network of second-order causes (Thomas: *causae particularia*) in relation to modern science. Thomism does affirm the integrity of the created order as maintained by God; Thomism also understands human

---

[13] Langdon Gilkey, "Cosmology, Ontology, and the Travail of Biblical Language," *The Journal of Religion* 41 (1961): 194-205, 198. Quoted in Russell, "Introduction," in *QC*, 7.

beings as psychosomatic unities. But the project is also challenging to traditional Catholic thought, in that Darwinian biology questions a straightforward directionality of evolution. A traditional concept of inner-worldly teleology seems not compatible with the openness of evolutionary development.

Stephen Happel has thus aimed to show that a Thomist position is also able to accommodate notions of chance and probability as well as strict laws. This openness, however, is paid for by a rather high degree of generality in the theological hypothesis of mediated divine actions: "Rocks cooperate [with God] as rocks, plants as plants, and dogs as dogs."[14] Another constructive response has been developed by Bill Stoeger who argues that a certain directionality, though in more open and flexible form, is in fact part of the trajectory of evolution, and can be substantiated on scientific grounds, if we combine the neo-Darwinian principles of selection and variation with self-organizational principles that result from nature's nested hierarchies.[15] These reappropriations of Thomism, however, cannot hide what remains a challenge for contemporary Catholic theology, namely the importance of biological evolution for understanding human mental capacities, including the human soul. As pointed out by George V. Coyne, the messages of Pope John Paul II on the evolution of humanity indeed mark important steps forward.[16] But it is probably fair to say that official Catholic doctrine has still not fully embraced an evolutionary understanding of the human person as a product of natural causes.

5 The Epistemological Commitments of the VO/CTNS Project

Since its beginning, the VO/CTNS project has catalyzed important moves in the official Roman Catholic teaching on Christian faith and modern science. The project has also persistently challenged various versions of the mainly Protestant two-language approaches. Hereby the project has kept alive the ambition that theology no less than science should have ambition of establishing theories open for cross-examination. By insisting on theology as capable of theoretical-empirical reasoning, the project may have done a considerable service to the credibility of theology at large, which should not be underrated. For sure, the project has not been to prove the "reality of God" or to induce the "efficacy of divine action" in the old style of natural theology.

---

[14] Stephen Happel, "Divine Providence and Instrumentality: Metaphors for Time in Self-Organizing Systems and Divine Action," in *CC*, 177-204, quotation 198; cf. my criticism in Niels Henrik Gregersen, "From Anthropic Design to Self-Organized Complexity," in *From Complexity to Life*, Niels Henrik Gregersen, ed. (New York: Oxford University Press, 2003), 230, no. 3.

[15] William Stoeger, "The Immanent Directionality of the Evolutionary Process, and its Relationship to Teleology," in *EMB*, 163-90.

[16] See the messages of the Pope, "Message to the Vatican Observatory Conference on Evolutionary and Molecular Biology" and "Message to the Pontifical Academy of Sciences," in *EMB* 1-9; see the Papal Statement in Ted Peters, ed., *Science and Theology: The New Consonance* (Boulder, Colo.: Westview Press, 1998), 149-52; and the assessment by George V. Coyne, "Evolution and the Human Person: The Pope in Dialogue," in *EMB*, 11-17.

Nor has the project claimed that it is possible to "trace" the active presence of God in nature, which would run contrary to the basic Christian assumption of the hiddenness of God's transformative presence in the world. As expressed by Robert J. Russell, the bold claims of the 'quantum proposal' should be understood under the proviso that the proposal "does *not* 'explain how God acts' or even constitute an argument *that* God acts."[17] That is, the proposal does not claim to possess a theory for "the causal joint" of divine action and natural events in the sense of tracking a cross-over from God to creation.[18] The point has rather been to develop distinct theological candidates for how the Christian affirmation of God's operative presence in the world of nature may *cohere with* specific theories of science and with more general philosophical thought models. As a spokesperson for using 'contextual coherence' as the most manageable criterion for evaluating interdisciplinary truth-claims,[19] I have been pleased to see the recurrent appeals to 'coherence' in the five volume series. For the sake of clarification, let my try to sort out the different degrees of coherence that may be attained in the dialogue between theology and the sciences. The formal range of possibilities might be from mere conceivability to plausibility up to a substantial coherence at the theory level of science and theology. Exactly what level of coherence can we reasonably expect to obtain when relating divine action and modern science?

First, both the quantum proposal and the chaos proposal point to *specific domains* of nature as primary locations for non-interventionist divine actions.[20] I shall therefore refer to these proposals as the *localized approaches* to divine action. Second, the 'quantum proposal' has a special epistemic status within the project. It makes the very specific claim that special divine actions can be conceived to take place at the microphysical level of quantum systems with macroscopic effects *within the theory structure of quantum mechanics* (when interpreted in the context of the 'orthodox' Copenhagen interpretation). That is, the theological quantum proposal presupposes a network of theoretical presuppositions, all of which, in principle, are open for discussion and refinement. These presuppositions are, among others, (i) the no-hidden variable theorem, that is, the understanding that quantum theory is the *fundamental* physical theory available to human observers, (ii) the adequacy of the 'orthodox' Copenhagen interpretation, which interprets nature

[17] Robert John Russell, "Divine Action and Quantum Mechanics: A Fresh Assessment," in *QM*, 295.

[18] Robert John Russell, "Is Nature Creation? Philosophical and Theological Implications of Physics and Cosmology from a Trinitarian Perspective," in *The Concept of Nature in Science & Theology*, Gregersen, Parsons, and Wassermann, eds., 94-124, 113; cf. idem, "Special Providence and Genetic Mutation: A New Defense of Theistic Evolution," in *EMB*, 216.

[19] Niels Henrik Gregersen, "A Contextual Coherence Theory for the Science-Theology Dialogue," in *Rethinking Theology and Science: Six Models for the Current Dialogue*, Niels Henrik Gregersen and J. Wentzel van Huyssteen, eds. (Grand Rapids, Mich.: Eerdmans, 1998), 181-231, esp. 189-98 and 226-8.

[20] Russell, "Special Providence and Genetic Mutation," 216: "More precisely, it [the quantum proposal] locates the domain in which that action—however mysterious it truly is in itself—may have an *effect* on the course of nature."

philosophically as being indeterministic at its very root, (iii) the theological suggestion that God works *together with nature* in determining concrete quantum events, which otherwise would be purely random, but does so *without* altering the deterministic wavefunction between the measurements (so to speak, behind the scenes) and *without* changing the over-all Gaussian statistics at the macro-level, and (iv) the scientific presupposition that concrete quantum events at the micro-level nonetheless can have, and do have, effects on the macro-world by way of physical and biological amplifiers *without* invoking quantum chaology.[21] As a matter of fact, however, no interpretation of quantum mechanics can avoid making metaphysical presuppositions.[22] The strength of the proposal is that it offers a theological interpretation which is specified by, and clarified by, the theory structure of quantum mechanics. In this sense, the proposal is "based on what we know about nature, assuming that quantum physics is the correct theory and that it can be interpreted as telling us that nature is ontologically indeterministic."[23]

Whereas the chaos proposal refers to the uncertainties of natural events under the assumption that non-equilibrium chaotic processes are not sufficiently described by the deterministic equations of mathematical chaos, the quantum proposal suggests that *God can act by determining the wave collapse within the uncertainties that are already known and described by quantum theory.*

The quantum way should thus be given the prize for the most specific proposal in terms of the intensity of contact between a theology of divine action and the theory level of basic physics. By the same token, it is also by far the most risky proposal, since the theological affirmations are made dependent upon Bohr and Heisenberg's specific interpretation of a quantum theory, a theory which is itself revisable. The proponents of the quantum proposal are aware of this riskiness, but have deliberately opted for the closest possible "traction" between science and theology, which inescapably involves revisable truth-candidates on both sides.

This more general method of the VO/CTNS series has been to combine a sense of *analogies* between theology and science and at the same time claim a *hierarchy of epistemic levels*, according to which the higher order theories (e.g. in biology, or theology) are constrained by, though not reducible to lower level theories of fundamental physics.[24] Following the methodological program

---

[21] See the concise clarifications by Russell, "Divine Action and Quantum Mechanics," 294-300.

[22] Unless, of course, one opts for a purely instrumentalist or empiricist interpretation of quantum mechanics, whereby one simply begs the question of the relevance of quantum theory for our worldview. Bas van Fraasen is a strong proponent of this view.

[23] Russell, "Divine Action and Quantum Mechanics," 295.

[24] See Robert John Russell, "Bodily Resurrection, Eschatology, and Scientific Cosmology", in *Resurrection: Theological and Scientific Assessments*, Ted Peters, Robert John Russell, and Michael Welker, eds. (Grand Rapids, Mich.: Eerdmans, 2002), 3-30, 10f. The hard issue is, of course, whether there is not only an epistemic autonomy of higher level theories from lower level theories, but also a process

already stipulated by Ian Barbour in *Issues in Science and Religion* (1966), the assumption has been that despite the differences in attitude between lived religion and the practices of science, there are nonetheless important similarities between science and second-order theological reflection.[25] In both cases, there is a *two-way interaction* between data and theory and between experience/experiments and interpretation; metaphysical interpretations are inescapable to both science and theology. Second, the *interpretative community* is essential to both science and theology, even though both scientists and religious scholars understand theories and models to disclose real features of the world, albeit only tentatively. And third, both scientific and theological theories make up *interconnected concepts* that are to be tested and evaluated in a holistic framework of coherence, comprehensiveness and experiential adequacy.[26]

In subsequent discussions, some have opted for reshaping theology as a genuine empirical research program. I take Nancey Murphy to be the strongest proponent for this move.[27] Others in the science-religion dialogue have underlined the importance of what I have called semantic explanations, according to which explanations in metaphysics and theology primarily are concerned about "making sense" of the world of nature as described by the sciences.[28] There might be, on the one hand, a kind of *Wahlverwantschaft* (Max Weber) or 'elective affinity' between a strongly empirical orientation and the search for a localized space for divine action (e.g. in quantum theory) and, on the other hand, an emphasis on theology as a semantic explanation of the world, and a stronger emphasis on God's involvement in emergent processes as a whole. As will appear, I believe that *the 'complexity path' for divine action makes up an independent case for God's operative presence, which does not rely on the quantum bottom-up proposal of divine action* (though the former might be supported by the latter).

---

autonomy of higher order processes in relation to lower processes, which is the claim I am going to make in the following.

[25] Ian Barbour, *Issues in Science and Religion* (New York: Harper and Row, 1966), esp. 264-70. In his early writings, however, Barbour did not always distinguish appropriately between lived religion as a first-order phenomenon, and theological theories as a second-order reflection on religious assumptions, see Christian Berg, *Das Werk Ian Barbours als Beitrag zur Bestimmung des verhältnisses von Theologie zu Naturwisseschaft und Technik*, unpublished Heidelberg Dissertation 2002, 108.

[26] See Barbour, *Issues in Science and Religion*, 264-70, esp. 267.

[27] Nancey Murphy, *Theology in the Age of Scientific Reasoning* (Ithaca, N.Y.: Cornell University Press, 1990), esp. chap. 6. Cf. my discussion in "A Contextual Coherence Theory for the Science-Theology Dialogue," 205-12.

[28] Philip Clayton, *Explanations from Physics to Theology* (New Haven, Conn.: Yale University Press, 1989), e.g. 36 and 46f (with reference Stephen Toulmin and John Passmore); cf. also idem, "Tracing the Lines: Constraints and Freedom from Quantum Physics to Theology," in *QM*, 211-34, which does not attempt to state any causal lines between divine action and quantum reality, but rather asks how quantum theory may constrain theological theories.

6 On the Status of the Distinction between General and Special Divine Action

Our concepts about what we mean by 'divine action' naturally inform our expectations about how God might bring about new states of nature that would not take place without God's constant, non-homogenous interaction with nature. There even seems to be a logical *pre* and logical *post* at work here. Logically, *the semantic explanation of God's operative presence precedes any causal explanations* about *how*, *where* and *why* God makes a causal difference in the world of ours. One the other hand, any semantic view of divine action will also want to know how God might be able to bring about all that which God is supposed to do. In this sense a semantic explanation of God's transformative presence cannot divorce itself from the causal issue that stands in the foreground in the divine action project.

The VO/CTNS project has from the outset taken a rather determinate stance concerning divine action. First, some extremes have been excluded from the center of interest: to be transcended on the left side is a *deism*, according to which divine activity is restricted to the initial phase of creation; to be avoided on the right side is an *occasionalism*, which sees God as the sole cause of natural events, while natural events only provide passive 'occasions' for divine action. On the deist assumption, the world is closed for divine interaction; on the occasionalist view, the sense of the integrity of nature is lost.[29] Second, the requirement has been made that a constructive theory of divine action must transcend the cavalier theory of divine action that simply claims that God is at work in and through the entire created order. The defect of this model is that it leaves no room for special divine action; moreover, it is unstable, so that it either slides back into occasionalism, or reduces divine action to a mere 'rubber stamp' on natural processes.[30]

But what are then the semantic presuppositions behind the notion of 'special,' 'mediated' and 'non-interventionist' divine action, and again between 'direct' and 'indirect' divine action? The distinction between SDA and GDA goes back at least to the Middle Ages. Traditionally, it was used to distinguish between God's sustaining of the created order as a whole, as opposed to God's salvific actions.[31] However, in its post-17th century usage SDA refers to acts of God that objectively change of the course of nature, which would have run otherwise according to the laws of nature, *if* God had not acted in this or that special way.

---

[29] Nancey Murphy, "Divine Action in the Natural Order: Buridan's Ass and Schrödinger's Cat," in *CC*, 332; cf. Russell, "Divine Action and Quantum Mechanics," 296f, no. 12; Tracy, "Creation, Providence, and Quantum Chance," 240. It should be noted that 'occasionalism' is used here in a rather broad sense that may well include many Biblical understandings of divine creativity (e.g. Ps. 104: 24-35); also Medieval and Reformation theologies usually understood 'earth' or 'matter' as intrinsically passive, while only the vivifying Spirit of God makes them alive. See below.

[30] Murphy, "Divine Action in the Natural Order," 332f.

[31] So, at least, in Bernhard of Clairvaux, *On Grace and Free Choice*, written in the first half of the 12th century.

The concept of SDA has proven to be useful. But if my observation so far is correct, it seems that the distinction between GDA and SDA is not quite as simple as it might appear. A first problem is that *the distinction between GDA and SDA presupposes contrafactual statements about what the causal effects of natural laws would have been, if God had not brought about a change of affairs by SDAs.* Accordingly, a concept of SDA presupposes that we have at hand a rather clear knowledge about what nature would have done, if God had not acted in these special and new ways. For instance, the claim that an interventionist miracle by God has taken place, is able to offer a much more convincing account about what would have happened, if the miracle had not occurred, than a theory of non-interventionist SDA! Eventually, an argument for SDA presupposes that we have available a rather safe picture of how nature works apart from SDA.

This brings us to the second problem: *the very concept of non-interventionist SDA (i.e. SDA within the constraint of the laws of nature) presupposes a strong ontological concept of the laws of nature in the first place.* This is not unfounded, in so far as a scientific realism rightly finds reason to believe that well-tested laws of physics (say, concerning gravity, relativity theory, quantum theory, bonding laws of chemistry) somehow correspond to universal laws of nature. But how far can we extend nomological universalism? What if only a few of our known laws of nature are genuinely universal (primarily those pertaining to fundamental physics), while others are *domain specific* (for example, genetics) and may even be late products of a contingent history of nature (for example, rules for chemical signal systems in different species of social systems)?[32] It can thus be argued that 'laws' in biology usually refers to invariant relations without needing to invoke a nomological necessity.[33] It is even debatable whether laws of physics apply in real-world situations outside the constrained domains of experimental set-ups. Nancy Cartwright, for one, has argued that science is more like a patchwork of explanations than like a hierarchy of covering-law explanations. The many faces of our scientific laws motivate us to be careful not to ontologize laws of nature prematurely, and not to understand all laws in a necessitarian scheme. However, *since a presupposed concept of the laws of nature serves as a foil for shaping the idea of non-interventionist SDA, it seems crucial for the Divine Action Project to clarify the concept of the laws of nature.*[34]

---

[32] The idea that laws are grounded on contingencies is rightly emphasized by Wolfhart Pannenberg, "Contingency and Natural Law," in *Essays on Theology and Science*, Ted Peters, ed. (Louisville, Ky.: Westminster/John Knox Press, 1993); 72-122, esp. 81-6.

[33] See Jim Woodward, "Law and Explanation in Biology: Invariance is the Kind of Stability that Matters," *Philosophy of Science* 68:1 (March 2001): 1-20.

[34] Cf. Nancey Murphy may escape this problem by pointing out that "the 'de-ontologizing' of the 'laws of nature' is a helpful move in understanding divine action," ("Divine Action in the Natural Order," 334). She is also careful to speak of the 'law-like' domains of nature rather than of the 'existence of laws of nature.' In a personal letter on May 28, 2004, Murphy has further clarified her view on laws of nature: "I see all laws of nature as descriptive (after the fact) rather than normative. This is first of all the case for quantum laws, but given the largely bottom-up structuring of the world, this

Against this background, I suggest we see the notion of a SDA as a helpful shorthand motivated by the need to overcome the uniformitarian view of divine action (the view that God always and everywhere does the same job). The concept of SDA articulates the Christian conviction that God acts in and around our lives in non-repetitive ways, and that God may actively respond to our thoughts and prayers. But do we need to suppose that the distinction between SDA and GDA reach into the mind of God? To put it bluntly, do we suppose that God knows when God is performing a GDA and when God is performing a SDA?

I do not find it easy to sustain a clear-cut distinction between SDA and GDA. It seems to me that there cannot be *two categories of divine action*, some *xx* merely general, some *yy* specific. All participants in the VO/CTNS project grant that every SDA is undergirded by GDA (such as God's sustaining of the physical world). SDA = GDA + SDA. But could we then perhaps see SDA and GDA as *two separate aspects* of one of the same compound divine action? On this account God is actually imagined to perform two actions at the same time, one purely 'general' or ordinary, and the other 'special' or extraordinary, similar to way we do something routinely (such as bicycling), while we are also doing something specific (such as shouting hello to a friend on the corner)? Evidently, this picture is all too anthropomorphic. It seems odd to claim that God's mind is divided between performing "standardized" action, in addition to which God also now and then performs "extra" actions. In short, the distinction between GDA and SDA belongs to our conceptual schemes, and has no ontological status.

I therefore suggest that we think the other way around: *What if we take the notion of SDA as fundamental rather than as something added on the top of GDA*? On this view, the singularity of divine activities has the ontological priority over against the notion of general divine actions. God's activity is always one and undivided, and yet complex and multifaceted in its manifestations. God is not first creating 'a being' of a thing (GDA), and then perhaps also determining its 'operations' (SDA). *Esse et operari unum sint.* To be a creature and to be capable of doing something (having a capacity) is one and the same thing. On this account, what we term as GDA is grounded in the divine self-consistency as self-giving love, but actualized ever anew as God's gracious activity from the midst of the material world itself. Speaking of God's general activity is an abstraction from God's singular acts, by which the world comes into being as a complex network of interacting events and processes, at once existing and in operation. At each moment, God creates both *ex nihilo* (out of nothing) and *ex ovo* (out of the potentialities of the past) *and* for the purpose of letting something new come into being (*ad novum*).

On this view, no divine action is related to 'physical entities in abstraction,' apart from the precise informational contexts, in which these physical entities play their roles. As soon as life and consciousness have taken hold on Earth (or any other planet), *God never creates a 'pure nature,' but*

undercuts any possible normative/*a priori* laws at higher levels. (In addition I now think that there may be regularities that result from patterns of top-down causation.) So my talking about the regime of law is only to describe the regularities we observe—this is the level where things look deterministic because of the regularities." Here as elsewhere, Murphy takes an empiricist rather a rationalistic approach.

always creates and recreates the whole embodied phenomena with all their God-given capacities: physical, biological, and social. [35] Just as SDA should be given priority over against a homogenized concept of GDA, so the multifaceted phenomenon should be given priority over physical laws.

On this theological account, God is and remains ontologically prior to the laws of nature. At the same time we can at once acknowledge the laws of nature that God actually has chosen for our world (and which science is uncovering for us) and yet be metaphysically open to the diverse canvas of laws that the different sciences present to us—metaphorically speaking—in their different shapes, sizes and colors.[36] *The uniformitarian view of a divine action, which always and everywhere repeats itself in accordance with a preestablished principle of nomological universality, is supplanted by a picture of a living God, who is constantly in a process of self-donation, both as ordering principle (Logos) and as operational presence (Spirit).*

Is there any precedence for such a view in theological tradition? I believe there is. Let me for convenience mention Paul Tillich's understanding of God as Being-Itself: God is "the power to be in and above all that exists."[37] If God is that which gives existence and operational capacities to any natural event, God cannot be divorced from the world of nature. Tillich is aware that he is here dependent both on the Thomistic understanding of God as *ipsum esse* (so 'being' is the only predicate that is shared univocally between God and created beings) and on Martin Luther's more actualistic understanding of God's radical in-being in nature. For Thomas as well as for Luther God is intimately and transformatively present in the world of nature. "God exists in the created things," as Thomas says.[38] In a context discussing how God can be "really" present in the bread and wine of the Eucharist, Luther affirms the intimacy of God's operational presence in the inner structures of the material world: "For it is God who, through his almighty power and right hand, creates, works in and contains all things...Therefore He must fully exist in each individual creature, in their intimacy as well as around them, through and through, below and above, in front and behind, since nothing can be more present or intimate in all creatures than God with His power."[39] God is not only "more intimate to

---

[35] With Arthur Peacocke and Philip Clayton I therefore prefer, in a metaphysical context, to talk about a multifaceted emergent monism rather than about a non-reductive physicalism, even though I recognize the strategy of choosing the latter term for making oneself understandable to a materialist audience.

[36] In a letter on May 28, 2004 in response to this article, Nancey Murphy writes: "I am in essential agreement with your rejection of the distinction between SDA and GDA, because God's sustenance of the universe via particular actions in all quantum events produces the regular features of nature. So SDA can only mean something like 'irregular' DA." She adds, however, that the concept of SDA can also be used to denote the expression of "a particular [divine] intention based on a personal relationship."

[37] Paul Tillich, *Systematic Theology*, Vol. 1 (Chicago: Chicago University Press [1953] 1978), 236.

[38] Thomas Aquinas, *Summa Theologiae*, Ia 8.

[39] Martin Luther, "That These Words, 'This Is My Body,' Still Stand Firm Against the Fanatics," in *Luther Works* (Saint Louis, Mo.: Concordia Publishing, 1958), vol. 37, 58. Original text in Martin Luther, *Werke. Kritische Gesamtausgabe* (Weimar: Hermann Böhlaus Nachfolger, 1883), vol. 23, 133-4. On this theme, see Niels Henrik

myself, than I am to myself," as Luther famously said, but this radical in-being of God in creation also applies to the material world.

Neither Thomas nor Luther can be accused of occasionalism.[40] Neither of them understood divine action in a haphazard manner, for God exactly makes nature capable of co-operating actively with God in producing the causal order. For as Thomas says, God activates the passive potentials (*potentia passiva*) of the creatures, and as Luther says, the divine blessing of the created is what makes natural growth possible: "What Moses calls a blessing (*benedictio*), the [natural] philosophers call fertility (*foecunditas*)."[41] In this manner both Thomas and Luther presupposed that God's activity is prior to the regularities of nature. The ontology of divine action determines the understanding of the laws of nature, rather than the other way around. I believe this remains the logical orientation, if we assume that God is creator of the law-like processes of nature in the first place. In effect, this means that *one can be a theological compatibilist (meaning that divine activity and natural causality coincide) without falling into the trap of uniformitarianism.* If one were a rock-bottom incompatibilist, nature could do something without God doing it (whereby nature is no longer thought of as created by God), or God could do something in creation without doing it mediated by created nature (whereby one would posit a creation outside of creation, which makes no sense). The only theological candidate for truth is to argue that nature is never devoid of divine activity: nature always equals divine-activity-in-and-through-nature.

What I find theologically attractive about the quantum proposal is exactly its possibility for reemphasizing God's operational presence in the most basic processes of nature known to us. Any spatial distinction between God and nature evaporates, for as expressed by Russell, "what we normally take as 'nature' is in reality the activity of 'God + nature'"; accordingly "we do not know what the world would be like *without* God's action."[42] It seems to me, however, that this insight is best articulated when God is conceived to be at work in determining the collapse of the wave function pervasively rather than only in particular cases. I here agree with Murphy's *votum*: "both doctrine and

---

Gregersen, "*Unio Creatoris et creaturae*: Martin Luther's Trinitarian View of Creation," in *Cracks in the Walls: Essays on Spirituality, Ecumenicity, and Ecclesiology*, Else Marie, Marie Wiberg Pedersen, and Johnnes Nissen, eds. (Frankfurt am Main: Peter Lang Verlag, 2005), 43-59.

[40] According to R. Specht, the term occasionalism should be reserved first and foremost to the post-Cartesian psychologists (like Cordemoy), who claimed—in contrast to Descartes—that also the human mind is merely passive and dependent upon God; secondly the term applies to the post-Cartesian critiques of Aristotelianism that we find exemplified by Malebranche and others, "Occasionalismus," *Historisches Wörterbuch der Phiolosophie* Vol. 6, 199x, 1090-91. Here the passivity of matter is used to deny even what the Thomists would call the *potentia passiva* of created reality.

[41] *Weimarer Ausgabe* 42, 40 or *Luther's Works* 1, 53. See further Niels Henrik Gregersen, "The Idea of Creation and the Theory of Autopoietic Processes," *Zygon* 33:3 (1998): 333-68, esp. 347-53.

[42] Robert John Russell, "Does the 'God who Acts' Really Act in Nature," in *Science and Theology*, Peters, ed., 77-102, 89.

logic suggest that if God acts at all, God is acting in everything that happens."[43] For in a sense, the result of God's action *is* nature.

Robert John Russell and Thomas Tracy, by contrast, sometimes hypothesize that God may choose to perform a SDA or refrain from it at any moment. The reason for this move is probably the scientific concern for not violating the overall probability distribution. On this issue, however, one can argue that a probabilistic law cannot really be violated within a finite time scale.[44] Nonetheless any scientific theory of a certain probability rate will become more and more implausible, if, say, a dice continues to give '6' more than 10,000 times in a row. One would begin to think whether the dice is not strongly loaded. The theological solution proposed above is to say that the actual outcomes of quantum events are 'ontologically' never 'purely' statistical, since there exists no *solo* nature, without God (Russell's point, above). *Divine action coincides with the actual outcomes in the actual universe that we inhabit; the outcomes would only be plainly random ('ontological indeterministic' in an ultimate sense), if God were not operationally present in the quantum world. On the view defended here, quantum systems are only 'ontological indeterministic' in a penultimate sense.* The point of the particularist version of quantum SDA could then be redeemed by stating that God's selection of quantum events does not always have the form of a unilateral control (as in classic-style theological determinism). Divine SDAs have a variety of forms, and our probabilistic laws reflect the way God has chosen to create the actual world. The only reason for making a distinction between special and general divine action is that God may act differently in different events (and thus may appear to be determinative in some situations, flexible in others).

Thus understood, God is always acting *as God* but *from within the world of creation.* God is at work "in" natural events as God, as creator (or as prime cause). On the view taken here, God cannot act as one factor among others at the level of secondary causes. However, it makes sense to ask for typical locations of God's renewing actions, although we can probably never disentangle what is divine and what is natural in an event.

On this view, there is *no causal joint* in terms of a third neutral meeting ground, nor any possibility of tracking the route from God to world.[45] Rather, the causal joint can only be the very creatures themselves—those who are gifted and burdened by God's giving existence and operational power.

7 The Many Faces of "Scientific Laws": Chaos, Complexity, and Divine Action

In the divine action series one finds a clear awareness of the distinction between the laws of nature, as they are in themselves, and our scientific laws

---

[43] Murphy, "Divine Action in the Natural Order," 330.

[44] As suggested by George Ellis in discussions.

[45] In process theology, this neutral ground is provided for by the 'eternal objects' and the 'creativity,' which is commonly shared between actual entities, whether divine or worldly.

that aim to grasp and formulate these laws.[46] The 'critical realism' supported by most scholars in the field of science and religion has served to keep this awareness alive. However, do we need to accept a nomological universality and a necessitarianism concerning the laws of nature (LN)—whatever they may be—by virtue of the very fact that classical physics formulates most laws in the form of deterministic equations?

It seems to me that proponents of the 'quantum way' sometimes play an unnecessary all-or-nothing game: Either it's ontological quantum indeterminacy or no indeterminacy at all.[47] It is thus assumed that the macroscopic world of nature—between the level of the quantum world and perhaps human consciousness—*must* (in light of current physics) or at least *should* (in an ideal future physics) be described as one tightly closed system. But is this view justified? Should one necessarily work in a worst case scenario of a fully deterministic classical system, with one exception for the particular case of quantum mechanics? I suggest not. It seems to me that nature is not necessarily deterministic, even though our equations are.[48] In addition, the universal laws of physics are not explanatorily relevant everywhere. At a closer look, modern science gives evidence of a variety of laws, principles and rules. Similar to the manner in which Christianity only has very few dogmas (in my view none but the Trinity, creation and incarnation), so does current physics only have relatively few universal laws (e.g. gravity, quantum mechanics, relativity) and some important meta-principles (e.g. the principle of the conservation of energy). Most science is eventually concerned about

---

[46] Especially in the important contributions by William Stoeger, "Contemporary Physical and the Ontological Status of the Laws of Nature," in *CC*, 209-34 and idem, "The Mind-Body Problem, the Laws of Nature, and Constitutive Relationships," in *NP*, 129-46, but also in the position taken by John Polkinghorne, "The Metaphysics of Divine Action," in *CC*, 153f.

[47] Nancey Murphy, "Divine Action in the Natural Order," 327, 329, 339-40 and 342: "The ontological reductionist thesis seems undeniable—macroscopic objects are *composed of* the entities of atomic and subatomic physics. This being the case, much (but not all) of the behavior of macro-level objects is determined by the behavior of their smallest constituents. Therefore, God's capacity to act at the macro-level mostly includes the ability to act upon the most basic constituents"; observe here, however, how the cautious statement in the antecedent ("much" (but not all)) is not reflected in the emphatic claim of the consequent ("Therefore"). Similarly Russell first argues that the quantum way of SDA should be located in a wider context of other approaches to divine action, such as top-down causality (via consciousness) and whole-part causation; yet he continues to argue that whole-part causation at the level of complex systems such as the ecological web "fails to be a candidate for noninterventionist divine action because of the underlying determinism of the processes involved, no matter how complex or inter-related they might be," ("Divine Action and Quantum Mechanics," 300, 301). In the same manner, Polkinghorne's 'chaos way' is excluded by reference to a putative implication of metaphysical determinism by virtue of deterministic equations, see Wildman and Russell, "Chaos," 82-6; similarly Drees, "Gaps for God?," 224-8. This is what I mean by the 'all-or-nothing game' played by the proponents of the quantum proposal.

[48] Popper, *The Open Universe* made a convincing case against scientific determinism against the background of classical physics, see my "Three Types of Indeterminacy," 166-9, no. 9.

identifying *singular causes and specific correlations,* which, by definition, do not posses the features of nomological universality. This becomes especially evident in complex biological and cultural systems.

I would thus argue that the variety of laws and principles needs to be more acknowledged in the future science and religion dialogue, and I believe that the divine action project in particular would benefit from reconsidering the status of laws and selection principles. Already much energy has been invested in exploring robust notions of top-down causality in discussion with current philosophy of science.[49] But more work has to be done in showing that much of modern science only consists of patchworks of more loosely connected laws. Moreover, many laws are purely descriptive, since they measure, in first-order empirical terms, natural capacities of domain-specific systems, the capacities of which only come forth under highly specific conditions.[50] Many of the 'laws' used to describe such systems (e.g. in medicine and physiology) have the form of *ceteris paribus* laws.[51] For example the regularity entailed in the sentence, "Aspirin cures headaches" (N. Cartwright's example) holds, 'all else being equal.'

The question I want to raise here is philosophical in nature, but, I believe, strongly linked to a specific understanding of what theologians can expect from 'scientific explanations.' I would thus like to see a further clarification on the status of laws, a clarification that is likely to influence future discussions on divine action, especially in the realm of evolved complex systems. Such investigations should seek to answer the following questions:

1. How far is it possible today, in the light of current philosophy of science, to be a *regularist* rather than a necessitarian concerning divine laws? On this account one gives ontological primacy to singular events and singular correlations (which are empirically testable) rather than to necessary causes and correlations (claimed

---

[49] Within the series, an alternative account of supervenience as context-dependent has been laid out by Nancey Murphy, "Supervenience and the Downward Efficacy of the Mental: A Nonreductive Physicalist Account of Human Action," in *NP* 147-64; Theo C. Meyering, "Mind Matters: Physicalism and the Autonomy of the Person," in *NP*, 165-80; cf. my own position in *The Human Person in Science and Theology* (Edinburgh: T & T Clark and Grand Rapids, Mich.: Eerdmans, 2000). These are examples of arguments in favor of a contextualism (in particular a holism) that argue for the insufficiency of a microphysical determination of mental events to brain events.

[50] Nancy Cartwright, *Nature's Capacities and their Measurement* (Oxford: Clarendon Press, 1989); idem, *The Dappled World: A Study of the Boundaries of Science* (Cambridge: Cambridge University Press, 1999), chap. 3. Cartwright argues that capacities are more fundamental than laws, which shows the Aristotelian thrust of her argument.

[51] See Peter Lipton, "All Else Being Equal," *Philosophy* Vol. 74 (1999): 155-68, for a plausible account of the difference between 'strict laws' and 'ceteris paribus laws.' Lipton argues for a dispositional view of natural capacities that is close to Cartwright's. A further step would be to show that that 'dispositions' in fact may change in the course of evolution. A strong case for this is autopoietic systems such as cells, immune systems, and brains, not to speak of semiotic systems such as language and cultural institutions.

to possess nomological universality). *Observe that on the regularist account there is no problem of conceiving non-interventionist SDA, since there are no necessitarian, preexisting laws to be violated by divine action.*[52] Even though this approach is a live option in current philosophy of science, it is probably seems difficult to be a pure regularist, though. Nonetheless, a great deal of current physics is *de facto* descriptive rather than prescriptive.[53]

2. What is the relation between descriptive laws that describe atemporal relations, and genuinely causal laws that specify temporal succession?[54] Think of the law T = pv, 'for a gas at equilibrium, temperature equals pressure times volume.' This law tells us nothing about the molecular friction, etc., that actually produces the heat. Or think of the logistic models of chaos theory which map the formative principles of some non-equilibrium systems, but do not trace the efficient causes of the empirical systems. Or think about the whole new field of 'computational complexity' that has not yet been discussed in the VO/CTNS series. Here the algorithmic models describe formative principles, but there is no reason to believe that a computer model of mountain formations (taking 15 seconds to run) discloses the real-world causes of mountain formation (that has taken millions of years).[55]

3. What if most of the "laws of nature" that we believe inhabit the world of nature, are only '*ceteris paribus* laws,' that is, laws that only have explanatory power "if everything is equal"—which is never the case? The issue is here to what extent the boundary conditions may begin to invade the notion of laws, so that we can no longer make a clear distinction between the 'system' and the ever-changing 'boundary conditions.'

---

[52] As rightly observed by Nicholas Saunders, *Divine Action and Modern Science* (Cambridge: Cambridge University Press), 62: "...if we make the theological assertion that SDA is a fundamental feature of the world, a broadly applied regularity approach will simply attempt to describe the world on the basis of the different singular events in it, and consequently subsume all instances of SDA."

[53] Cf. Ronald N. Giere, *Science without Laws* (Chicago: Chicago University Press, 1999).

[54] Cf. C. A. Hooker, "Laws, Natural," in *Routledge Encyclopedia of Philosophy*, Vol. 5 (London: Routledge, 1998), 470-5, 471: "There is no single or simple relationship of laws to causes (of law statements to causal statements)," and "Philosophical theories of science have often ignored or simplified the distinctions among, and rich diversity of, scientific laws."

[55] See Niels Henrik Gregersen, "Complexity: What is at Stake for Religious Reflection," in *The Significance of Complexity: Approaching a Complex World Through Science, Theology and the Humanities*, Kees van Kooten Niekerk and Hans Buhl, eds. (Aldershot: Ashgate, 2004), 135-66, esp. 142-3.

4. Provided that *ceteris paribus* laws play a major role in the actual pursuit of science, one might begin to look for an ontology that steers a middle course between isolated singular events (the in-and-out regulatarian approach) and universal laws (the necessitarian approach). A very appealing candidate here is the various dispositional theories of laws, referring to 'propensities' or 'capacities.' The dispositional understanding of laws transcends a merely regulatarian account, in so far as they claim that causal powers reside in capacities, which can often be empirically measured. This view may also be metaphysically appealing, because capacities may increase and decrease over time, according to their use and functional success of their use. Especially for temporal theists, the concept of causal capacities as prior to laws may offer a new basis for a theological theory of how God may facilitate and respond to the autopoietic capacities of self-evolving creatures. On the dispositional view of natural laws of development, the feature of autopoiesis, or natural self-productivity, would stand in the center of a theory of divine action.

Even though new aspects of the question of divine action in a world of natural causes might be worthwhile pursuing, any future discussion will do well to consult the insights and dense scholarship provided by the VO/CTNS project on divine action. It is no hype to say that this project is absolutely unparalleled in its depth, and historic in its dimensions. The five volume series will no doubt remain the prime resource for future scholarship on this perennial question.

SOME REFLECTIONS ON
"SCIENTIFIC PERSPECTIVES ON DIVINE ACTION"

Arthur Peacocke

1 Introduction

Insurers used to describe inexplicable, unpredictable, unexpected events as 'acts of God' and if these were favorable to their purposes or health, individuals would often describe them as 'miracles.' The increasing success of the sciences in accounting for physical events has led to confusion and obscurity around the whole question of how and whether God 'acts' in the world. To this must be added the moral dilemma that, if God can and does change events in the world, why is evil allowed to flourish? All of which arises from the notion of 'acts' of God in which God is supposed to ensure that certain *particular* events, or patterns of events, occur when otherwise they would not have done so but for God specifically willing them. But such special divine action[1] is not the only way in which God has been conceived of as interacting with the world. Christian (and Jewish and Islamic) theology has affirmed that God gives existence to all-that-is by an act of God's own will and as the expression of God's own nature. This is what the classical Christian doctrine of *creatio ex nihilo* was about. The epic of evolution has renewed an emphasis on God as the eternal Creator sustaining in existence processes which are endowed by God with an inherent capacity of generating new forms and so with possessing a derived creativity. Classically, monotheists have regarded this as the sustaining activity of God. In the past this was conceived of in somewhat too static terms, seeing God rather like the figure of Atlas supporting the terrestrial globe. However, in view of the evolutionary character of natural processes, both cosmic and biological, this now has to be endowed with much more dynamic

---

[1] The possible forms of divine action were a perennial focus in the conferences (q.v., *inter alia*: Robert John Russell, "Introduction," in *QC*, 1-32; Keith Ward, "God as a Principle of Cosmological Change," in *QC*, 261; Ted Peters, "The Trinity In and Beyond Time," in *QC*, 275; Willem B. Drees, "A Case Against Temporal Critical Realism?: Consequences of Quantum Cosmology for Theology," in *QC*, 349; Robert John Russell, "Introduction," in *CC*, 1-13; Wesley J. Wildman and Robert John Russell, "Chaos: A Mathematical Introduction with Philosophical Reflections," in *CC*, 85; Denis Edwards, "The Discovery of Chaos and the Retrieval of the Trinity," in *CC*, 163 and 172-4; Stephen Happel, "Divine Providence and Instrumentality: Metaphors for Time in Self-Organizing Systems and Divine Action," in *CC*, 181; Jürgen Moltmann, "Reflections on Chaos and God's Interaction with the World from a Trinitarian Perspective," in *CC*, 208; Willem B. Drees, "Gaps for God?," in *CC*, 236; William R. Stoeger, S. J., "Describing God's Action in the World in Light of Scientific Knowledge of Reality," in *CC*, 244 and 248; Paul Davies, "Teleology Without Teleology: Purpose through Emergent Complexity," in *EMB*, 154; William R. Stoeger, S. J., "The Immanent Directionality of the Evolutionary Process, and its Relationship to Teleology," in *EMB*, 165; Denis Edwards, "Original Sin and Saving Grace in Evolutionary Context," in *EMB*, 389; John F. Haught, "Darwin's Gift to Theology," in *EMB*, 394).

imagery. God gives existence to each instance of spacetime with all forms of matter-energy themselves dynamically and continuously and creatively being metamorphosed into new entities, forms and patterns. These latter have included ourselves—all human beings and their societies and history. From this human perspective, the sustaining creative interaction of God with the world has often been called God's 'general providence.' The scientific vista we now possess reinforces and enriches these two aspects of God's creating and sustaining interactions with the world.

This is not the case when we consider the possibility of *particular* events, or patterns of events, being other than they would 'naturally' have been because God intended them to be different. Such possibilities have often been denoted as the 'special' providence in God's interaction with the world—the outcomes of special divine action. This category can be extended to include 'miracles' regarded as events not conforming to natural regularities believed to be well established. Indeed the followers of the monotheist faiths, the 'children of Abraham'—Jews, Muslims and Christians—shape their lives and religious practices on the general belief that God does indeed influence people and events. Private devotions and public liturgies in these religious traditions include much else of much more spiritual significance (for example: thanksgiving, adoration, contemplation, meditation). Yet they certainly involve an inexpugnable element of petitionary prayer which is based on the belief that God can make particular events happen, if God so wills.

This belief in special divine action has for over three centuries been largely colored by two presuppositions which were taken to be either as validated by the world-view of the natural sciences or, at least, as being implications of it. It would not be too strong to aver that these are two ghosts that still haunt some, certainly popular, theological thinking even of recent decades. I refer to:

1. the understanding of the natural world basically as a mechanism, controlled by inviolable 'laws of nature,' deterministic and therefore, 'in principle' at least, predictable; and

2. the dualist assumption that human beings consist of two entities and that these represent two different ontological orders of reality, sometimes denoted respectively as 'matter' and 'spirit/mind,' whose relationship, admittedly problematical, mysterious even, is regarded as exemplified in the human experience of being an agent.

On presupposition (1), it is not possible to conceive of God's interaction with the world, other than his general and perpetual sustaining of it in existence, except in terms of God 'intervening' (not the most felicitous word, many thought, in this context[2]) in the natural course of events. God is

---

[2] The possibility of such divine 'intervention' was a major theme of the conferences (q.v., *inter alia*: W. Alston, "Divine Action, Human Freedom, and the Laws of Nature," in *QC*, 186, 190, and 199; Robert John Russell, "Finite Creation without a Beginning: The Doctrine of Creation in Relation to Big Bang and Quantum

presupposed to bring about results in the world which are other than they would have been had the law-determined processes, which the sciences reveal, taken their usual course. However, the successes of the sciences in unraveling the intricate, often complex, yet rationally beautifully articulated, web of relationships between structures, processes and entities in the world have made it increasingly problematic to regard God as 'intervening' in the world to bring about events that are not in accordance with these divinely created patterns and regularities that the sciences increasingly unravel. The very idea of an 'intervening' God appears to presuppose that God is in some sense 'outside' the created world and has in some way, not specified, to come back into it to achieve his purposes. A mechanistic view of the world inevitably engenders, as it did in the eighteenth century, a deistic view of God and of his relation to it. Yet the recognition, through our new knowledge of cosmic and biological evolution, that the processes of creation are continuous has led to that renewed emphasis on the presence and immanence of God within the processes of the created order. The very notion of God as the faithful source of rationality and regularity in the created order appears to be undermined if one simultaneously wishes to depict his action as both sustaining the 'laws of nature' that express his divine will for creation and at the same time intervening to act in ways abrogating these very laws—almost as if God had had second thoughts about whether he can achieve his purposes in what is divinely created!

Indeed for most scientifically-educated monotheists, their very belief in the existence and nature of the Creator God depends on that rationality and regularity of the world. The transcendence of God, God's essential otherness and distinct kind of being from everything else,[3] always allows in principle the theoretical possibility that God *could* act to overrule the very regularities to which God has given existence. However, setting aside the immense and real moral issues about why God does not intervene to prevent rampant evil, the proposal of such divine intervention gives rise more fundamentally to an incoherence in our understanding of God's nature. For it suggests an arbitrary and magic-making Agent far removed from the concept of the One who created and is creating the world which science reveals. That world appears

Cosmologies," in *QC*, 323-4; George F. R. Ellis, "The Theology of the Anthropic Principle," in *QC*, 389 and 397; John Polkinghorne, "The Laws of Nature and the Laws of Physics," in *QC*, 433; Wildman and Russell, "Chaos," 85; Stoeger, "Describing God's Action in the World in Light of Scientific Knowledge of Reality," 244, 248-9, and 260; George F. R. Ellis, "Ordinary and Extraordinary Divine Action: The Nexus of Interaction," in *CC*, 382-4; George V. Coyne, S. J., "Evolution and the Human Person: The Pope in Dialogue," in *EMB*, 17; Davies, "Teleology Without Teleology: Purpose through Emergent Complexity," 152 and 159; Robert John Russell, "Special Providence and Genetic Mutation: A New Defense of Theistic Evolution," in *EMB*, 199 and 216; Michael A. Arbib, "Towards a Neuroscience of the Person," in *NP*, 100; Ted Peters, "Resurrection of the Very Embodied Soul?," in *NP*, 324-5; George F. R. Ellis, "Intimations of Transcendence: Relations of the Mind and God," in *NP*, 470; Thomas F. Tracy, "Creation, Providence, and Quantum Chance," in *QM*, 237 and 242; Robert John Russell, "Divine Action and Quantum Mechanics: A Fresh Assessment," in *QM*, 295-6).

[3] In particular, God's omnipotence, the inherent capacity to do all that it is logically possible to do.

increasingly convincingly as closed[4] to causal interventions from outside of the kind that classical philosophical theism postulated (e.g., in the idea of a 'miracle' as a breaking of the laws of nature).[5]

Furthermore, one has to recognize, with Hume, that adequate historical evidence for such a contravention of the originally divinely established laws of nature could never, in practice, be forthcoming. *Ex hypothesi* the 'laws' are themselves statements of both observed regularities in sequences of events, as Hume would have seen them, and also revisable expressions of underlying fundamental relationships between terms depicting realities, as a critical realist interpretation of modern science would endorse.[6] One would need vastly more

---

[4] Whether or not there is, in fact, any degree of openness and flexibility in the course of natural events, as opposed to determinism, was one of the key questions of the conferences (q.v., *inter alia*: C. J. Isham, "Quantum Theories of the Creation of the Universe," in *QC*, 64; Paul Davies, "The Intelligibility of Nature," in *QC*, 156; William R. Stoeger, S. J., "Contemporary Physics and the Ontological Status of the Laws of Nature," in *QC*, 233-4; Ward, "God as a Principle of Cosmological Explanation," 256 and 260-1; Polkinghorne, "The Laws of Nature and the Laws of Physics," 442; Russell, "Introduction," in *CC*, 7; Wildman and Russell, "Chaos," 82-3; Stoeger, "The Immanent Directionality of the Evolutionary Process, and its Relationship to Teleology," 168; Russell, "Special Providence and Genetic Mutation," 202; Philip Hefner, "Biocultural Evolution: A Clue to the Meaning of Nature," in *EMB*, 332; Thomas F. Tracy, "Evolution, Divine Action, and the Problem of Evil," in *EMB*, 518; Nancey Murphy, "Supervenience and the Downward Efficacy of the Mental: A Nonreductive Physicalist Account of Human Action," in *NP*, 147; Philip Clayton, "Neuroscience, the Person, and God: An Emergentist Account," in *NP*, 209; Ian G. Barbour, "Neuroscience, Artificial Intelligence, and Human Nature: Theological and Philosophical Reflections," in *NP*, 261; William R. Stoeger, S. J., "Epistemological and Ontological Issues Arising from Quantum Theory," in *QM*, 89; Jeremy Butterfield, "Some Worlds of Quantum Theory," in *QM*, 112; Tracy, "Creation, Providence, and Quantum Chance," 237 and 243; Russell, "Divine Action and Quantum Mechanics," 298).

[5] Even if one conceives of these 'interventions' as rare, as made only for significant purposes such as, say, the education of humanity in God's ways or for the revelation of God's purposes, one still faces the question of whether it is coherent to think of God's action in the world in the light of other insights into the nature of God, for example, God's willingness to allow human beings to possess free will.

[6] The understanding of the 'laws of nature' that I find the most acceptable is that expounded by William Stoeger, "Contemporary Physics and the Ontological Status of the Laws of Nature":

"[T]hese laws and theories are actually "models," or approximate descriptions, albeit very accurate and detailed ones" (212).

"[U]nified theories and laws certainly do more and more adequately describe and, even at a certain level, account for what we observe...they do not seem to give us more than detailed and reliable *models* of certain aspects of nature and of its regularities" (214).

"[T]heories and models and the laws they encompass have a definite basis in reality as we observe and experiment with it. But to go further and maintain that they constitute the underlying pattern or plan of physical reality as it is in itself is thoroughly unjustified. The most we can say is that there are regularities and interrelationships in reality as it is in itself—a fundamental order—which are imperfectly reflected in our models and laws" (221).

evidence for the occurrence of any event supposed to have contravened such regularities or 'laws of nature' than for one not thought to be doing so—and it is of the nature of our fragmentary historical (even contemporary) evidence that this cannot be forthcoming. Any favorable assessment that the historical evidence could indeed be interpreted as evidence for a divine 'intervention' would clearly be sensitively dependent on the degree to which the witnesses and/or assessors believed *a priori* in the possibility of such 'intervention' ever occurring.

Scientists who are theists infer to there being a Creator God as the best explanation of the very existence of the world and of its inbuilt rationality. For such a theist it is incoherent to accept the pre-supposition that God intervenes in the created processes of the world, in the divinely created fabric of existence, of which human beings are an integral and emergent part. A God who intervenes could only and rightly be regarded by them, and by all who well-advisedly adopt a scientific perspective on the world, as being a kind of semi-magical arbitrary Great Fixer or occasional Meddler in the divinely created, natural and historical networks of causes and effects.

Hence, in contrast to these popular religious perceptions, already mentioned, a 'presumption of naturalism'—no supernatural causes, no *intervening* God—prevails in the present cultural milieu in which the monotheistic religions, especially Christianity in the West, operate. That milieu is dominated by the success of the sciences in explaining not only physical events but also human psychological and social ones—the whole epic of evolution from the 'hot big bang' to humanity has become intelligible in scientific terms.

This presumption in favor of intelligible, scientific explanations is reinforced by its methodological necessity in our investigations of the world and by the emergentist monist position which I myself espouse. On this view, all-that-is (the 'world') is made up of whatever physicists conclude are the basic building blocks of matter. There is 'nothing else' in the world in one

---

"[T]he laws we discover are essentially descriptive—not prescriptive—and only weakly explanatory and revelatory of the grounds of possibility and of necessity, and of the relationships which flow from them...In the biological sciences in particular—whether the functional disciplines or the evolutionary disciplines—laws are rarely spoken of seriously, and then only in an analogous sense" (221).

"[A] very strong case can be made for saying that such theories—because of their success have in practice been elevated to the position of laws—closely model in great qualitative and quantitative detail the fundamental patterns of order, causal influence and constraint we observe at different levels of the physical and chemical world. (222)...Thus these theories do have a very strong basis, in observed reality" (223).

"[T]he laws [of nature] cannot be said to be the source of the behavior nor of the constraints on behavior—they model or describe it" (224).

"[T]here seems to be little support for the position that the law is the cause of the regularity observed or that it forces physical entities to behave in the way they do. It is rather a very detailed and specific description of that regularity and of its fundamental character" (225).

"[T]hey [the laws of nature] provide intermediate, highly detailed descriptions which closely link phenomena which originally seemed unconnected, but they do not completely explain why the reality is that way rather than some other way" (225).

sense, nevertheless there is 'more to be said' than such a statement seems to imply. There are hierarchies of complexity constituted of those fundamental entities and these complexes display emergent properties which can have causal efficacy on lower levels. So, although they represent higher-level realities than those fundamental ones, there are no grounds for any kinds of supernaturalism, non-natural causal agents, vital forces (the ghosts of discarded vitalisms), any of the 'fields' modern occultisms postulate or even for mind/body or spirit/body duality in human beings.

It is this basically monistic, but many-leveled and so emergentist, world with which God must be seen to be interacting. If so, then *how* are we to conceive of God's special divine action, of a form of God's interaction with the world that influences, steers or even directs events to be other than they would have been had not God particularly willed them to be so? The presumption of naturalism has tightened into the realization that the causal nexus of the world is increasingly perceived as closed.[5] There would seem to be no way for God to affect events other than by direct intervention in causal chains or providing new environing structural causes. This remains in principle, of course, a possibility since God could bring about events in the world by simply overriding the divinely-created relationships and regularities. As stated above, that is always a theoretical, though incoherent, possibility for the monotheist, but our present understanding of the world increases to the point of unattainability the onus on those who believe in such special providence and/or miracles to obtain convincing historical evidence for them. The more irregular the event, the better the evidence must be. Moreover, it is not sufficient for an event to be inexplicable by current science, for science has continuously closed the gaps in our knowledge and understanding of the world. Any 'God of the gaps' is vulnerable to being squeezed out by increasing knowledge as is widely recognized by theologians.

Given this cultural and intellectual impasse in our ability to conceive of how God could interact with the world through special providence and/or miracles, it is not surprising that this has become the key issue in the quickening pace of the dialogue between science and theology in the last two decades. At the spearhead of this attempt to relate our knowledge of the natural world to received theological beliefs, especially in regard to this issue, have been the biennial research symposia instigated since 1987 by the Vatican Observatory (VO) with the cooperation of the Center for Theology and the Natural Sciences in Berkeley, California (CTNS). The scientists, theologians and philosophers (often embodied in the same individuals) have, beginning with *PPT*, produced since then a succession of state-of-the-art volumes on scientific perspectives on divine action. These have been focused on different areas of the sciences: quantum cosmology and the laws of nature; chaos and complexity; evolutionary and molecular biology; neurosciences and the person; and quantum mechanics. It cannot be pretended that consensus has yet been obtained but the nature of the problems and the strengths and weaknesses of various proposals have been and continue to be thoroughly examined. Meanwhile popular Christianity continues to affirm the miraculous nature not only of certain putative events recorded in the Bible, but even some of those in everyday life. It often does so without recognizing the incoherence and insupportability of such beliefs in a cultural milieu more critically informed of the nature of the world than in the previous generation—in fact, light years from the presuppositions of the biblical (and Koranic) texts. At present this is an especially acute issue for Christianity which has borne the brunt of the critique of the Enlightenment of its sacred sources and of the effect of the widening perspectives through science of the origin and course of nature, especially of humanity. Other major religions have yet to experience the equivalent awakening to rational criticism.

In what follows I will summarize in a broad way where I think those discussions have led and state my own position on these controversial issues. I shall not conceal my conviction that certain routes of exploration, albeit followed by very able investigators, have proved to be cul-de-sacs and I shall point to the lines of inquiry that I judge still to be fruitful.

Part of the underlying problem in such studies is the general, and often vague, assumption that 'science' has somehow ensured that events in the world are predictable and we must first look at this background issue.

2 Predictability and Causality

At the beginning of the 17th century, John Donne (in his *Anatomie of the World*) lamented the collapse of the medieval synthesis: "Tis all in pieces, all coherence gone"—but, after that century, nothing could stem the rising tide of an individualism in which the self surveyed the world as subject over against object. This way of viewing the world involved a process of abstraction in which the entities, structures and processes of the world were broken down into their constituent units. These parts were conceived as wholes in themselves, whose law-like relations it was the task of the "new philosophy" to discover. It may be depicted, somewhat over-succinctly, as the asking of, firstly, "What's there?"; then, "What are the relations between what is there?"; and finally, "What are the laws describing these relations?" To implement this aim a *methodologically* reductionist approach was essential, especially when studying the complexities of matter and of living organisms. The natural world came to be described as a world of entities involved in lawlike relations which determined the course of events in time and so allow predictability.

The success of these procedures has continued to the present day, in spite of the revolution necessitated by the advent of quantum theory in our understanding of how and what we know about the sub-atomic world. For at the larger-scale level that is the focus of most of the sciences from chemistry to population genetics, the unpredictabilities of quantum events at the sub-atomic level are usually either ironed out in the statistics of the behavior of large populations of small entities or can be neglected because of the size of the entities involved or both. Predictability was expected in such macroscopic systems and, by and large, it became possible after due scientific investigation. However, it has turned out that science, being the art of the soluble, has, by and large, until recently been concentrating on those phenomena most amenable to such interpretations.

The world is, almost notoriously, in a state of continuous flux. As Heraclitus said in the 5th century BCE, "Nobody can step twice into the same river." It has, not surprisingly, been one of the major preoccupations of the sciences ever since to understand the changes that occur at all levels of the natural world. Science has asked "What is going on?" and "How did these entities and structures we now observe get here and come to be the way they are?" The object of our curiosity is both causal explanation of past changes in order to understand the present and also prediction of the future course of events, of changes in the entities and structures with which we are concerned.

The notions of explanation of the past and present and predictability of the future are closely interlocked with the concept of causality. For detection of a causal sequence in which, say, A causes B, which causes C, and so on, is frequently taken to be an explanation of the present in terms of the past. It is also predictive of the future, insofar as observation of A gives one grounds for inferring that B and C will follow as time elapses, since the original A-B-C... sequence was itself a succession in time. It has been widely recognized that causality in scientific accounts of natural sequences of events can only reliably be attributed when there have been discovered some underlying relationships of an intelligible kind between the successive forms of the entities involved. The fundamental concern of the sciences is with the explanation of change and so with predictability and causality. It transpires that different kinds of natural systems display various degrees of predictability and that the corresponding accounts of causality are therefore different.

Science began to gain its great ascendancy in western culture through the succession of intellectual pioneers in mathematics, mechanics and astronomy which led to the triumph of the Newtonian system with its explanation not only of many of the relationships in many terrestrial systems but, more particularly, of planetary motions in the solar system. Knowledge of both the governing laws and of the values of the variables describing initial conditions apparently allowed complete predictability with respect to these particular variables. This led, not surprisingly considering the sheer intellectual power and beauty of the Newtonian scheme, to the domination of this criterion of predictability in the perception of what science should, at its best, always aim to provide—even though single-level systems such as those studied in both terrestrial and celestial mechanics are comparatively rare. It also reinforced the notion that science proceeded, indeed should proceed, by breaking down the world in general, and any investigated system in particular, into their constituent entities. So it led to a view of the world as mechanical, deterministic and predictable. The concept of causality in such systems can be broadly subsumed into that of intelligible and mathematical relations with their implication of the existence of something analogous to an underlying mechanism that generates these relationships. Furthermore certain properties of a total assembly can sometimes be predicted in more complex systems. For example, the gas laws are not vitiated by our lack of knowledge of the direction and velocities of individual molecules.

It is well known that the predictability of events at the atomic and sub-atomic level has been radically undermined by the realization that there is a fundamental indeterminacy in the measurement of certain key quantities in quantum mechanical systems. It arises from there being only a probabilistic knowledge of the outcome of the 'collapse of the wave function' that occurs when measurements are made. This introduces an inherent limitation, in some respects though not all, in the predictability of the future states of such systems. A related example of such inherent unpredictability occurs in a collection of radioactive atoms. It is never possible to predict at what instant the nucleus of any particular atom will disintegrate—all that can be known is the *probability* of it breaking up in a given time interval. This exemplifies the current state of quantum theory which allows only for the dependence on each other of the probable measured values of certain variables and so for a looser form of causal coupling at this micro-level than had been taken for granted in classical physics. But note that causality, as such, is not eliminated.

However there are also Newtonian systems which are deterministic yet unpredictable at the micro-level of description. This has been a time-bomb ticking away from at least as long ago as the 1900s under the edifice of the deterministic, and so predictable, paradigm of what constitutes the world-view of science. The French mathematician Henri Poincaré then pointed out that the ability of the (essentially Newtonian) theory of dynamical systems to make predictions depends on knowing not only the rules for describing how the system will change with time, but also on knowing the initial conditions of the system. Predictability often proved to be extremely sensitive to the accuracy of our knowledge of the variables characterizing those initial conditions. Thus it can be shown that even in assemblies of bodies obeying Newtonian mechanics there is a real limit to the period during which the micro-level description of the system can continue to be specified, that is, there is a limit to predictability at this level. This limit has been called the horizon of "eventual unpredictability."[7] There is no unlimited predictability by us because of our inability ever to determine sufficiently precisely the initial conditions, in spite of the deterministic character of Newton's laws.

For example, in a game of billiards suppose that, after the first shot, the balls are sent in a continuous series of collisions, that there are a very large number of balls and that collisions occur with a negligible loss of energy. If the average distance between the balls is ten times their radius, then it can be shown that an error of one in the 1000th decimal place in the angle of the first impact means that all predictability is lost after 1000 collisions. Clearly, *infinite* initial accuracy is needed for the total predictability through infinite time. The uncertainty of the directions of movement grows with each impact as the originally minute uncertainty becomes amplified. So, although the system is deterministic in principle—the constituent entities obey Newtonian mechanics—it is never totally predictable *in practice*.

Moreover, it is not predictable for another reason, for even if the error in our knowledge of the angle of the first impact were zero, unpredictability still enters because no such system can ever be isolated completely from the effects of everything else in the universe—such as gravity and, of course, the mechanical and thermal interactions with its immediate surroundings.

Furthermore, attempts to specify more and more finely the initial conditions will eventually, at the quantum level, come up against the barrier of the measurement problem to our knowledge of key variables characterizing the initial conditions even in these "Newtonian" systems. So new questions arise. Does this limitation on our knowledge of these variables pertaining to individual units (whether atoms, molecules or billiard balls) in an assembly reduce the period of time within which the trajectory of any individual unit can

---

[7] Wildman and Russell, "Chaos," 76.

be traced? Is there an ultimate upper limit to the predictability horizon set by the irreducible quantum "fuzziness" in the values of those key initial conditions to which the eventual states of these systems are so sensitive? Does "eventual unpredictability" also prevail with respect to the values of those same parameters which characterized the initial conditions and to which quantum uncertainty can apply? Many physicists think so.

3 Possible Loci of Special Divine Action

The possibility of the occurrence of special divine action which is not 'intervention,' in the sense of disrupting the regularities which some sciences attempt to codify as 'laws of nature,' has been held by some authors[8] to imply that there are *intrinsic* gaps in the causal web of natural events not covered by deterministic inevitabilities—that is, there exist points of under-determination of the shaping of the course of subsequent events in the natural world (including the mental processes of human beings). In this perception there is an inherent flexibility built into the course of natural events, especially at higher levels of complexity. If this were so, they would constitute a plausible locus for special divine action, so it was argued, in particular with respect to 'chaotic' or quantum-mechanical systems. These proposals were a recurrent theme at all the VO/CTNS conferences, together with that of whole-part influences,[9] or 'top-down' causation, operating on the world-as-a whole and perhaps in and on other levels of complexity.

3.1 'Chaotic' Systems

One of the striking developments in science in recent years has been the increasing recognition that many relatively macroscopic systems—physical, chemical, biological and indeed neurological—can in practice become unpredictable in their macroscopically observable behavior. This is so even when the course of events is governed by equations which are deterministic in their consequences so that the final states are determined. Basically the unpredictability *by us* arises because two states of the systems in question which differ only slightly in their initial conditions eventually generate radically different subsequent states and we are unable to measure the significant differences in those initial conditions. These are generally, and somewhat misleadingly, called 'chaotic' systems. One particular equation (the so-called 'logistic' equation) has proved to be significant for a number of natural systems (e.g., predator-prey patterns, yearly variation in insect and

---

[8] Proponents included *inter alia* John Polkinghorne, "The Metaphysics of Divine Action," in *CC*, 151ff.; Russell, "Divine Action and Quantum Mechanics," *QM*, 295; Tracy, "Creation, Providence, and Quantum Chance," 250; and the notion was disputed by Drees, "Gaps for God," 223-237.

[9] *Inter alia*: Arthur Peacocke, "God's Interaction with the World: The Implications of Deterministic 'Chaos' and of Interconnected and Interdependent Complexity," in *CC*, 272-6 and 282; idem, "The Sound of Sheer Silence: How Does God Communicate with Humanity?," in *NP*, 220-9 and 235-7; Murphy, "Supervenience and the Downward Efficacy of the Mental," 154-7; and Ellis, "Ordinary and Extraordinary Divine Action," 362-3.

other populations, and physical systems too). In some other systems there can also occur an amplification of a fluctuation of the values of particular variables (e.g., pressure, concentration of a key substance, etc.) so that the state of the system as a whole undergoes a marked transition to a new regime of patterns of these variables. The state of such systems is then critically dependent on the initial conditions which prevail within the key, transitory fluctuation which is subsequently amplified. A well-known example is the "butterfly effect" of Edward Lorenz whereby a butterfly disturbing the air here today could affect what weather occurs on the other side of the world in a month's time through amplifications cascading through a chain of complex interactions. Another is the transition to turbulent flow in liquids at certain combinations of speed of flow and external conditions. Yet another is the appearance, consequent upon localized fluctuations in reactant concentrations, of spatial and temporal patterns of concentration of the reactants in otherwise homogenous systems when these involve autocatalytic steps—as is often the case, significantly, in key biochemical processes in living organisms. It is now realized that the time-sequence of the value of key parameters of such complex dynamical systems can take many forms: limit cycles; regular oscillations in time and space; and flipping between two alternative allowed states.

In the real world most systems do not conserve energy: they are usually dissipative ones through which energy and matter flow, and so are also "open." Such systems can often give rise to the kind of sequence just mentioned. Moreover, recent physics has also led to the recognition that, in such changeovers to temporal and spatial patterns of system behavior, we have examples of "self-organization." The point is that new patterns of the constituents of the system in space and time become established when a key parameter passes a critical value.

Explicit awareness of all this is only relatively recent in science and necessitates a re-assessment of the potentialities of the stuff of the world, in which pattern formation had previously been thought to be confined only to the large-scale, static, equilibrium state. In these recently examined systems, matter displays its potential to be self-organizing and thereby to bring into existence new forms entirely by the operation of forces and the manifestation of properties we already understand. "Through amplification of small fluctuations it [nature] can provide natural systems with access to novelty."[10]

How do the notions of causality and predictability relate to our new awareness of these phenomena? (We shall discuss this, for the time being, without taking account of quantum uncertainties.) Causality, as usually understood, is clearly evidenced in the systems just discussed. Nevertheless the identification of the causal chain now has to be extended to include unobservable fluctuations at the micro-level whose effects in certain systems may extend through the whole system so as to produce effects which extend over a spatial range many orders of magnitude greater.

The equations governing all these systems are deterministic which means that *if* the initial conditions were known with infinite precision, prediction would be valid into the infinite future. But it is of the nature of our knowledge

---

[10] J. P. Crutchfield, J. D. Farmer, N. H. Packard and R. S. Shaw, "Chaos," *Scientific American* (December 1986): 48.

of the real numbers used to represent initial conditions that they have an infinite decimal representation and we can know only their representation up to a certain limit. Hence there will always be, for systems whose states are sensitive to the values of the parameters describing their initial conditions, an "eventual unpredictability" *by us* of their future states beyond a certain point. The "butterfly effect," the amplification of micro-fluctuations, turns out to be but one example of this. In such cases the provoking fluctuations would anyway be inaccessible to us experimentally. Note, however, that—after much discussion—it has become clear that although these states are unpredictable *by us* they are not intrinsically (more technically, 'ontologically') indeterminate, in the way that quantities dependent on quantum states and sensitive to measurement indeterminacy are taken to be, in the prevailing 'orthodox' interpretation of physicists. Nevertheless, this whole scientific development does show that there can exist states of particular systems that are extremely close in energy, yet differ in their pattern or organization, and so in 'information content.'

In spite of the excitement generated by this recently won awareness of the character of the systems described above, the basis of the unpredictability in practice of their overall states is, after all, no different from that of the eventual unpredictability at the micro-level of the description of 'classical' Newtonian systems. However, the world appears to us less and less to possess the predictability that had been the presupposition of much theological reflection on God's interaction with the world since Newton. We now observe it to possess a degree of openness and flexibility within a law-like framework, so that certain developments are genuinely unpredictable by us on the basis of any conceivable science. We have good reasons for saying, from the relevant science and mathematics, that this unpredictability will, in practice, continue.

The history of the relation between the natural sciences and the Christian religion affords many instances of such gaps in human ability to give causal explanations—that is, instances of unpredictability—being exploited by theists as evidence of the presence and activity of God who thereby filled the explanatory gap. But now, as discussed in the preface to this section (3), we have to take account of, as it were, permanent gaps in our ability to predict events in the natural world. Should we propose a "God of the (to us) *uncloseable* gaps?" There would then be no possibility of such a God being squeezed out by increases in scientific knowledge. This raises two questions of theological import: (1) "Does *God* know the outcome of these situations/systems that are unpredictable by us?"; and (2) "Does God act within such situations/systems to effect the divine will?"

With respect to (1),[11] an omniscient God may be presumed to know, not only all the relevant, deterministic laws that apply to any system (its inherent regularities), but also the relevant initial values of the determining variables to

[11] Excluding quantum theory considerations and assuming that the future does not already exist in any framework of reference *for* God *to* know, for which I have argued elsewhere; see *Theology for a Scientific Age* (London: SCM Press and Minneapolis: Fortress Press, 1993), 128-34; see also J. R. Lucas, "The Temporality of God," in *QC*, 235-46, *contra* the classical Boethian view that God has timelessly, immediate, direct apprehension all events that to us are past, present and future.

the degree of precision required to predict its state at any future time, however far ahead, and also the effects of any external influences from anywhere else in the universe, however small. So, for those systems whose future states are sensitive to the initial conditions, there would be no eventual unpredictability for an infinite, omniscient God, even though there is such a limiting horizon for finite human beings because of the nature of our knowledge of real numbers and because of ineluctable observational limitations. Divine omniscience must, for example, be conceived to be such that *God* would know and be able to track the minutiae of those fluctuations in dissipative systems which are unpredictable and unobservable *by us* and whose amplification leads at the large-scale level to one particular outcome rather than another—but still unpredictable *by us*. God would just *know* all that it is logically conceivable to know about the systems (initial conditions, controlling equations, etc.), indeed knows them as they deterministically are, and so knows what they will be—even if the future does not yet exist for him to know with direct immediacy.

This is an affirmative answer to (1). Could we then go on to postulate that God might choose to influence events in such systems by changing those initial conditions so as to bring about a different macroscopic consequence conforming to the divine will and purposes? This would be also to answer question (2) affirmatively. God would then be conceived of as acting, as it were, "within" the (to us) flexibility of these unpredictable situations in a way that, in principle, we could never detect. Such a mode of divine action would always be consistent with our scientific knowledge of the situation. In the significant case of those dissipative systems whose macro-states arise from the amplification of fluctuations at the micro-level, God would have to be conceived of, presumably by an input of energy/matter, as actually manipulating micro-events in these initiating fluctuations in the natural world in order to produce the results at the large-scale level that God wills.

Such a conception of God's action in these, to us, unpredictable situations would then be no different in principle from that of God *intervening* in the order of nature with all the problems that that evokes for a rationally coherent belief in God as the Creator of that order. The only difference in this proposal from that of earlier ones postulating divine intervention would be that, given our recent recognition of the actual unpredictability, on our part, of many natural systems, God's intervention would always be hidden from us.

At first sight, then, this introduction of an unpredictability, open-endedness and flexibility into our picture of the natural world seemed to help us to propose a possible location for where God might act in the world in now uncloseable "gaps." However the above considerations indicate that such divine action would be simply a kind of that divine intervention which we were striving to avoid postulating.[12]

---

[12] Note, too, that this analysis continues to assume that God can know all it is logically possible to know about natural events, that God is omniscient and so *does* know the outcome of deterministic natural situations which are unpredictable by us.

In this context, John Polkinghorne, stimulated by his own analysis of chaotic systems, has made an interesting metaphysical proposal[13] which, although itself not deducible from the character of chaotic systems, as such, could well be valid and so significant in its own right, hard though it will be to establish. According to this proposal the deterministic character of many physical systems (including chaotic ones) is but a "downwardly emergent approximation to a more subtle and supple physical reality" and "for a chaotic system, its strange attractor represents the envelope of possibility within which its future motion will be contained."[14] For the discussion of these systems has served to emphasize how, in the limit represented by the 'strange attractors,'[15] chaotic systems may become very close in energy but different in pattern (and so in information content). Polkinghorne proposes that, if in the (infinite?) limit the difference became zero, they could possibly become capable of being the subject of special divine action by the input of 'active information.' However, the energy differences between adjacent paths through the attractor could become so small that they would come into the range of Heisenberg energy uncertainties[16]—and at present, there is no theory available to deal with such 'quantum chaos' and therefore no firm scientific basis for pursuing this as a possible basis for special divine action.[17]

Moreover, there are in the view of many of us,[18] serious flaws in arguing from the unpredictability of chaotic systems, as such, to an ontological openness in natural processes at the higher levels of complexity and organization. On Polkinghorne's views concerning 'downward emergence' and the laws of physics being "but an asymptotic approximation to a more subtle (and more supple) whole," Thomas Tracy[19] cogently commented: "If this view were vindicated, then chaos theory could be regarded as an instance of it. But is hard to see why we should run the inference in the other direction, that is, from chaos theory to a more holistic view. For chaos theory...is quite prepared to explain unpredictability entirely as a function of its deterministic equations."

---

[13] For his exposition and fuller references, see Polkinghorne, "The Laws of Nature and the Laws of Physics," 439; idem, "The Metaphysics of Divine Action," 147-56; and idem, "Physical Processes, Quantum Events, and Divine Agency," in *QM*, 189.

[14] Polkinghorne, "The Metaphysics of Divine Action," 153.

[15] The limit as trajectories become infinitely close together is called the 'strange attractor' in the mathematical theory of chaos.

[16] A problem pointed out by N. Saunders, *Divine Action and Modern Science* (Cambridge: Cambridge University Press, 2002), 94-150, and recognized by Polkinghorne, "Physical Processes, Quantum Events, and Divine Agency," 189, no. 9.

[17] Because in some chaotic systems, involving irreducibly finite numbers of (say) molecules or biological species, the limit of infinite complexity (zero energy difference between trajectories) cannot in principle be reached, Saunders has also argued that Polkinghorne's proposal for special divine action by active information input into chaotic systems requires a quantum chaos model.

[18] For example, Nancey Murphy, "Divine Action in the Natural Order: Buridan's Ass and Schrödinger's Cat," in *CC*, 327-8.

[19] Thomas Tracy, "Particular Providence and the God of the Gaps," in *CC*, 314.

Nevertheless, although it seems that it cannot be inferred from chaos theory, his hypothesis of openendedness and flexibility at higher levels of complexity along with the apparent and well-established determinism of lower level systems (including chaotic ones) as being the consequence of a confining and restriction at those levels ('downward emergence') is an attractive one, for it might open up a metaphysical path for integrating the flexibility and indeterminate character of mental experience in relation to brain states and other non-reducible features of personal and social, and indeed historical, existence which at present persist in remaining bafflingly incongruous with their physical matrix and basis.

All of this leads us to the conclusion that this newly-won awareness of the unpredictability inherent in chaotic systems does not, of itself, help directly to illuminate the "causal joint" of where God acts in the world—much as it alters our understanding of what is going on. This route for understanding special divine action based on our scientific perceptions has therefore proven to be a cul-de-sac while raising interesting speculations about the possibility of flexibility and openendedness at higher levels of complexity and organization. The exercise has demonstrated how open *for us* are the outcomes of many scientifically understood processes and that we are wise not to assume our total ability always to be able to predict in situations where there are many alternative states of structures and sequences of events that are very close in energy but which differ in pattern, and so in their content of 'information.'

3.2 Quantum Events

Another route in the exploration towards an understanding of divine action in light of the sciences which was followed in the VO/CTNS conferences is the possibility that the indeterminacy of the outcomes of measurements at the quantum level might provide another possible 'uncloseable gap' in the causal chains of nature. This indeterminacy is regarded as inherent, ontological and basic by most physicists. They do not believe there are 'hidden variables' to be discovered to render quantum events deterministic in the classical way. In such an uncloseable gap at the quantum level, God, it is proposed, could be affecting the outcomes of particular events consistently with the laws of the relevant science (quantum mechanics in this case) and accomplish this unbeknown to human observers. While some protagonists of this view (all carefully nuance their proposals) have argued for God being active in all quantum events,[20] others postulate divine action only in some,[21] or primarily in brain events[22]—and most prefer those locations for such special divine action whose consequences are amplified to larger-scale levels, for example by 'chaotic' processes, and/or in dissipative systems. This last kind of proposal is particularly attractive, seductive even, insofar as it can be propounded as the

---

[20] For example, Murphy, "Divine Action in the Natural Order," 343; and Russell, "Special Providence and Genetic Mutation," 214; idem, "Divine Action and Quantum Mechanics," 316-7.

[21] For example, Tracy, "Creation, Providence, and Quantum Chance," 235-58.

[22] For example, George F. R. Ellis, "Quantum Theory and the Macroscopic World," in *QM*, 259-91.

means whereby God might affect the course of biological evolution and even of human thinking—by divine action at the quantum level of mutations in DNA and on neuronal synapses, respectively. So it is not surprising that this hypothesis has attracted weighty supporters.

Those making this kind of proposal all accept, with most physicists, the inherent, ontological indeterminacy of the outcomes of particular quantum events—more precisely, of the outcomes of measurements at the quantum level. They recognize that there can be only a probabilistic advance knowledge of certain key parameters of a quantum mechanical system which would result from measurements made upon it. The state of such a system, according to the most widely accepted view, is represented by a superposition of wave-functions before the measurement and that this 'collapses' into a single one after it. *Which* one it will collapse into, *which* state the system will then be in, can be known in advance only probabilistically and no other more precise, advance knowledge is possible—in principle, most physicists would add. That is the force of the adverb 'ontologically' when it precedes 'indeterminate' and is applied to such situations as we shall see below.

Those who argue for direct divine involvement in *all* such quantum events, in all such wavefunction collapses—such is the basic underpinning of all natural events at the quantum level—cannot, it seems to me, avoid implicitly supporting divine responsibility for *all* events. This verges on the theological view called 'occasionalism' according to which God is the sole cause of all events and which entails notorious problems concerning evil and free will, amongst many others. For, if God's action is thought to be mediated through involvement in *all* quantum events then, because of the ubiquity of such events and their underpinning of almost all other events, it is hard not to infer that God is at least a contributory cause of all particular events. So the most defensible form of this hypothesis it seems to me, is that which proposes that God influences directly the outcomes of only *some* quantum events, in particular those whose outcomes can be amplified to bring about specific effects at larger-scale levels.

All involved in this discussion, whether or not in agreement with the proposals just outlined, accept that God upholds and sustains, gives continuous existence to, those processes and events in the natural world for which quantum mechanics is the current best (and indeed highly successful) interpretation. That is not the issue. What is under discussion is whether or not more direct and particular divine action needs to be postulated at this basic structural level of known nature.

If the probabilistic (that is, stochastic) 'laws' of quantum theory governing the outcomes of measurements of quantum events only *describe* the behavior of quantum systems and their ontological status is not affirmed, then any action of God in changing events in such systems could reasonably be construed as not interventionist in the theologically desired sense expounded above. (This is the view attributed to Robert Russell and Nancey Murphy by Wesley Wildman at the Capstone Conference which this volume reports[23]). Most physicists assume that the outcomes of measurements on quantum-mechanical systems really *are* in fact ontologically indeterminate, within the

[23] The 'Capstone' Conference held at Castel Gandolfo, September 7-13, 2003.

restrictions of the deterministic equations governing the development of the state of the system in time.[24] The ontological laws expressed in these equations govern only the probabilities of the outcome of measurements. Since God can know[25] only what it is logically possible to know (what is meant by God being omniscient) and that is confined to the probabilities of the outcome of any measurement, God cannot—logically cannot—know definitively the *precise* outcome of any particular measurement. Furthermore, if God were to alter one such event in a particular way then it seems to follow that, for the overall probabilistic relationships which govern the quantum events to be obeyed, many others—perhaps absurdly many over a long period—would also have to be changed, even if the departing from the statistical law could not be detected by observation in any practicable period. This is therefore no tidy, neat way to solve the problem.

There are other critical issues about the 'quantum divine action' proposals which arise when one asks *how* God can actually influence the outcomes of quantum events—more precisely, the outcomes of measurements on quantum-mechanical systems, some of which could then be amplified to the macroscopic level of observable events. For example, would such actions be divine 'interventions' of the kind we were trying to avoid? The answer to this question turns, it was urged,[26] on the extent to which the laws prescribing probabilities for the measurement events apply only in the sense (a) of constraining large ensembles of events in statistical fashion *or* (b) of applying to each particular measurement event in the sense of constraining even individual measurement events. If the former, weaker sense (a) of the laws being 'ontological' holds, any divine action in altering the outcome of a particular event[27] could be construed as 'non-intervening,' since the law for

---

[24] The trajectory of the wavefunction, which is interpreted as being related to the probable value of certain physical parameters, is controlled by the deterministic Schrödinger equation.

[25] See n.12 in connection with this statement.

[26] In particular, by Wesley Wildman at the 'Capstone' Conference.

[27] Saunders argues in *Divine Action and Modern Science* that there are only four broad ways in which God might conceivably influence the outcomes of measurements on quantum-mechanical systems 'quantum divine action': (i) God alters the wave-function between measurements; (ii) God makes God's own measurement on a given system; (iii) God alters the probability of obtaining a particular result; and (iv) God controls the outcome of measurements.

Saunders's careful analysis shows that (i) is not only a highly interventionist action but that, at the point of measurement, there would be no guarantee that the result intended by God would be obtained. So does God then, as it were, 'switch off' indeterminacy to get the desired result? Moreover the time evolution of such a system is entirely deterministic, governed by the Schrödinger equation, and if it is still to be valid cannot allow the introduction of an entirely new component wave function. So (i) seems unlikely.

Possibility (ii) fares no better fate. If God 'makes a measurement', presumably and questionably via an observable part of creation, then, like any other measurement on the system, the outcome is governed by the probabilities of which the un-measured state is already compounded. So it is not possible for God to achieve any *particular* intended result. The result of a measurement is not determined—only its probabilities.

the ensemble would not, immediately and at that point, be abrogated—for *each* event is anyway not sufficiently caused. It is so construed, it would appear, by many (at least at the Capstone Conference[25]) of those who propose special divine action of this form at the quantum level. However, because the ensemble *is* an ensemble of individual events, I myself cannot see how this proposal avoids having implicitly to affirm a divine 'effect' at the level of the *individual* events which constitute the ensemble subject to the law in order to render those statistical laws operative and this appears to make the proposal (a) vulnerable to the same interpretation—as involving divine intervention—as does the assumption of (b). Only a minority of physicists, it was affirmed at the Capstone Conference, adopt the second ontological approach (b) to quantum probabilities (i.e., the probabilities applying to *each* particular measurement) and on this view any alteration by divine action of the probability of the outcome of a quantum measurement certainly appears to be intervening. I myself can see no difference between this and proposing that God alters the outcome of any other law-described regularity or sequence of events which the proponents of quantum divine action would unhesitatingly describe as 'intervention.'[28]

The various forms of the proposal of quantum divine action all rest on still much-disputed metaphysical interpretations of quantum theory, and even on variations within these (e.g., different forms of ontological assertions *re*

Suggestion (iii) involves God altering the probabilities prevailing hitherto in a measurement so that the divinely intended result is more likely. Its probability will range between certainty (probability one) and impossibility (probability zero). This proposal assumes, problematically, that in some sense the probabilities exist as features of the system in question before the measurement. They *describe* the nature of physical reality. Hence God would, on this proposal, be altering the nature of reality prior to a measurement and this involves intermittent and interventionist action on God's part.

Suggestion (iv) also involves God being involved in the process of measurement. God simply, as it were, sidesteps the probabilities predicted by normal quantum mechanics and just controls the outcomes of measurement. This involves a contrary assumption to that implied by (iii), namely that the probabilities *follow* from the measurements and not *vice versa.* This would imply, as (iii), that divine action is intermittent—because if God acted directly to control the outcomes of *all* such measurements, then God would be conceived of as arbitrarily making sure they fit the probabilities prescribed by quantum mechanics (and that would be a very 'occasionalist' proposal).

From his analysis Saunders concludes, I think rightly, that (ii) combined with (iv), or (iv) alone, prove to be the most plausible proposals for 'quantum divine action', with only some events, some measurements, being the direct action of God. Russell supports his own form of proposal (iv) as saying "that God acts with nature to bring about the outcomes of particular measurements consistent with the probabilities given before the event occurs" ("Divine Action and Quantum Mechanics," 296, no.11). (Note that the proposals of quantum divine action often need to rely on amplification of quantum events and the capacity of 'chaotic' processes to do this. However, this capacity is itself highly problematic in the light of the unresolved problems in relating quantum mechanics to chaos theory.)

[28] The exponents of 'quantum divine action' at the Capstone Conference in fact appeared to assume (a) and not (b), although this has not previously always been affirmed explicitly.

quantum laws), and—given also that the ghost of 'interventionism' is not convincingly exorcised—I judge that this is not the most fruitful path to follow at the present to understand special divine action in a non-interventionist way.

Let us now consider what I think is a more promising, if only partial, path in our exploration towards understanding the interaction of God with the world, one which the sciences of the last decades of the second millennium have opened up.

3.3 Whole-Part Influence

Causality in complex systems made up of units at various levels of interlocking organization can best be understood as a two-way process. There is clearly a 'bottom-up' effect of the constituent parts on the properties and behavior of the whole complex. However, real features of the total system-as-a-whole are frequently an influence upon what happens to the units (which may themselves be complex) at lower levels. The units behave as they do *because* they are part of these particular systems. What happens to the component units is the *joint* effect of their own properties, explicable in terms of the lower-level science appropriate to them, and also the properties of the system-as-a-whole which result from its particular form of organization. No lower-level regularities/laws are abrogated thereby. When that higher level can also be understood only in terms not reducible to lower-level ones, then new realities having causal efficacy can be said to have emerged at the higher levels.

The world-as-a-whole may be regarded as a kind of overall System-of-systems for its very different (e.g., quantum, biological, cosmological) component systems are interconnected and interdependent across space and time with, of course, wide variations in the degree of coupling. There will therefore be an influence on the component unit systems, at all levels, of the states and patterns of this overall world-System, and of its succession of states and patterns. Moreover God, by God's own very nature as omniscient, is the One Who has an unsurpassed awareness of such states and patterns of the world-System in all its interconnectedness and interdependence. These would be totally and luminously clear to an omniscient God in all their ramifications and degrees of coupling across space and time. For God is present to, and constitutes the circumambient Reality of all-that-is—that is what is implied by the strong emphasis on the immanence of God as a result of our scientific apprehensions of creation through the continuities of cosmic and biological evolution.

I have proposed[29] that such theological insights informed by these new scientific perspectives, might provide a resource for exploring into how we are to conceive of God interacting with the world. I am *not* postulating that the

---

[29] Originally in 1987, for references see Peacocke, "God's Interaction with the World," 263, no.1. For fuller accounts of mine, see ibid., 272-4, 282-3 and also idem, "The Sound of Sheer Silence," 220-1, 225-9, and 235-240, where I respond to the comments of Tracy, "Particular Providence and the God of the Gaps," 306, no.39. Peacocke, "The Sound of Sheer Silence," 223, refers in particular to those of Drees, "Gaps for God," 233-7.

world is, as it were "God's body" for, although I do think the world may best be regarded as 'in' God (pan*en*theism[30]), God's Being is distinct from all created beings in a way that we are not distinct from our bodies. Although the world is not organized like a human body, it is nevertheless a "System," indeed a "System-of-systems," for all-that-is displays real, if variable, interconnectedness and interdependence over time and space. So we shall continue to speak of the 'world-System' without relying, at this stage, upon any analogy with the mind-brain-body relation or with personal agency.

If God interacts with the world-System as a totality then God, by affecting its overall state, could be envisaged as being able to exercise influence upon events in the myriad sub-levels of existence of which it is made without abrogating the laws and regularities that specifically apply to them. Moreover God would be doing this without 'intervening' within the supposed 'gaps' provided by the in-principle, inherent unpredictabilities noted earlier. *Particular* events could occur in the world and be what they are because God intends them to be so, without any contravention of the laws of physics, biology, psychology, or whatever is the pertinent science *for the level in question*—as whole-part influences operate for the many individual, constituent systems of the world.

This model is based on the recognition that an omniscient God uniquely knows, over all frameworks of reference of time and space, everything that it is possible to know about the state(s) of the world-System, including the interconnectedness and interdependence of the world's entities, structures and processes. By analogy with the operation of whole-part influence in real systems, the suggestion is that, because the 'ontological gap' between the world and God is located simply *everywhere* in space and time, God could affect holistically the state of the world-System. Thence, mediated by the whole-part influences of the world-System on its constituents, God could cause particular patterns of events to occur which would express God's intentions. These latter would not otherwise have happened had God not so specifically intended.

Any such interaction of God with the world-System would be initially with it as a whole. This initial interaction would be expected to be followed by a kind of 'trickle down' effect as each level affected by the particular divine intention then has an influence on lower levels and so on down the hierarchies of complexity to the level at which God intends to effect a particular purpose. We have already seen how in 'chaotic' systems, especially dissipative ones, there can be states differing in pattern and organization (and so in 'information content'), yet very close in energy. This provides a flexible route for the transmission of divinely influencing 'information' within the world-System as a whole down to particular systems within that whole. Especially sensitive recipients could well be those of individual human-brains-in-human-bodies and could be the means whereby God is experienced in acts of meditation and worship—as well as recognized as 'special providence' in events judged to be

[30] For recent critical analyses and expositions of this position, see Philip Clayton and Arthur Peacocke, eds., *In Whom We Live and Have Our Being: Panentheistic Reflections on God's Presence in a Scientific World* (Grand Rapids, Mich.: Eerdmans, 2004).

responses to such human acts. If such divine responses were so transmitted then they would often, but not necessarily always, be indirect, elusive, not easily discerned and could well take a long time by human perceptions—and this corresponds to actual religious experience.[31] This action of God on the world is to be distinguished from God's universal creative action in that particular intentions of God for *particular* patterns of events to occur would be effected thereby.

The ontological 'interface' at which God must be deemed to be influencing the world is, on this model, being located in that which occurs between God and the totality of the world-System and this, from a panentheistic perception, is 'within' God's own self. What passes across this 'interface' may perhaps, as already hinted, be appropriately conceived of as a 'flow of information' without energy transfer (as would necessarily accompany it in such flows within the world-System). But one has to recognize that there will always be a distinction, and so a gulf, between the very nature of God and that of all created entities, structures and processes (the notorious 'ontological gap at the causal joint' of Austin Farrer[32]). Hence this model can only attempt to postulate intelligibly the 'location' and tentative character of the initial effect of God on the world-System seen, as it were, from our side of the boundary. Whether or not this analogical use of the notion of information flow proves helpful in this context, we do need some way of indicating that the effect of God at this, and so at all, levels is that of pattern-shaping in its most general sense, without abrogating lower-level regularities/laws. I am encouraged in this kind of exploration by the recognition that the concept of the *Logos*, the Word, of God is usually taken to refer to God's self-expression *in* the world as God's creative patterning of it.

The model is propounded to be coherent with the monist concept that all concrete particulars in the world-System are composed only of basic physical entities; and with the conviction that the world-System is causally closed. There are no dualistic, no vitalistic, no supernatural levels through which God might be supposed to exercising *special* divine activity. In this model, the proposed kind of interactions of God with the world-System would not, according to panentheism, be from 'outside' but, as it were, from 'inside' it. For the world-System is regarded as being 'in God'. This seems to be a fruitful way of combining God's ultimate otherness with God's ability to interface holistically with the world-System. This model of God's interaction with the world as including a whole-part influence has proved, in my view, to be a promising path to take in our exploration from science towards an understanding of God's special providence and has indeed been adopted by

---

[31] For further discussion of the role of human persons as particular recipients of divinely communicated 'information' see: Arthur Peacocke, "Emergence, Mind and Divine Action: the Hierarchy of the Sciences in Relation to Body-Brain-Mind" in *The Re-emergence of Emergence*, Philip Clayton and Paul Davies, eds. (Oxford: Oxford University Press, 2006), chap.12.

[32] Austin Farrer, *Faith and Speculation* (London: A & C Black, 1967), 66.

others,[33] though often in combination with other 'bottom-up' proposals, such as those involving chaotic systems and/or quantum events.

3.4 God as 'Personal Agent' in the World

I hope the model as described so far has a degree of plausibility in depending on an analogy only with complex natural systems in general and on the way whole-part influence operates in them. It is, however, clearly too impersonal to do justice to the *personal* character of many (but not all) of the profoundest human experiences of God. So there is little doubt that it needs to be rendered more cogent by the recognition that, among natural systems, the instance *par excellence* of whole-part influence in a complex system is that of personal agency. Indeed we could not avoid above speaking of God's 'intentions' and implying that, like human persons, God had purposes to be implemented in the world. For if God is going to affect events and patterns of event in the world, then we cannot avoid attributing the personal predicates of intentions and purposes to God—inadequate and easily misunderstood as they are. So we have to say that though God is ineffable and ultimately unknowable in essence, yet God 'is at least personal' and personal language attributed to God is less misleading than saying nothing!

That being so, we can now legitimately turn to the exemplification of whole-part influence in the mind-brain-body relation as a resource for modeling God's interaction with the world. When we do so the ascendancy of the 'personal' as a category for explicating the wholeness of human agency comes to the fore and the traditional, indeed biblical, model of God as in some sense a 'personal' agent in the world is rehabilitated. It is re-established here in a quite different metaphysical, non-dualist framework from that of much traditional theology but now coherently with that understanding of what is actually in and going on in the world which the sciences provide. Accounts of religious experience are, of course, deeply suffused with the language of personal interaction with God and at this point our more philosophical and theological explorations towards God begin to make contact with the common experiences of believers in God.

When we were using non-human systems in their whole-part relationships as a model for God's relation to the world in 'special providence' we resorted to the idea of a 'flow of information' as being a helpful pointer to what might be conceived as crossing the 'ontological gap' between God and the world-as-a-whole. But now as we turn to more personal categories to explicate this relation and interchange, it is natural to interpret a 'flow of information' between God and the world, including humanity, in terms of the 'communication' that occurs between persons—rather in the way that a flow of 'information,' in the technical engineering sense, transmutes, say in a telephone call, in the human brain into 'information' in the ordinary sense of the word, so that 'communication' occurs between persons. Thus whatever else may be involved in God's personal interaction with the world, communication must be involved and this raises the question of to whom God

[33] For example by Murphy, "Supervenience and the Downward Efficacy of the Mental," 147-64; and Ellis, "Ordinary and Extraordinary Divine Action," 359-95.

might be communicating. Indeed, there would not have been the intense and worthwhile investigations at the VO/CTNS conferences on "Scientific Perspectives on Divine Action" if had not been the case that humanity distinctively and, it appears, uniquely has regarded itself as the recipient of communication from that Ultimate Reality, named in English as 'God!'

4 A Postscript

Nicholas Saunders at the conclusion of his recent survey of *Divine Action and Modern Science*[18] asks "Would it be correct to argue on the basis of the forgoing critique that the prospects for supporting anything like the 'traditional understanding' of God's activity in the world are extremely bleak?" and responds: "To a large extent the answer to this question must be yes. *In fact it is no real exaggeration to state that contemporary theology is in crisis*...such a wide range of doctrine is dependent on a coherent account of God's action in the world, and we simply do not have anything other than bold assertions and a belief that SDA [special divine action] takes place."[34]

To many this may seem too pessimistic for the VO/CTNS conferences and their resulting publications have superbly faced up to this challenge that, as Saunders puts it, "much of the traditional account of God's activity cannot hold up against our understanding of modern science" as "we" (presumably, that is, Christian and the other Abrahamic theists) continue to wish "to assert that God is active in the physical world."[35] The proponents of the current relevance of quantum events and chaotic systems to solving the problem of special divine action seem to me to be more confident and optimistic than I am that they are useful paths to pursue. From my perspective on divine action the incorporation of whole-part divine influence, linked with a form of pan*en*theistic understanding,[36] continues to seem to me to be a viable, and philosophically tenable,[37] current approach worth developing further with all the radical shifts in received Christian theology it will inevitably entail.

---

[34] Saunders, *Divine Action and Modern Science*, 215 (emphasis in the original).

[35] Ibid., 216.

[36] For further discussion of this, not uncontroversial idea, see Clayton and Peacocke, eds., *In Whom We Live and Have Our Being.*

[37] For Saunders, too, apparently (Saunders, *Divine Action and Modern Science*, 214).

CONCEIVING DIVINE ACTION IN A DYNAMIC UNIVERSE

William R. Stoeger, S. J.

1 Introduction

In the Vatican Observatory (VO) and the Center for Theology and the Natural Sciences (CTNS) divine action program there has been significant progress in framing the question of divine action within the context of conclusions from the various natural sciences, as well as in modeling both God's creative action in nature and God's special action in history in ways coherent with the understanding of nature we have from the sciences and acceptable from the point of Christian theological teaching and tradition.[1] In this paper I shall synthesize the main features of the conception of divine action which has emerged in this focused interaction between theology and the natural sciences, and then refine it further in light of clarification or development of several other issues which, in my judgment, were not adequately emphasized, addressed or resolved during the program. In doing this, I shall also briefly comment on some issues featured in the program which I shall not incorporate in my portrait of divine action—either because I do not consider them central to my synthesis, or because I do not agree with the way they have been invoked in models of divine action.

The key distinction regarding divine action which emerged from the program and was employed extensively during it is that between God's universal creative action in nature and God's special action in history in favor of particular persons or groups of persons. The former has been described dominantly in terms of traditional *creatio ex nihilo* / *creatio continua* primary-causality models, and understood essentially as an ontological—not temporal—origination and dependence. Obviously, this does not rule out a temporal origin of created reality—a temporal origin, either real or metaphorical, would symbolize and even emphasize ontological contingency.[2] But it is not the primary content of the concept. The other side of this notion of creation is that the natural sciences describe the secondary causes, the regularities, relationships, processes and structures—the laws of nature—through which the universe and nature unfold in all their richness. God *continues* to act as creator in and through these causal channels, sustaining

[1]Some researchers consider that God's special action also can occur in nature—that is, without any reference to God's relationship with persons or groups of persons. I do not think this is the case—I believe a compelling argument can be given to show that God's action in nature is creative. In fact, as we indicate at the end of this paper, it may be that all divine action is essential creative, including what we categorize as special. These issues will be treated more fully later on.

[2] Cf. Robert John Russell, "Finite Creation without a Beginning: The Doctrine of Creation in Relation to Big Bang and Quantum Cosmologies," in *QC*, 293-329; idem, "*T*=0: Is It Theological Significant?," in *Religion and Science: History, Method, Dialogue*, W. Mark Richardson and Wesley Wildman, eds. (New York and London: Routledge, 1996), 201-24.

them in their existence and their efficacy. In fact, the principal way in which God continues to act, and manifest God's self, is through the dynamisms of "the laws of nature," *not* outside them. Exactly what this comprises and implies requires further elaboration, analysis and interpretation. From one point of view, *modulo* careful qualifications concerning what God is responsible for and what God is not, this can include all that occurs in history—all that human beings as individuals and as communities accomplish. The VO/CTNS participants seem to be in agreement that the laws of nature investigated by the sciences require an ontological grounding which they themselves cannot provide, and that they are the channel of God's continuing creative action in the universe. As a group, however, they were not ready to accept that this includes the ambiguities and tragedies of human history.

In section 2 we shall examine with some care certain aspects of the *creatio ex nihilo*—primary causality model of creative divine action—especially the transcendental character of the "divine action" involved, its fundamental resistance to any conceptualization, and the protocols which must therefore be enforced in analogically describing it via concepts like "cause," "act," or "ground." I have provided this detailed summary, because, from a philosophical and theological point of view it is, in my opinion, *creatio ex nihilo* and the ideas closely connected with it that provide a fundamental basis for properly understanding God's action in the world—both God's universal, creative action, and God's special action. Even those approaches which distance themselves from *creatio ex nihilo* or from the closely related primary/secondary causal distinction generally adopt some of its essential features, such as the fundamental ontological dependence of reality upon God and the unique character of the divine causal role. Secondly, as I shall point out later, one of the important influences of our scientific knowledge of nature is to reinforce and nuance this articulation of divine action. In fact, as I summarize its important features, the impact of the natural sciences, though implicit, will be evident. And thirdly, in my judgment, *creatio ex nihilo*, the issues connected with it and the consequences flowing from it were not adequately engaged in the course of the VO/CTNS series. Such engagement—particularly from a philosophical point of view—should be on our agenda for the future.

In section 3 will go on to review some of the insights which have emerged regarding how "the laws of nature" are to be most helpfully and accurately conceived, how God's universal creative action is effected and manifested in and through them, and take a brief look at divine action under the metaphors of the influence of the whole on its parts and the action of top-level causes on lower-level processes and entities. These considerations will provide the background and the foundation for examining some of the central issues regarding God's special divine action in history.

That is the more problematic side of our basic categorization of divine action. Given the restrictions which our understanding of the natural sciences seem to impose on the influences of any purely immaterial causal agency, internal or external to the material world, how can we adequately understand and model such special divine action? Leaving aside those suggested solutions which are simply interpretational, psychological or metaphorical, there have been a number of suggestions which allow God to act in special ways, in addition to God's universal creative action. Prominent among the possibilities which have been discussed and developed in the course of the VO/CTNS program have been:

1. God's special acts within the windows of opportunity provided by quantum indeterminacy;[3]
2. postulating that "the laws of nature" as they actually function in reality are much richer and deeper than what we have modeled in the natural and the human sciences, thus allowing for God's special acts without their violation;[4]
3. conceiving God's special acts as flowing from the "top-down" or "whole-part" causalities exercised by God toward creation;[5]
4. denying that the natural sciences enforce a determinism—even at the classical macroscopic level—that rules out God's special acts in history.[6]

Categories (2) to (4) point in roughly the same direction, and possibly can be fruitfully pursued together, once their differing points of departure and underlying metaphors are recognized. In accounting for God's perceived special acts within history, besides the issue of intervening in or violating the established "laws of nature," there is the theological requirement that God not be identified as simply another secondary cause, filling in gaps or directly functioning in ways that seem to be always accomplished by created causes or agents. Thus, there has been a strong inclination by all participants to link God's special action more closely with God's universal creative role. We shall return to this issue later.

But first we shall deal with several crucial issues whose resolution, along with how we conceive God's universal creative action and the laws of nature, strongly influences any conclusions we attempt to draw in this area. In section 4, we shall discuss the meaning and limits of determinism in nature, whether or not it holds, or can be shown to hold—and in what sense—and whether special divine action requires windows of indeterminacy in order to operate (the compatibilism/incompatibilism divide). And then finally in section 5, relying on the conclusions we have drawn along the way, we shall, taking our cue

---

[3] See Robert John Russell, "Divine Action and Quantum Mechanics: A Fresh Assessment," in *QM*, 293-328, and references therein; see also Wesley J. Wildman, "The Divine Action Project, 1988-2003," *Theology and Science*, Vol. 2, No. 1 (2004): 31-75, and references therein.

[4] See William R. Stoeger, S. J., "Contemporary Physics and the Ontological Status of the Laws of Nature," in *QC*, 209-34; idem, "The Mind-Brain Problem, the Laws of Nature and Constitutive Relationships," in *NP*, 129-46, and references therein; John Polkinghorne, "The Laws of Nature and the Laws of Physics," in *QC*, 437-48, and references therein.

[5] Arthur R. Peacocke, "The Sound of Sheer Silence: How Does God Communicate with Humanity?," in *NP*, 215-42, and references therein.

[6] William Alston, "Divine Action, Human Freedom, and the Laws of Nature," in *QC*, 185-207; Keith Ward, "God as a Principle of Cosmological Explanation," in *QC*, 247-62.

from Tom Tracy's ideas,[7] venture a portrayal of special divine action as a manifestation or mode of God's universal creative action, properly conceived.

2 God as Creator—God's Relation to the World

The traditional Judeo-Christian-Islamic expression of God's relation to the world is in terms of creation—God as the Creator and Sustainer of all that is, freely endowing all that is not God with existence, order and dynamism—that is, "the laws of nature." Or we can turn this around, and say that creation is the relation of all that is to the Creator.[8] From God's side it is a self-communication, an act of efficient causality (but not limited to efficient causality), producing creatures (i.e. things whose being is not that of God's being). From our side it is the relation of ultimate dependence.[9] This has almost always been articulated as *creatio ex nihilo*, creation from absolutely nothing—God producing or establishing reality without any prior raw material, energy, or laws of nature except, of course, God. As has been indicated often in the science-theology dialogue, this formulation is complemented by the idea of *creatio continua*, continuing creation, which stresses that the divine "act" of creation continues as long as anything that is not God exists. The basic content of *creatio ex nihilo* and *creatio continua* is the absolute dependence of all that is not God on God. God is the ontological origin of everything that exists—the universe, the world and all that they contain. This may or may not involve a chronological origin—"a first moment of creation"—but that is a secondary issue. There have been many different precisions, variations and modifications of this basic approach in the history of philosophy and philosophical theology. We shall not even attempt to discuss these. Instead we shall connect this fundamental model of creation with other key concepts, discuss how these concepts are to be qualified and interpreted in light of the impossibility of saying anything adequately descriptive of God and God's action, and briefly examine some of the problems with this approach and some of the alternatives. All this has been said many times before.[10] Here, I am summarizing it very briefly, as an introduction to some important closely related points which are rarely made, and which help us interpret and apply these insights much more flexibly and insightfully. In particular, I want to suggest how this articulation

---

[7] Thomas F. Tracy, "Creation, Providence, and Quantum Chance," in *QM*, 238ff.

[8] See Catherine Mowry LaCugna, *God for Us: The Trinity and Christian Life* (San Francisco: HarperSanFrancisco, 1993), 160; as Robert J. Russell, private communication, has said, "To be a creation *is* to be in relation to God as Creator and to be entirely constituted *by* that relation."

[9] LaCugna, *God for Us*, 60.

[10] For substantial but brief summary treatments of this constellation of traditional ideas on creation, see Langdon Gilkey, "Creation, Being, and Nonbeing," in *God and Creation: An Ecumenical Symposium*, David B. Burrell and Bernard McGinn, eds. (Notre Dame, Ind.: University of Notre Dame Press, 1990), 226-41; LaCugna, *God for Us*, 158-67; William R. Stoeger, S. J., "The Origin of the Universe in Science and Religion," in *Cosmos, Bios, Theos: Scientists Reflect on Science, God, and the Origins of the Universe, Life and Homo Sapiens*, Henry Margenau and Roy A. Varghese, eds. (La Salle, Ill.: Open Court, 1992), 254-69.

of God's action can be usefully refined and modulated within the careful description and analyses of other related experiences and phenomena—from the natural sciences, from philosophy, from theology.

Creatio ex nihilo is intimately related to the well-known "primary-secondary cause" distinction and classification. According to this, God, who is the Creator, is the primary, first, or ultimate "cause" of everything, and as such is uncaused. In that sense, God's nature is "to be"—to exist, and God is the necessary, ultimate "being," pure act—on which all else depends, and from which all else flows. All other causes are secondary causes, and ultimately dependent on the primary cause. God as "the primary cause" is unlike any other cause. As "primary cause"—as Creator—God is effective, but super-effective in some ways, and markedly under-effective in others. God's creative action, as we have already indicated, does not need any pre-existing material to mold, or fashion. It ultimately effects, forms, sustains and coordinates all the laws of nature—the regularities, relationships and processes which embody the complexity and intricacy of nature on its many different levels of organization. But, in doing that, it endows them with their own independent characters and efficacies, leaving them free to operate within their proper contexts. God is the ultimate or foundational "cause," then, of order as well as of existence. But, at the same time, God as primary cause is hidden, is not isolatable, cannot be picked out as a recognizable "cause" completely distinct from other, scientifically accessible causes.

God as primary cause *is* distinct, though, from all secondary causes and entities, but it would be incorrect to say that God is "outside" them. Unlike any other cause, because God is transcendent (subject to no barriers or constraints), God is radically immanent (interior) to all that is—but in a highly differentiated way, according to the character of each process, relationship or object. God, then, is not an entity, like other entities—God is more like a verb, a continuing action in which everything else participates, but participates according to its own individuality. God's primary causality does not substitute for nor interfere with nor countermand the integrity and adequacy of the (secondary) causal structures of nature or of history—despite being their ultimate foundation and source.

Thus, the use of the word "cause" for God's creative action requires extensive qualification, or rather explanation. We are really using this word to describe a reality that it does not fit, and cannot even come close to describing adequately. But it is the best we have, as far as applicable concepts go. As we struggle to articulate divine "causation" more precisely and more accurately, we shall let a number of other helpful concepts and disclaimers interact with "primary causality" in order to temper its inadequacy. But we shall never completely succeed. Still it is important to continue—in order to recognize the transcendental character (the radically inconceivable character) of the ontological horizon or ground careful reflection on our experience reveals.

These reflections underscore the metaphorical and mythical character of our language about God, especially when we speak of God as an actor, or as a cause. In fact, it is precisely for this reason that the primary-secondary causal model was developed—to emphasize that when we speak of God as a cause, or as subject of an act, it is in a way which is unlike any other cause or act, transcending what we can describe or articulate. This negative theological disclaimer must always be applied to whatever we end up saying about God's action. At the same time, this should not be taken to the extreme of holding that all analogical language about God is unhelpful or misleading—or that absolutely nothing at all can be asserted about God.

In this connection, I strongly suspect that the category of "relationship" is more fundamental than that of "act" or "cause." As I have argued elsewhere,[11] a more detailed and precise philosophy and theology of the different interconnected levels of relationality within creation, and of creation with God—within a dynamic Trinitarian framework—would improve our understanding of God's action in the world. This understanding would be based on a careful elaboration of how the interrelationships within nature, which are partially described by the sciences, are intimately linked with the creative relationship of God with the world, and therefore with the inner relationships of the Trinity which constitute God as God.

In characterizing *creatio ex nihilo* and the causality it represents, using somewhat traditional philosophical terms (the main point, of course, is not what terms we use, but what we are attempting to signify by them), we have done so referring to the context of our scientific knowledge of the world and the universe. In fact, the conclusions of the natural sciences, while not able to say anything directly about God or God's action, provide strong indirect support for these nuanced ways of describing God's action. It is clear, for instance, particularly from cosmology,[12] that cosmology and physics will never be able to describe or model the ultimate cause of the universe, nor of the laws of nature—that which effects the transition from absolute nothingness to existence and that which grounds the order of what exists. From that realization, it is a quick step to recognizing that whatever ultimately effects and maintains existence and order must be a cause unlike any other we know of. Furthermore, the findings of all the natural sciences strongly reinforce the formational and functional integrity[13] and relative autonomy, of nature. This, in turn, helps us articulate more concretely the transcendent immanence of God's action. There is no need or room for God to act in nature as a created cause[14]—but God as Creator is present and active in each process and in the whole network of processes and relationships. Thus, we can also say, in a way we could not have before the advent of the natural sciences, that God's universal creative action, though unique, is also realized in a highly differentiated and evolving way throughout nature. This evolutionary,

---

[11] William R. Stoeger, S. J., "God and Time: The Action and Life of the Triune God in the World," *Theology Today* 55, No. 3 (October 1998): 365-88.

[12] See, for instance, C. J. Isham, "Creation of the Universe as a Quantum Process," in *PPT*, 375-408; idem, "Quantum Theories of the Creation of the Universe," in *QC*, 49-89; William R. Stoeger, S. J., "Contemporary Cosmology and Its Implications for the Science-Religion Dialogue," in *PPT*, 219-47; and Russell, "Finite Creation without a Beginning."

[13] See Howard J. Van Till, *The Fourth Day: What the Bible and the Heavens Are Telling Us About Creation* (Grand Rapids, Mich.: Eerdmans, 1986).

[14] No participant in the divine action project has intentionally treated God as a created cause. However, my concern is that we may not have completely avoided doing so in our modeling and discussions of special divine action.

emergent and unfinished character of creation revealed by the sciences serves to emphasize the continuing character of God's action through the regularities, processes and relationships God sustains. From a scientific perspective we can point also to the central importance of relationality in the hierarchical structuring of physical and biological reality—this, too, is consonant and supportive of what we know of God and of God's activity from theology.[15] Finally, linked with evolutionary character of nature, we also notice the transience of entities and groups of entities within creation, serving the continuation and evolution of the whole. This, too, is consonant with what we as Christians have come to understand about God's presence and action within creation.[16] Along with the properties of divine creative action which these key characteristics of nature, as the sciences reveal them, reflect, modify and reinforce is that it cannot be simply an efficient cause. It must also, at the same time, endow creation with potentiality, form and purpose. This is manifested particularly evident in the highly differentiated relationality and evolutionary character of reality.[17]

Returning to *creatio ex nihilo* itself, some of the other important things to point out about this model of God's universal creative action follow from our discussion so far. These are not characteristics of all versions of the model. I am accentuating them, because in my opinion they are characteristics of the strongest and least inadequate version. First, God's creative act is not necessary—it is not an automatic "emanation" from the divine Creator, as Platonists and Neo-Platonists would have it. It is, rather, a free decision or choice of God to create. Secondly, the action is in a sense one action—with many manifestations, which are all eternally present to God. From the creature's perspective it is obviously continuing in time—perduring in time. This can be at least intuitively rendered somewhat intelligible by emphasizing once again that God is not an entity, but rather "an activity." As such, from the point of view of "the divine activity," there is no other activity or act, nor is there any intrinsic limit—beginning or ending—or other type of restriction to this activity,[18] and there is thus no time, in that sense. There are, however, relationships—relationships within God, God's self, (Trinitarian relationships, according to the Christian view) and relationships between God and the world. From the point of view of these relationships, we might say that there is a type of time—we call it "eternity"—the ultimate connectedness or "togetherness of

---

[15] For a further discussion of these characteristics, see William R. Stoeger, S. J., "Cosmology and a Theology of Creation," in *Interdisciplinary Perspectives on Cosmology and Biological Evolution*, Hilary D. Regan and Mark Worthing, eds. (Adelaide, Australia: Australian Theological Forum, 2002), 128-45.

[16] This is particularly clear if we consider God's creative action continuing and being completed by the new creation effected by Christ in His Life, Death, Resurrection and Sending of the Spirit and our participation in that.

[17] See Paul Davies, "Teleology without Teleology: Purpose through Emergent Complexity," in *EMB*, 151-62; and William R. Stoeger, S. J., "The Immanent Directionality of the Evolutionary Process, and Its Relationship to Teleology," in *EMB*, 163-90.

[18] See Joseph A. Bracken, *The Divine Matrix: Creativity as Link between East and West* (Maryknoll, N.Y.: Orbis Books, 1995), 11-37, 128-40.

the temporal modes" within creation itself.[19] This "eternity" finds its foundation and condition of possibility in the enhanced, ultimate "togetherness" of activity and relationship within. the Creator—the Creator's "eternity." So far, we have reflected on what Bernard McGinn refers to as the four basic constituents of the idea of creation: dependence, distinction, decision and duration.[20] These characteristics are intimately inter-related, which is obvious upon careful reflection. We shall not elaborate these relationships here, beyond what we have already implied.

But now there are some further nuances required in understanding these features of creation. Continuing with our enumeration, thirdly, according to almost all proponents of this approach, God's act of creation does not change or perfect God in God's essence. God would still be God without having created anything outside of God, and God would not be any less loving, good or complete by not having done so. Creation manifests all these qualities of God, but does not constitute them for God. Fourthly, though God is completely distinct from creation, and creation is utterly dependent upon God, God is deeply immanent within creation—closer and more intimately present to creation than creation is to itself. This is the transcendent immanence of the Creator—as we have briefly said above, the transcendence of God enables God's immanence. God, though radically distinct from Creation is profoundly present and active within it—in highly differentiated ways. And, though present and active, also, in a very definite sense, hidden and vulnerable within it, too.

Fifthly and finally, God's creative action involves God's practical knowledge of creation in a very peculiar and transcendent way.[21] Essentially, God's knowledge as creator—though working through secondary causes—is like that of the artisan, articulated or expressed in the act of creation itself and is fundamentally of particulars—individuals—rather than of abstract universals, though, at the same time, God's knowledge as Creator includes such knowledge. This, along with God's free choice in creation, makes God's creative action personal—present, oriented, and active within each particular with reverence for what it is in itself. How then does God know uncreated, unactualized possibilities? Only as what could have existed but does not, since God in God's eternity is present to all that existed, exists or will exist.[22] It is difficult to appreciate how both of these types of expression of particular divine creative knowledge are possible in light of the fact that God enables the autonomy and integrity of the laws of nature—seeming to delegate or parcel out divine creative decisions and power to blind, insensitive processes. But that would be to forget God as Creator continues to be profoundly immanent in and ontologically supportive of all those processes and secondary causes *all the time.* As Creator, "God knows what God is doing," not propositionally or

---

[19] Robert Cummings Neville, *Eternity and Time's Flow* (Albany, N.Y.: SUNY Press, 1993), 268; see also Stoeger, "God and Time."

[20] Bernard McGinn, "Do Christian Platonists Really Believe in Creation?," in *God and Creation*, Burrell and McGinn, eds., 208-9.

[21] See David B. Burrell, *Knowing the Unknowable God: Ibn-Sina, Maimonides, Aquinas* (Notre Dame, Ind.: University of Notre Dame Press, 1986), 71-91.

[22] Ibid., 99-104.

speculatively but practically and intimately, in endowing each individual object and entity with participation in God's being, even though this is achieved, as God intends, "through the integrity of natural causes."[23]

The drawbacks and inadequacies of this type of traditional approach to God's creative action are well known, and have been discussed frequently in the literature. As John O'Donnell helpfully summarizes, this concept of God—God as absolute, immutable, without potentiality, outside of time, and unaffected by what God creates—is considered by many to be both "metaphysically incoherent" and religiously objectionable.[24] It seems to be "metaphysically incoherent," because it is difficult to reconcile this portrayal of God's transcendent character with what it also asserts regarding God's immanence and knowledge. It seems that a God who really knows each object and is immanent in each, must, at least in some way, contain temporality and must be affected by that knowledge, especially if creation is also relatively autonomous and free to be itself, as this approach also emphasizes. It is religiously objectionable, because it seems to rule out any incorporation of genuine love and vulnerability, and thus seems to contradict what is proclaimed by Christian, Islamic and Jewish faith. It seems to rule out any self-communication of God, which is essential to what the theological concept of God implies.[25]

Langdon Gilkey[26] emphasizes the same failures, but goes on to analyze perceptively the root of the problem. As many, many theologians and philosophers have pointed out, and as we have already emphasized, God simply cannot be described or conceptualized—or objectified as other existents or beings are. God is beyond—transcends—all categories and language. When we use a concept to describe God we can do so only "symbolically" or "analogically." Essentially, God is intrinsic mystery, inexhaustible richness which is constantly being revealed but whose depths can never be adequately plumbed nor reached. What we have often done, however, is mistakenly to translate this transcendence and intrinsic mystery into "the absoluteness of God's being." The finitude of being we experience and understand has been naively extrapolated—"transcended"—into an absolute degree of the very sort of being we know and experience. Thus, as Gilkey says, "God's nature becomes defined by its unconditional and absolute character rather than by its mystery, and the dialectical nature of the relation of that mystery to the being that is God is lost—as the continual relatedness of God to finite being is also sacrificed."[27] Thus, we end up deluding ourselves into thinking we have adequately described God by this move, but in so doing we have dismissed the radically mysterious character of the divine reality, and "overaffirmed" the

---

[23] Ibid., 98-9.

[24] John J. O'Donnell, S. J., *Trinity and Temporality: The Christian Doctrine of God in the Light of Process Theology and the Theology of Hope* (Oxford: Oxford University Press, 1983), 17-21.

[25] Robert W. Jenson, *Unbaptized God: The Basic Flaw in Ecumenical Theology* (Minneapolis: Fortress Press, 1992), 138-9.

[26] Gilkey, "Creation, Being, and Nonbeing," 231-6.

[27] Ibid., 232.

being of God in terms of finite being.[28] Instead, we need to preserve the intrinsic mystery of God as fundamental, and insure that any affirmation of God's being or God's relationality is dialectically couched. These must be dialectically expressed simply because any words or concepts are inadequate to the mystery of God's being. Thus, in agreement with Gilkey, we are not called to reject the constellation of intricately inter-related metaphysical concepts and analyses which constitute the richer formulations of traditional *creatio ex nihilo* theology of creation. Rather, we are invited to radically revise it, or rather situate it dialectically as one well-honed but imperfect articulation of an important aspect of God's being and activity within the full, paradoxical, rich and incomprehensible mystery that our experiences of transcendence and divine revelation disclose.

Theologians and theologically oriented philosophers have done this in a variety of ways. Gilkey himself, for instance,[29] takes the basic traditional insights and re-configures them in a dialectical divine polarity of being and nonbeing. Bracken[30] transforms the basic traditional insights in terms of a metaphysically enriched Whiteheadian-like Creativity, a divine activity or process and not an entity, of which God is the primordial exemplar and in which all that exists participates. Others, like Karl Rahner,[31] Catherine LaCugna,[32] Colin Gunton,[33] and Nicholas Lash,[34] do something similar but more radically inter-relational by emphasizing the Trinitarian character of God, and of God's relationship to all that is not God.[35] Jenson[36] links the full realization of God as Trinity with Incarnation and all that flows from it in time.

Before concluding our reflections on God's relationship as Creator with the world in light of traditional and contemporary attempts to portray it, I wish to offer two further avenues of reflection and a final suggestion. The reflections return us to the fundamental issue of the radical mystery that God is, and to our complete inability to conceptualize God or God's activity. The first has to do with the realization that, despite this radical mystery and transcendence of God, we *do* have manifestations—revelations, self-communications—of God in persons and in experiences addressed to us which we can appropriate. God is disclosed to us in creation. And, as we have indicated above, the characteristics of creation manifested to us by the natural

[28] Ibid., 234, 236.

[29] Ibid., 236-40.

[30] Bracken, *The Divine Matrix*, 128-40.

[31] Karl Rahner, *The Trinity* (New York: Herder and Herder, 1970), 120; idem, *Foundations of Christian Faith: An Introduction to the Idea of Christianity* (New York: The Seabury Press, 1978), see esp. 133-7, 178-203.

[32] LaCugna, *God for Us*, 158-66.

[33] Colin E. Gunton, *The One, the Three and the Many: God, Creation and the Culture of Modernity* (Cambridge: Cambridge University Press, 1993), 248.

[34] Nicholas Lash, *Believing Three Ways in One God: A Reading of the Apostles' Creed* (Notre Dame, Ind.: University of Notre Dame Press, 1993), 136.

[35] Bracken also attempts to bring in the Trinitarian dimension in an essential way in his thinking, linking it with his notion of Creativity.

[36] Jenson, *Unbaptized God*, 138-45.

sciences are in harmony with, and even indirectly support, many of the fundamental qualities of God's creative action. Thus, in conceptualizing God and God's action, though no portrayal is adequate and the reality is far beyond all portrayals, we have some basis for determining the suitability of articulations about God and God's action. *Some portrayals of God and God's relationship to the world are much less inadequate and incorrect than others.*

The second reflection is simply that, as many have indicated, there are fundamental aspects of reality which we cannot conceptualize—which are transcendental and mysterious precisely because they are so fundamental, so essential, so rich, so pervasive that they cannot be defined or adequately described. Essentially, these are the characteristics of reality which cannot be isolated and studied separately from anything else, the intrinsic dynamism of reality, its existence, and the source of its order, all of which are deeply interconnected. We can point to them, but we cannot describe them or model their origin. And then there are the horizons within which we find and know all that we know, and within which we make choices and decisions, as well as what we can only refer to as "the conditions of possibility" for what we experience as existing, and for our knowledge of what exists. These are transcendentals, and defy any adequate categorical treatment.[37] And yet they are essential to an understanding of experience and reality as we know it. These are intimately connected with the mystery that we often refer to as "God," and with God's universal creative and sustaining grounding of all that is.

And now to my suggestion! As we have already stressed, none of our conceptual modeling comes anywhere near an adequate description of God the Creator, nor therefore of God's relationship to the world. We do the best we can with the concepts we have available to provide an impressionistic portrait of God and of God's creative action which avoids the more obvious difficulties. For example, we say that somehow God the Creator is the ground or basis of the existence and dynamism of all that is, but at the same time God is unlike any other cause, is not an entity like other entities, but rather more of an all-pervasive activity or process in which all else participates, and, though present everywhere cannot be isolated as either a separate cause or force. We are using these concepts and images like an impressionistic artist uses colors, juxtaposing and interrelating them in unusual and provocative ways in order to disclose a profound reality and convey a transcendent experience which is beyond what any carefully precise or fully logical use of language is capable. Our portrait is not by any means adequate either, but it possesses an openness and expresses a radical mystery which would have been missed, or at least flattened, by any description with clear and precise conceptual lines. Allowing a riot of images and concepts to modify and qualify one another in this way—with the help of philosophical and theological analysis—is probably the only way of optimizing our characterization of God and of God's action. Reliance on a single conceptual framework will never yield the nuances and subtleties a careful opened-ended collage of images and concepts is capable of. There

---

[37] See, for instance, James J. Bacik, *Apologetics and the Eclipse of Mystery: Mystagogy according to Karl Rahner* (Notre Dame, Ind.: University of Notre Dame Press, 1980), 166.

seem to be strong indications that the basic gist of *creatio ex nihilo* along with primary/secondary causality considerations is fundamental to any characterization of God's creative action. But, at the same time, the strongly relational and dynamic insights of both process and Trinitarian frameworks are essential.

3 The Laws of Nature and Top-Down Action

In the introduction to this paper we tagged six general approaches to God's special action which have been proposed in the VO/CTNS divine action program:

1. appealing to God's direct action within the indeterminacies offered by quantum theory (according to the most prominent interpretation of it—the Copenhagen interpretation);
2. considering that the laws of nature as they actually function in reality significantly transcend what we have been able to model;
3. conceiving special divine action as originating from the "top-down" or "whole-part" influence or relationship of God towards all that is not God;
4. opening nature at higher levels of process and complexity to God's intentional active information input (Polkinghorne);
5. employing the somewhat similar but less well-defined process idea of God presenting new possibilities to nature and inviting it to respond—eliciting but not determining novelty (Barbour, Birch, and Haught);
6. showing simply that there is no determinism which rules out God's special acts.

I shall not directly evaluate these six proposals. Instead I shall focus on several core issues which are essential to any such evaluation, and which have not been fully resolved.

In this paper I shall not discuss category (1), which has been dealt with in detail by many other contributors to this volume. There has been and is considerable controversy and disagreement concerning how these proposals are to be understood. The detailed critical review by Wildman offers a specific attempt at clarifying how they should be understood.[38]

The most serious issue in all these approaches really is how a completely immaterial "cause" or "entity" can act to influence a material configuration or entity in a way which is not that of a "primary cause." Thus, for example, there is no real problem in allowing for free human activity—barring a rigid physical determinism—since that involves an embodied agent, and important restrictions such as the conservation of energy would still be in force. The challenge is to determine whether or not and how, and to model how, the divine presence can act directly in nature and in human situations without "violating" the laws of nature. Such violations are very problematic in view of the infrequency and lack of substantiation of such patterns of intervention, and

[38] See Wildman, "The Divine Action Project."

of issues relating to the problem of evil. Even of more concern, they threaten to cast God's role in the world into that of another secondary cause. One way around these constraints and problems is to relativize the determinisms seemingly demanded by the natural sciences by broadening our conceptions of "the laws of nature, such as in category (2) above. Strictly speaking, even leaving to one side the indeterminacies at the quantum level, it is clear, as Thomas Tracy emphasizes, that "the natural sciences do not require (on methodological grounds) or establish (on evidential grounds) an exceptionless causal determinism, though neither do they rule out such a metaphysical interpretation."[39]

What are "the laws of nature"? As I have discussed elsewhere,[40] "the laws of nature" *are* the regularities, processes, structures and relationships which we find in reality. There are really two meanings we can attach to this phrase. They can refer to the regularities, processes, structures and relationships in nature as we imperfectly and provisionally describe them and understand them—"our laws of nature." Or they can refer to all the regularities, processes, structures and relationships as they actually function within reality, whether or not we recognize them or understand them. Obviously, "our laws of nature" are much more limited and uncertain than the full range of "the laws" in themselves, and only describe what occurs. They do not prescribe or enforce behavior. This distinction[41] is very important, because, not only are there regularities, processes and relationships which the sciences have not yet discovered nor adequately modeled, there are also such processes and relationships which are in principle beyond the competencies of the natural sciences to investigate and to model. These would include those which reflect radically personal, metaphysical and theological (ultimate, transcendent and perceived revelatory), special and particular aspects of reality. These have been referred to by Albert Borgmann as "the deictic," those features of our experience which can be pointed out but not subsumed under general patterns and laws, but which endow our lives with meaning, value and orientation.[42]

Thus, what we may consider to be "violations" of "our laws of nature," might not be when understood in the light of the full range of the relationships and processes within reality as they actually function—particularly the way in which certain "higher" regularities and relationships having to do with the personal, the social, and the transcendent may in some cases override, modify, or subsume the laws of nature we model in physics, chemistry and biology. It seems very clear, from the point of view of the contemporary natural sciences

---

[39] Tracy, "Creation, Providence, Quantum Chance," 237.

[40] Stoeger, "Contemporary Physics and the Ontological Status of the Laws of Nature"; Polkinghorne, "The Laws of Nature and the Laws of Physics"; William R. Stoeger, S.J., "Science, the Laws of Nature, and Divine Action," in *Interdisciplinary Perspectives on Cosmology and Biological Evolution*, Regan and Worthing, eds., 117-27.

[41] Recently, some years after first insisting on this fundamental distinction, I found that Mario Bunge had already pointed it out in similar terms in his book *Causality and Modern Science*, 3rd rev. ed. (New York: Dover Publications, Inc., 1979), 249ff.

[42] Albert Borgmann, *Technology and the Character of Contemporary Life* (Chicago: University of Chicago Press, 1984).

and from our philosophical and theological reflections on them, that God acts primarily through the laws of nature as they actually function, including those which constitute evolutionary processes and including as well those which pertain to personal relationships and our transcendent relationship to God as our ground of existence and order. God's creation of human beings, or of any particular species of organism, for instance, has been effected through the processes of physics, chemistry, and biology—according to the regularities and processes which comprise evolution—and not apart from those. When it comes to God's special action, however, it is helpful to recognize that it is always framed in virtue of God's personal relationship to individuals or to groups of individuals open to God. Thus, it seems that the "higher laws of nature" which pertain to these relationships are always essential for properly understanding such divine action in history.[43]

Thus, as Keith Ward emphasizes in a way which, I believe, reflects the thinking of most of the VO/CTNS participants,

> The laws of the physical universe should not be set in opposition to God, as inviolable general rules which God, as an external power, would have to violate in order to act. The laws give the structure within which the purposes of God are brought to realization; and at every point there is the possibility of a creatively free initiative or response from creatures and from God. The existence of a general rational structure in nature in no way inhibits the possibility of discontinuous emergent events which disclose the underlying character of the divine presence and prefigure the consummation of value which is the goal of creation. It is better to construe miracles as such transformations of the physical to disclose its spiritual foundation and goal than to think of them as violations of inflexible and purposeless laws of nature.[44]

It is certainly true that these "creatively free initiatives or responses" from creatures and from God are constrained by the laws of nature in definite ways. In fact they are expressed through "the laws of nature." But, at the same time, Nature is open to such initiatives at all levels—through indeterminancies and openness to chance at lower levels, through natural selection, symbiotic relationships and amplification of feedback mechanisms at intermediate levels of complexity, and through goal directed behavior, personal decisions and relationships among conscious beings at higher levels. It is vital in this regard to recognize and to describe in greater detail those relevant phenomena which transcend the present competencies of the natural sciences—for example, mental reasoning and complex human decision-making processes and the role of inter-personal and social relationships in living and working. It is also

---

[43] As I have indicated in the introduction, I hold that God's special action, insofar as we can distinguish it from God's universal creative action, is initiated in virtue of God's relationships with persons or groups of persons (it may involve natural processes or events, certainly, but in order to express or enhance the personal relationships which are their foci). Otherwise, God's universal creative action—in the highly differentiated forms it assumes via *creatio continua*—is sufficient. In fact, it may be, as we shall see later, that all special divine action can be subsumed under divine creative action properly understood. Does this eliminate "special providence" from natural processes? It depends on how you define that term.

[44] Keith Ward, "God as a Principle of Cosmological Explanation," in *QC*, 260.

crucial to understand more fully how the different levels of the laws of nature are connected—principally but not exclusively through different types and levels of reducibility and emergence. Finally, the interplay of internal and external constitutive relationships must always be considered,[45] particularly in cases in which semantic or pragmatic—as distinct from purely syntactic—information is being generated and is playing a key role.

Turning to the operation of top-down—and whole part—causality, we really need to articulate more clearly how such concepts or images express the special divine acts of salvation history in ways which mesh with and qualify the apparent restrictions imposed by the findings of the natural sciences, and provide enough dynamism and novelty to constitute these special acts themselves. These top-down and whole-part metaphors focus our considerations on important "external" or transcendently immanent relationships which have been there all along and which are either presumed by or hidden from the natural sciences. They only manifest themselves locally when the lower-level entities or parts open themselves to the potentialities contained therein. It is likely, furthermore, that such a characterization is simply an alternative model of what is being expressed when we invoke "laws of nature" beyond those which are within the competency of the sciences to describe. Though some suggestions along these lines point towards a strong panentheism, such as Chris Clarke's notion of "èntrainment,"[46] they serve as clear examples of such all-embracing causality. He observes:

> My experience of the divine is of a guidance that is immanent, in being part of the concrete flow of events around me, and transcendent in the sense of not being contained within any given contextual framework, of being always greater than my current horizons. It thus makes sense for me to think of the divine as analogous to an ideal outermost system, beyond any imaginable context and of divine action as being top-down action of will through entrainment that coordinates and informs all the individual acts of will that it contains.

This is a model of a very strong type of connection between God's universal creative action and God's particular "actions" in history, in which all creation is essentially found "within" God, with its dynamisms ultimately derived from and coordinated by God. But how can this be understood in light of the natural sciences? Obviously, all—or at least a large part—of the coordination and informing of individual acts by an immanent God must be done by the laws of nature as they actually function, those we have modeled and more or less understood through the sciences, and those we have not, or cannot, understand through what they reveal. And just as obviously, this top-down causation must be qualified by what we have already pointed out about God's universal creative action in section 2, and by our pervasive experience of the formational and functional integrity[47] and the relative autonomy of nature.

---

[45] William R. Stoeger, S. J., "The Mind-Brain Problem, the Laws of Nature, and Constitutive Relationships," in *NP*, 129-46.

[46] Chris Clarke, "The Histories Interpretation of Quantum Theory and the Problem of Human/Divine Action," in *QM*, 176-8.

[47] Van Till, *The Fourth Day*.

4 The Specter of Determinism

A central issue with regard to special divine action is determinism: Do God's special acts in history require indeterminism in the laws of nature as they really function? Or can special divine action, instead, occur even though the laws of nature at a given level are deterministic? An affirmative answer to this latter question characterizes what has come to be known as compatibilism—God's special acts are compatible with determinism. Requiring windows of indeterminacy in nature for God's acts, in contrast, is the incompatibilist stance. One strong motivation for assuming the incompatibilist position is to achieve "traction" on the issue of divine action relative to testable conclusions of the natural sciences.[48] That is, the incompatibilist stance, according to its proponents, enables one to rule out types of special divine action on the basis of our scientific understanding of the laws of nature—God may only act in a special, direct way in the "causal gaps" opened by indeterminacies. Otherwise, God's special action would conflict with the causal processes within nature. Detailed discussion of these and related issues are found in Wesley Wildman's[49] and Philip Clayton's[50] contributions to this volume.

Here we shall confine our discussion to the meaning and extent of determinism as it applies to the laws of nature as they actually function, and to a consideration of whether God's special action requires indeterminism. How strongly do the natural sciences compel us to push special divine acts into the windows of indeterminacy in the natural order? As we shall see, it is very difficult to justify a strict determinism in the fabric of nature, even setting aside quantum indeterminacy. However, at the same time, whether or not freedom from strict determinism is required for special divine action depends very much on how such action is conceived or modeled. If it is modeled to mimic the secondary causes with which we are familiar, then it certainly does require freedom from strict determinism. But, if, on the other hand, it is conceived in some way as an influence which works in and through such secondary causes, as a mode of God's universal creative action, for instance, then it does not. This particular characterization of special divine action is the focus of section 5, immediately below.

But what is determinism? As we have implied above, it is basically a mathematical concept. A system is deterministic if its state or configuration at any time automatically leads to (*determines*) a single definite state or configuration at some later time. That is, once we set the state of the system at a certain time, the system will inevitably end up in a definite state at any specified later time—if there is no other outside influence impinging upon it. Or if we know the state of system with exact precision at one time, we can, if we know the equations governing the system, tell what the precise state of the system will be at any later time (this, of course, cannot be done in a chaotic system, which *is* deterministic, simply because it exhibits extreme sensitivity

[48] Philip Clayton, "Toward a Theory of Divine Action That Has Traction," in this volume, and references therein.

[49] Wildman, "The Divine Action Project," in this volume.

[50] Philip Clayton, "Toward a Theory of Divine Action."

to initial conditions and we cannot know the initial state of the system with the necessary precision.[51]

Now, what is clear from this characterization of determinism is that it is system-specific. That is, we can only apply the term to a definite well defined system, which must be isolated from other systems. If it is not isolated, then we must examine the smallest larger completely isolated system of which the original system is a part—to see if that system is deterministic. Furthermore, determinism is in most cases level-specific, that is, finding out that systems on one level of organization are deterministic (at the level of molecules, say) does not imply either that the less complex systems on more fundamental levels which constitute the system on the given level, or the more complex systems on higher levels, which are constituted by the systems on the given level, are deterministic. This will only be true if causal reducibility[52] holds in the latter case, and if the more fundamental lower-level system is deterministic in its own right (it may not be!) in the former case. Furthermore, as one can already see, it is very, very difficult—if not impossible—to discern in practice whether a given real system is deterministic or not. Idealized cases are relatively easy. But it is extremely difficult to test whether or not a given real system of interest is isolated, or to specify the larger system it is part of which is isolated—so that its deterministic character can be checked.

When we consider the observable universe itself, as it really is and not just how it has been theoretically modeled in a simple way, discerning whether or not it is deterministic is fraught with difficulty. Consider the case of most interest, the Earth and all its subsystems. First and most fundamentally, there is the quantum substrate, which, according to the most accepted interpretation, induces an irreducibly indeterministic element into the macroscopic world. But, leaving this to one side, we are also aware that there are innumerable levels of complexity which are intertwined. If we look at any single-level subsystem it is virtually impossible to isolate a system which we could then study to find out whether or not it is deterministic. An idealized *model* of it might be—such as a deterministic chaotic model—but that does not mean that *the actual system* is. Furthermore, how would we make such a conclusion, that the determinism of the model is also a characteristic of the real system? Secondly, determinism on any level does not, as we have already seen, trickle up or down levels—unless it is already present on the lower level, and unless causal reductionism holds. The emergent properties of higher levels (see section 3 above) may, and probably often do, introduce new irreducible causal elements into the equation, most notably the self-conscious intentionality of human beings.

---

[51] See, for instance, Robert John Russell, "Introduction," in *CC*, 1-31, and references therein.

[52] Nancey Murphy, "Supervenience and the Downward Efficacy of the Mental: A Nonreductive Physicalist Account of Human Action," in *NP*, 146-164; see also William R. Stoeger, "Reductionism and Emergence: Implications for the Interaction of Theology with the Natural Sciences," in *Evolution and Emergence: Systems, Organisms, Persons*, Nancey Murphy and William R. Stoeger, eds. (Oxford: Oxford University Press, 2007), 229-247.

Our conclusion from this basic discussion of determinism is simply that it cannot be applied to multi-level organization of nature as a whole, even though it may in limited cases be applied to one or other sub-system on a given level. Since it certainly does not characterize the organization nor the dynamics on most levels, including the quantum level, and since causal reducibility can be compelling shown not to hold,[53] strict determinism is not a characteristic of nature as a whole. This is further supported by careful analysis of the different types of emergent systems and processes which develop as the universe evolves. These involve causal influences which are radically new, often depending on feedback from the macroscopic to the microscopic domain and, in more advanced cases, generation and transmission of semantic/pragmatic information (as in DNA/protein systems), and irreducible to causal influences at lower levels. At even higher levels of organization we have the eruption of intentional, goal-directed action and behavior, some of it freely initiated by conscious agents, who can anticipate the consequences of their actions. Although such choices are influenced and conditioned by environment and history, they are not determined by them.

Of course, as we have already pointed out earlier, the natural sciences themselves as such do not in any way require or establish "an exceptionless causal determinism," either methodologically or evidentially.[54] However, at the same time, we want to avoid modeling or considering God's actions as simply extraordinary exceptions to the secondary causal fabric that characterizes nature—as another secondary cause hidden in the interstices of nature which unpredictably manifests itself in events which fall outside the usual patterns of natural regularities and processes. One way of doing this, as we have already seen, is to recognize that there are influential relationships which transcend what the natural sciences are able to investigate and understand on their own terms—which we have identified as the personal, the social, the transcendent, the spiritual, and the particular as particular. These also belong to "the laws of nature" as they actually function, even though they cannot be adequately investigated or modeled by the natural sciences. And these influential relationships almost always transcend—rise above—the constraints which seem to enforce determinism on a given level, though they are often conditioned by them.

Because of these considerations, there is a deep ambiguity in the issue of compatibilism vs. incompatibilism. If one's reference is to a thoroughgoing strict determinism which pervades all levels of organization and relationship, including those which transcend the natural sciences, then, of course, it seems clear that direct special divine action would be impossible, unless compatibilism holds. (We should recognize at this point that very few serious scholars any longer espouse such an all-embracing determinism.) But, in any case, it then seems very difficult, if not impossible, to understand such direct

---

[53] See, for instance, Steven Rose, *Lifelines* (Oxford: Oxford University Press, 1998), esp. chap. 4; Martinez J. Hewlett, "True to Life? Biological Models of Origin and Evolution," in *Evolution and Emergence: Systems, Organisms, Persons*, Nancey Murphy and William R. Stoeger, eds. (Oxford: Oxford University Press, 2007), 158-172; Murphy, "Supervenience and the Downward Efficacy of the Mental."

[54] Tracy, "Creation, Providence, and Quantum Chance," 237.

special divine action in a coherent fashion. What specific feature of reality and of God's relationship to it would render compatibilism a coherent and acceptable position? Thus, it seems correct to venture that special divine action, however it is conceived, does require "causal" openness to God at some level—an active relationship to the divine—which cannot be reduced simply to a naive grounding of the secondary causes of the material world.[55] This does not mean, however, that such divine influence competes with or countermands natural processes or relationships, though it may subsume them, marshal them in a definite way, or subsume them, in virtue of other higher relationships or processes—as the biological often does to the physical and the chemical, and personal to the biological and physical. In fact, as we have already emphasized, God's relationship with nature and with us is articulated precisely through the laws of nature—those we understand and have modeled, and those we have not. The principal difficulty arises when we terminate or absolutize "the laws of nature" according to what we presently understand and have modeled, without taking into consideration deeper or ultimate levels of relationship which escape adequate generalization in terms of physical, chemical or biological regularities.

If, on the other hand, our reference is to a determinism on one or other level, or even on several levels, apropos of our observations about physical determinism, then, of course, we may still be compatibilist—since that limited determinism is still compatible with divine action which engages conscious agents at levels which transcend those at which determinism is effective. This distinction, it seems to me, enables us to avoid some of the confusions which sometimes afflict these discussions. Another way of putting this is to say that the determinacies of, say, macroscopic physics are relative and not absolute—that is, relative to the levels at which the laws of physics and chemistry dominate, but not ruling out the operation of causal relationships, such as the mental and the personal, which transcend those described by physics and chemistry.[56]

Ultimately, as I have recently pointed out,[57] determinism is really a "red herring" in this particular discussion. What really seems to distinguish the two schools of thought is whether or not direct special divine action involves God's

---

[55] Here I am *not* attempting to argue definitively that special divine action requires indeterminism at whatever level it operates. What I am discussing is the meaning and the limited reach of determinism, which apparently strongly restricts the application of the usual compatibilist/incompatibilist distinction. This seems to indicate, but does not prove, that there is indeterminism at many levels within nature. Furthermore, this does not demonstrate one way or the other whether such indeterminism is necessary for God's special action—just that the issue itself may be less relevant than we first believed.

[56] Another way of conceiving this occurring is by distinguishing between models and the phenomena modeled: While the models we use may themselves be deterministic, there may be other important aspects of the realities being modeled which escape the determinism of the model. Obviously, it would be very important to lay out definite evidence of such transcendence of lower-level determinisms.

[57] William R. Stoeger, S. J., "The Divine Action Project: Reflections on the Compatibilism/Incompatibilism Divide," *Theology and Science*, Vol. 2, No. 2 (2004): 192-6.

operation as a secondary cause—as we have carefully distinguished this from, and related it to, God's primary causal role. Incompatiblists, though usually not intending to, effectively give God such a secondary causal role in divine special action; compatiblists carefully try to avoid any move in that direction. Of course, distinguishing these roles with precision is subject to further terminological and conceptual challenges—how do we precisely distinguish the range of God's universal creative action from what would be a secondary causal function? It is not at all clear how this distinction should be made.

5 Special Divine Action as a Mode of Divine Creative Action

Much of what we have been saying about special divine action leads me to suspect that it should not be separated from God's universal creative action, but rather considered as a particular manifestation or mode of that divine creative action, more broadly conceived. Thomas Tracy[58] points to another complementary way (other than the insights provided by "higher laws of nature" and top-down or whole-part causal metaphors). Let us briefly reflect on his suggestion and see where it leads.

The basic approach is to recognize that God's special actions within history are in virtue of God's universal role as the Creator of history. As Tracy explicates this, the key component is that "God's creative action includes the continuous 'giving of being' to the created world in its entirety," enabling each being and system of beings within creation to function and develop according to its own capacities and dynamisms, including those of conscious freely deciding and acting persons and communities of persons. It is by what material beings and systems of beings at all levels accomplish through the operation of their God-given potentialities that God continues to act creatively in the world—God acting in and through secondary causes, as we have already seen. Thus, ultimately, all that happens within the created world can be considered an act of God—in the sense of an act that God ultimately supports and allows.

But how does this approach provide a way of better understanding "God's special salvific acts" within history as manifestations of God's fundamental continuing act as Creator of all that is? Obviously, not all—and not even the vast majority of developments or actions worked by secondary causes within nature and history—would by any stretch of the imagination be considered "God's special salvific acts," even though they are ultimately supported and allowed ontologically by God's universal creative action. Most are just the result of the unconscious operation of the regularities, processes, and relationships with which God has endowed nature. And many others, initiated by human beings, and possibly by other self-reflective, intelligent entities, are undoubtedly directly opposed to the basic purposes and desires God has for creation, and to the fulfillment creation, and the various integrated systems within creation, yearn for. However, despite self-serving or evil intentions, and even the destructive character, of certain free acts, God as creator has established the potentialities and avenues within creation which enable them, and continues to sustain all that is freely chosen and executed in them, as well as those freely chosen acts, personal and communal patterns of behavior, and

[58] Tracy, "Creation, Providence, and Quantum Chance," 238ff.

cultural/political/economic structures which are in tune with God's ultimate purposes and are essentially life-giving. Furthermore, God also knows all there is to know about how the causal influences of all these events and situations fit together—even though God does not directly determine them—and how they succeed or fail to reach their divinely intended goals.

In light of this, we might suppose either (1) that God, as Creator, orchestrates all "the laws of nature" (not just those we understand and have modeled) and the initial and boundary conditions that are involved in their concrete operations, in order to insure that creation reaches its overall goal, without in any way limiting the autonomies and freedoms that have developed within creation on many different levels, or (2) that certain events or sequences of events initiated by secondary causal agents (either those which are freely and consciously choosing, or those which are not) are turning points within creation, or within history, and thus are specially revelatory of God's immanent creative presence. As such, these are in deeper harmony with God's intended purposes and with the essential structures and relationships already established with creation itself. In fact, these two alternatives are not exclusive of one another. Both may be operating in concert.

In the second case, these events or sequences of events would indeed be "God's special salvific acts," even though they were not direct acts of God (i.e., not without the intermediary of a secondary cause). Nor would they lack a sufficient cause within the manifold of dynamisms and potentialities of creation or of history, presupposing the universal primary creative action of God. They would, nonetheless, be special, salvific and revelatory, precisely because they are clear expressions of what God intends, and/or fulfill God's purposes and intentions in a particularly unambiguous or exemplary fashion. In most cases, this will be because the human, or other conscious, agents will have intuited or discerned from his or her contemplation and experience of nature or history what those intentions, purposes and goals are, and will have appropriated those and opened him or herself to all the dynamisms within creation oriented in that direction. This would normally not be the result of a single act of intuition or discernment on the part of the individual or the community, but the result of an ongoing quest to understand and to live out who we are and what nature is.

This overall approach, it seems to me, has definite advantages. It connects directly with the richly differentiated, transcendently immanent presence and action of the Creator God within creation[59] and with God's radically kenotic, deeply effective but hidden availability within nature.[60] Furthermore, it emphasizes that what is fundamental is not so much God's action, or actions, but rather God's ongoing relationship with creation. Again, the divine creative relationship is highly differentiated with respect to each entity and system within the universe—and God's action flows from the character of that

---

[59] William R. Stoeger, S. J., "Describing God's Action in the World in Light of Scientific Knowledge of Reality," in *CC*, 256-8.

[60] George F. R. Ellis, "The Theology of the Anthropic Principle," in *QC*, 377-99; idem, "Quantum Theory and the Macroscopic World," in *QM*, 285ff., and references therein.

relationship. In fact, it can only be understood in terms of God's relationship with creatures and communities of creatures.

This way of conceiving God's special salvific acts in terms of God's overall creative action does not eliminate the need, so to speak, for God to establish the primordial laws of nature in such a way that the divine purposes will be fulfilled in some way, despite God's surrendering detailed control of creation to the autonomies and freedoms of created causal agents—to the regularities, processes, relationships and emerging systems we refer to as "the laws of nature." As Ellis has so aptly described it, the overarching divine program is based upon and calls forth "self-sacrificing love."[61] This requires, apparently, the establishment of a broad range of evolving relationships at many different organizational levels, some of which are well modeled by the natural sciences and others which are not at all describable in those terms. Among these latter are those constituting the interpersonal, the social, and the cultural, and most importantly those effecting and expressing the profound ontological connections between God and the various entities and communities which make up creation—including those involving what Christians have come to recognize as the Incarnation, and all that flows from it. Thus, in this view the Incarnation, and the Death and Resurrection of Jesus, should not be seen primarily as salvific acts initiated by the divine to rescue creation gone wrong, but as primordially intended by God as the eventual destiny and completion of creation in God. All that creation involves is directed in one way or another toward this fulfillment, and has been established to facilitate this long-range, creative-salvific plan—subject to the autonomous evolutionary unfolding of its components.

6 Conclusion

I have focused in this discussion on several aspects central to clarifying our understanding of divine action in light of what the natural sciences are gradually revealing about reality which I judge demand more attention and development than they received in the VO/CTNS Project itself. I have not fulfilled that demand, but merely attempted to point our efforts in directions I consider worthwhile.

The principal need I began to address was for a more detailed and thorough understanding—an understanding involving an interaction with the natural sciences—of *creatio ex nihilo*, and *creatio continua*, along with the closely related issues of primary and secondary causality, the inadequacy of language about God, analogy, God's knowledge of creation, as well as the shortcomings of these models and the possible ways of improving them. Whatever the limitations and frustrations of these models and the points of view they represent, we seem to be drawn back to them, even though the language we use may be different.

Other important issues I briefly revisited were: the meaning and scope of the laws of nature and of top-down, and whole-part, causality as applicable to special divine action; the meaning of determinism and its relevance for discussions of special divine action; and the attractive avenue opened up by

---

[61] Ellis, "The Theology of the Anthropic Principle."

Tom Tracy towards developing a more fundamental understanding of special divine actions as richly differentiated modes or expressions of God's universal creative action.

The mystery of reality and of the Creator are infinitely rich and inexhaustible. We shall never adequately understand it. But I hope that these reflections, and those of my colleagues in this endeavor, serve to deepen our appreciation and wonder at what unfolds before us.

SPECIAL DIVINE ACTION AND THE LAWS OF NATURE

Thomas F. Tracy

The affirmation that God acts in the world is an obviously prominent element in the Biblical religious traditions. It is at the center of the dramatic scriptural narratives that have shaped Jewish and Christian theological reflection for millennia, and it is embedded in the liturgy, preaching, and prayer of the communities that live out these traditions today. This is not to deny that there have been deep disagreements within and between the theistic traditions about just what we should say that God has done and is doing. The Biblical texts themselves speak with a diversity of voices, and so have generated a long history of theological debate; it is fair to say that religious traditions are best described as ongoing (and often acrimonious) discussions rather than as the transmission of an unchanging doctrinal content. Nonetheless, within the Christian faith there has been broad agreement that God acts in the world to advance the purposes for which created things were called into being. Many of the central doctrines of classical Christianity involve claims about divine action (e.g., about God's redemptive presence and action in the life of Christ), and these beliefs have sponsored theological interpretations of history according to which God is at work in events both great and small, from the rise and fall of empires to the circuitous course and intimate struggles of an individual life (as Augustine, for example, suggests in *The City of God* and the *Confessions* respectively). Theologians who abandon the idea of particular divine action in the world commit themselves to a profound revision of this belief structure.

The theological centrality of this idea is matched, however, by its conceptual difficulty. Some of the most prominent modern puzzles about the concept of particular divine action arise in connection with natural science. How might talk about divine action in the world be brought into dialogue with the powerful explanatory theories emerging within the various sciences? A decade-long research project, conducted jointly by the Center for Theology and the Natural Science (CTNS) and by the Vatican Observatory (VO), put this question at the center of its inquiry. This remarkable undertaking combined a consistent theological focus on divine action with careful attention to a series of topics from scientific sub-fields. The result was an accumulating conversation that made substantial progress in clarifying key issues, mapping alternative approaches, and advancing a number of constructive proposals. A project of this scope, involving so many voices from diverse fields, weaves an extraordinarily rich skein of discussion, and I will comment here on just one important thread running through it. At the center of the discussion there has been a debate about an emerging theological program, a program that offers both a diagnosis of the principal problem about divine action and a strategy for addressing it. I will outline briefly what I take this theological program to be, note some objections and alternatives to it, comment on some of the proposals that have emerged as ways of advancing it, and offer some tentative remarks on what remains to be done.

1 A Theological Program

In order to describe this theological program with some precision, we need be clear at the outset about the idea of "special," or "particular," divine action, since it plays so crucial a role in the discussion. There are at least three senses in which an event may be singled out as a special action of God.

First, an event may be distinguished from others by virtue of its *epistemic role*, that is, its disclosive, or revelatory, importance. Particular events may become the occasion through which individuals and communities recognize with special clarity the presence and purposes of God. It need not be the case that God acts in these events in a way that is different from God's universal action in every event. What marks them out as special is rather that they reveal and exemplify *for us* the direction of God's work in history. Even if, for example, the escape of the Hebrew people across the shallows of the Red Sea involved only the ordinary processes of nature, this event may reveal for this community God's liberating purposes in history. I will refer to events that are distinguished solely by their epistemic role as "*subjectively special divine actions*."

Second, an event may be distinguished from others by virtue of its *role in realizing and advancing God's purposes in the world.* Once again, the event need not be brought about by God in any distinctive way; we can suppose that God acts in this special event in just the way God acts everywhere (as the creator and sustainer of a system of natural causes and free human agents). Yet this event may constitute a significant turning point in the course of events that moves God's purposes toward fulfillment. This will be a fact about the event, and not just about our response to it. The escape of Hebrew people, on this view, not only discloses God's intentions to us, it also advances them in an especially significant way. I will refer to such events as "*materially, or functionally, special divine actions*."[1]

Third, an event may be distinguished from others by virtue of the way in which God brings it about, that is, by its *causal history*. God may act within the ongoing course of the world's history to generate a development that would not have occurred had God not so acted. The causal history of such an event will include, along with all its finite causal antecedents, this special divine act. It is important to note that God's agency in such an event should not be exclusively identified with the special act that modifies the world's history. In the context of a classical conception of God as creator *ex nihilo*, we

[1] Terminological difficulties crop up here (and they run the risk of literally giving useful distinctions 'a bad name'). First, I have elsewhere referred to this second sense of special divine action as "causally special divine action," since these events can be distinguished by their *causal role* in the developing course of history that flows *from* them. I have switched here to the term "materially, or functionally, special" in order to avoid confusion with the third sense of special divine action, which distinguishes events by the special *causal history* that leads *to* them. Second, the idea of materially special action does not appear within the typology of positions on divine action developed by Robert Russell. (See the "Introduction," in *QM*, ii-vi.) That typology contrasts uniform divine action and objectively special divine action, and materially special divine action would simply be an additional subtype of uniform action.

must say that (1) God acts directly as the absolute ontological ground of every entity and event, and (2) God acts indirectly by means of the operation of created causes and agents. This third sense in which we may speak of special divine action adds the further claim that (3) God acts directly within or upon the order of created causes to bring about an event that finite causes alone are not sufficient to produce. An event of this sort may or may not evoke in us a recognition of God's working (i.e. it may not be subjectively special), and it may or may not represent an especially significant achievement or turning point in the course of events (i.e., it may not be materially special), but even if it remains hidden in the minutia of history, it constitutes a particular divine action in the world. I will call these events "*objectively special divine actions*."

It is special divine action in this third sense that has presented so many problems for modern theology. The claim that God acts in nature and history at particular times and places has seemed unsustainable to a substantial number of theologians. There are many reasons for this, including concerns internal to theology about the consistency of God's creative purposes and about God's justice in the face of the problem of evil. It is also clear, however, that many misgivings about the idea of particular divine action have at their center a perceived collision between this way of thinking about God and scientific descriptions of the world as an intelligible order of natural causes. One ongoing research program within the VO/CTNS conference series offers a diagnosis of how this putative conflict arises and pursues a corresponding strategy for how it can be avoided.[2]

Let me comment first on the diagnosis. The conflict, many of us have argued, is *not* an unavoidable result of developments in the contemporary natural sciences, but rather is a consequence of uncritically adopting a familiar but faulty philosophy of science. This widespread view of the sciences is reductionist and determinist; that is, it endorses some version of the claim that explanations offered at higher levels in the hierarchy of structure described by the various sciences can be reexpressed without remainder in terms of the lower levels, and it holds that explanations at the most basic level (and perhaps also at higher levels) are or will be causally complete.[3] In a deterministic natural system every event will have sufficient causal conditions in the events that precede it. The entire history of the system backward and forward in time could in principle be deduced from knowledge of its laws and a complete description of its state at any moment.

---

[2] As will become apparent in the discussion that follows, I have in mind here views developed by, for example, Philip Clayton, George Ellis, Nancey Murphy, John Polkinghorne, Robert Russell, and, in a rather different way, Arthur Peacocke. My own essays also fit the pattern I will describe.

[3] Many essays in the conference series have discussed reductionism in its various forms. For helpful ways of sorting out the many different senses of "reduction" see, among others, Nancey Murphy, "Supervenience and the Nonreducibility of Ethics to Biology," in *EMB*, 463-90; idem, "Supervenience and the Downward Efficacy of the Mental: A Nonreductive Physicalist Account of Human Action," in *NP*, 147-64; William Stoeger, S. J., "The Mind-Brain Problem, the Laws of Nature, and Constitutive Relationships," in *NP*, 129-46; Theo Meyering, "Mind Matters: Physicalism and the Autonomy of the Person," in *NP*, 165-77.

Given this understanding of the sciences, the consequences for a theology of divine action are clear. God can shape the direction of events in the world in just two ways: by determining the laws of nature and boundary conditions in the act of creation or by intervening in the world's history to interrupt the order of natural causes and redirect the course of events. God must either do the whole job at the outset (as it were), or step in later on to perform miracles. Further, the scientific modes of understanding that confront us with these theological alternatives also render problematic the idea of miraculous divine intervention. Among other difficulties, there are substantial evidential challenges facing any specific claim that a miracle has occurred (as Hume famously argued). So it is not surprising that first deists and then liberal theologians, as they sought to articulate the Christian faith in this intellectual context, retreated from talk about God acting to affect the course of events in the world, and offered instead various interpretations of this language according to which it actually expresses beliefs about God's overall purposes in creation. This involves interpreting the idea of particular providence entirely in terms of general providence; in creating the world, God establishes (and sustains) its possibilities, dynamics, and limits in a way that provides for the eventual fulfillment of God's purposes. On this account God does not so much *act in* history as *enact* history as a whole, and an event in the world can be identified as a special act of God only in the first and/or second of the three senses noted above.[4] This has consequences, of course, for virtually every major doctrinal topic in theology.

This diagnosis of the central challenges facing talk about divine action suggests a strategy of reply. If there are good reasons to reject deterministic interpretations of the natural sciences, then a third alternative may open up for understanding divine action. Perhaps God has created a world that includes events with necessary but not sufficient causal conditions in nature. In this case, the created order will be an intrinsically flexible interweaving of law and chance, open to new possibilities not rigidly prescribed by its past. God as creator sets the terms of this open cosmic history, and acts as the sustaining ground of every event within it. God might also continuously shape the unfolding course of events in such a world by determining some or all of what is otherwise left underdetermined within the order of created causes. Divine action of this sort would have particular consequences; God's activity in the world will turn events in new directions, setting in motion novel lines of development. The events to which God contributes in this way would be objectively special divine acts, but they would *not* involve a miraculous intervention. God's action here does not shoulder aside natural processes, interrupting an otherwise closed causal series, because (*ex hypothesi*) the events in question do not have sufficient natural causes. Instead, God actualizes of one of the possibilities rooted in, but not already selected by, the past history of the universe.

This intriguing idea can be successfully elaborated and defended, however, only if it can find backing in a plausible interpretation of the relevant natural sciences. It is easy enough to see what in general will be needed; the

---

[4] I have discussed this strategy in more detail in "Particular Providence and the God of the Gaps," in *CC*, 289-324.

difficult challenge is to determine whether developments in any of the sciences hold promise of meeting these needs. A scientific theory well-suited to being enlisted in this theological program would meet three conditions. First, it would permit, even if it does not require, an indeterministic interpretation; that is, it would describe a system in which at least some events can be understood to have necessary but not sufficient causal conditions in prior states of the system. These events will not merely be unpredictable for us (epistemic chance), but also causally underdetermined (ontological chance). Second, it would describe a natural system in which these underdetermined events can make the right sort of difference in subsequent causal processes. If the only effect of such events is to accumulate in probabilistic patterns that constitute higher-level regularities, then they will not provide a means by which particular changes could be made in the causal history of the world. There will need to be processes in nature that can amplify the effects of ontological chance. Third, it would locate underdetermined events within an intelligible natural structure. Although their physical explanation will be incomplete, these events will be an integral part of the order of nature, rather than merely a random interruption of it. If all three of these conditions are met by a theory within one of the sciences, then we would have a promising candidate for theological interpretation of the sort we have sketched. One prominent (though by no means the sole) research agenda in the VO/CTNS project on divine action has been to explore whether any of the scientific fields chosen for study include theories that meet these theologically informed desiderata.

2 Objections: Should We Turn to the Sciences to Find "Room" for Divine Action?

This program for relating theology and science faces challenges of two general sorts: first, objections can be raised to the way the program itself is conceived, and second, difficulties may crop up for any particular use of the sciences in carrying out this program. I want to start with challenges to the program itself, since these issues hold the potential to strike at its theological foundations. The theological side of the religion-science dialogue must, of course, be given the same careful attention as the scientific side, and it is vitally important to acknowledge the context of theological debate within which this project is undertaken. I will then turn in section 3 to specific proposals that enlist the sciences in advancing this theological agenda.

2.1 Contrasting Divine and Created Action in a Zero-Sum Game

It might be objected that the effort to identify openings in the causal structures of nature for non-interventionist objective divine action reflects an underlying theological error. The alleged mistake is to assume what we might call a "contrastive," or "zero-sum," picture of divine and created agency.[5] The basic

---

[5] Kathyrn Tanner offers a critique of "contrastive" understandings of God's transcendence in *God and Creation in Christian Theology* (Oxford: Basil Blackwell, 1988), chap. 2. William Placher makes use of this idea in *The Domestication of Transcendence: How Modern Thinking about God Went Wrong* (Louisville, Ky.:

idea here is that God's activity and that of created things stand as contrasting alternatives. If God acts, the creature must be passive, and if the creature is to act, then God must forbear. It has sometimes been said, for example, that in creating a world, God must at least partially withdraw from it, restricting divine activity so that creatures may exercise the powers that God has given them. The problem then arises about how to get God back into the world in a way that does not displace natural causes and agents. William Placher, in a book provocatively titled *The Domestication of Transcendence*, argues that this way of understanding God and the world is one of the characteristic marks of modern theology. "At the beginning of the modern era...theologians and philosophers began to worry about just where to put God in the universe. Debates about miracles and about grace and free will dominated the theology of the seventeenth and eighteenth centuries, and both those debates involved asking which things God, as opposed to someone or something else, did."[6] Placher contends that this puzzlement about the boundaries between divine and created agencies resulted from the loss of an earlier tradition that appreciated more deeply the theological principle that God's action transcends all the forms of agency with which we are familiar. This enabled the earlier tradition (which Placher finds in Aquinas, Luther, and Calvin) to recognize that God is active in all events. "Since God was not one agent among others, but operated at a different level of agency, it made no sense to ask which things God had done and which things had been done by someone or something else."[7]

It could be argued that it is precisely the modern truncated understanding of God's agency that leads to the search for openness in structures of nature. If we suppose that God acts only when natural causes and agents do not, then the rise of determinism in early modern science must have the effect of pushing God out of the world. The only solution, given a zero-sum understanding of divine action, is to look to the sciences for ways to pry the world back open again. "In our own day, various intellectual schemes which seek to make room for the agency of creatures or which find theological significance for divine action in terms of the 'ontological openness' of quantum mechanics and chaos theory fail to recognize the profound metaphysical point that divine causality transcends any other category of causality."[8]

Should we conclude that the theological program I have described is premised upon a failure to appreciate the transcendent character of God's agency? I think not. This objection assumes that if we properly acknowledge the uniqueness of the divine agency we will also see that it cannot in principle be juxtaposed with created agency in an "either/or" alternative; to suppose that divine action is incompatible with *any* form of creaturely action is to

---

Westminster John Knox Press, 1996). Ted Peters has often criticized the idea that divine self-limitation is a precondition for creation, most recently in *Anticipating Omega: Science, Faith, and Our Ultimate Future* (Gottingen, Germany: Vandenhoeck & Ruprecht, 2006), pp. 21-22.

[6] Placher, *The Domestication of Transcendence*, 128.

[7] Ibid.

[8] William E. Carroll, "Aquinas on Creation and the Metaphysical Foundations of Science," *Sapientia* 54 (1999): 69-91. Also see idem, "Thomas Aquinas and Big Bang Cosmology," *Sapientia* 53 (1998): 73-95.

misunderstand the nature of divine action. In one respect this is right, and in another it is wrong. The objection is right in pointing out that God and creatures not only may but *must* (on a classical understanding of God as creator *ex nihilo*) act together in every event, though they do so in fundamentally different ways. The Thomistic distinction between primary and secondary causes provides one historically important conceptual device for expressing this point. God as creator gives being to creatures, and does so at every moment throughout the creature's history. This divine creative action does not cause a change in the creature, but rather brings it about that creature *is* at all; unless God does this, there is no creature to change. Created things, on the other hand, cause changes in other created things. So both God and creatures act in every change that takes place in nature. "The same effect is not attributed to a natural cause and to divine power in such a way that it is partly done by God, and partly by the natural agent: rather, it is wholly done by both, according to a different way, just as the same effect is wholly attributed to the instrument and also wholly to the principal agent."[9] God does not have to withdraw in order for creatures to act; on the contrary, God must act (as creator/sustainer *ex nihilo*) in order for creatures to act (as efficient causes in nature). It is precisely through God's presence to creatures as their absolute ontological ground that they are able to exist and to operate as they do. In thinking about this continuous creative relation, it clearly is a mistake to adopt a zero-sum account.

This does not fully resolve questions about the relation of divine and created agency, however. We have supposed that God confers upon creatures active powers of their own. For just this reason there will be some forms of divine action that can be contrasted with creaturely action. For example, Aquinas insists that God could produce directly any effect that ordinarily is produced by a created cause.[10] If God were to act in this way, it would be correct to say that God *rather than* the creature produced the effect. The non-contrastive creative relationship, therefore, does not rule out all trade-offs between divine and created agency; whether God's activity ever stands in partial contrast to that of creatures will depend on the capacities for action that God grants to creatures and on how God acts in relation to these created agencies. Two cases are of particular importance to us here. In the first, God acts directly to bring about an effect that created causes alone are *insufficient* to produce (in contrast to the Aquinas example); this is objectively special divine action. In the second, God empowers the finite free agent to undertake self-determining actions. Both cases are complex, and require careful consideration.

2.1.1 Objectively Special Divine Action and Created Causes

If an event is singled out as a special act of God in the subjective sense or the material/functional sense, then the divine and created causes of the event need not stand as alternatives to each other. This event will be brought about

[9] Thomas Aquinas, *Summa Contra Gentiles*, III, 70, 8.

[10] "It is erroneous to say that God cannot Himself produce all the determinate effects which are produced by any created causes." ST I.105.2

through God's universal action as the primary creative cause who acts by means of the order of natural causes. The third, and strongest, sense of particular divine action takes a further step, however. Here God acts not only by sustaining the existence of secondary causes, but also by directly bringing about particular changes in the world. This is to say that God acts among, and not only by means of, the causes and agents that God has created, and this entails that there will be a trade-off between divine and created agency; creatures, exercising the causal powers that God upholds in them, would not be sufficient to produce this effect without God's special action.

At this point, a theological debate is joined. In order to avoid treating divine and created agents as alternatives in this special case, one might contend that God does not act among secondary causes. It would be difficult to make the case that God *cannot* act in this way; if God is understood as the creator *ex nihilo*, then surely God may choose to act within the creation as well as at its ontological foundation (as we just saw that Aquinas affirms). It is more promising to contend that while God can act in this way, God chooses not to do so. One's position on this question involves judgments on a network of theological issues. It might be argued, for example, that if we think of God as acting within the stream of historical events to bring about particular effects, then we are left with a diminished and excessively anthropomorphic conception of divine agency. It is often said that this treats God as an additional secondary causal power at work in the world, the "biggest" and most impressive no doubt, but a player on the same level as the forces of nature. Note, however, that the most familiar way of pressing this objection can readily be met by the theological program we are considering. If God is understood as creator *ex nihilo* who acts in and with every finite cause, then God will never be *just* one cause in the world among others, even if God chooses *also* to act in this way among secondary causes. Nonetheless, the idea that God directly brings about particular effects in the world may seem to call in question God's wisdom in creation; why would God make a world that requires this ongoing divine contribution or, as it is sometime put dismissively, this constant "tweaking?" And if God can and does act in this way, then why is God not more effective in preventing or blunting the various evils that we see around us?

These and related concerns provide theological motivation for attempting to frame an account of divine action strictly in terms of God's universal activity as the ground of all created causes and agents. I have argued elsewhere that this approach can go a considerable distance toward giving an account of particular providence; the debate, of course, is over whether it can go far enough.[11] On such an account, God will act directly in every operation of nature as its metaphysical ground, and God will act indirectly in every change that is brought about through the causal powers of created things. Particular events can be identified as subjectively special acts of God in the limited sense that they reveal God's purposes to us, and perhaps also as materially/functionally special acts of God, by virtue of their particular role in realizing and advancing God's purposes in the world's history. These special divine

---

[11] See my "Creation, Providence, and Quantum Chance," in *QM*, sect. 2.

acts, however, will bear the same relation to God's agency as do all other events.

Can a theology of this sort, which interprets God's redemptive activity in history entirely in terms of the creative act that establishes and sustains the structures of nature, do full justice to the dynamic relationship depicted in the texts and invoked in the practices of historic Christian communities? The God portrayed in these traditions is one whose ongoing activity makes a difference in the course of the world's history, a God who calls, promises, judges, renews, and acts in innumerable other particular ways so that the divine purposes in creation will be advanced and fulfilled. Biblical, theological, and liturgical language of this sort suggests that God acts *in response* to human actions. This is one of the central theological concerns driving the effort to articulate a viable conception of particular providence in the third sense, and it is one to which we ought to give explicit attention. If God acts exclusively as the absolute ontological ground of all events, and never acts directly to affect the course of history, can we say that God responds to the dramas of human history, to the struggles the Church, to the cries of the oppressed, to the restless human heart? It appears that the creator's "response" to the actions of creatures will need to be built into the plan of creation from the start. It is possible to frame an account of responsive divine action in these terms, though the success of this enterprise turns upon some very difficult issues in the doctrine of God (e.g., about God's knowledge of future contingent events and about God's relation to time). The shape of such a position is also determined in crucial ways by how we handle the second general case in which divine and created agencies may partially contrast with one another: free action.

2.1.2 Divine Agency and Free Human Action

This topic is both central to the enterprise of thinking about divine and created agencies, and profoundly difficult. On one view of the nature of free action, there will be no problem about the relation of divine and human agency. If we give a compatibilist account, which holds that free intentional action is compatible with (the right sort of) causal determination, then divine action and human free action will perfectly coincide. In this case, God could build into the plan of creation a response to free human acts because every event, including those free acts, is determined by God's will. As some Calvinists said, God does not just foreknow human actions, God foreordains from eternity both those acts and God's response to them. On the other hand, we might adopt an incompatibilist and "libertarian" view, which holds that free intentional action requires causal underdetermination (as a necessary but not sufficient condition).[12] According to this view, God establishes and sustains the capacity

---

12 The distinction between compatibilism and incompatibilism has been borrowed from the discussion of human freedom and used to structure alternatives in considering God's relation to the world in action. There are some illuminating parallels here, but there are also important differences that we ignore at our conceptual peril.

We can usefully contrast compatibilist and incompatibilist accounts of objectively special divine action (OSDA). The latter hold that OSDA is incompatible with deterministic causal closure of the created world; if God acts to alter the course of

of finite agents for free action, and God permits the exercise of this capacity in a history of relationships with other persons and with God. Human beings contribute creatively to making the selves we become, and since this includes the capacity to act against God's purposes for us, we are simultaneously entangled in our own unmaking as well. This understanding of freedom complicates attempts to "front load" into the act of creation God's responses to specific human choices; it appears that responsive divine action will affect the course of events in the world in ways that were not already written into the program of history.[13]

---

events once it is underway, then there will be a departure from perfect natural determination (either because God interrupts the operation of natural law or because God has created a natural order that includes underdetermination as part of its given structure). A compatibilist regarding OSDA denies this. In this essay (and elsewhere) I have been arguing for incompatibilism in *this* limited sense.

Thus far, the issue is well-defined. If, however, we talk about compatibilist versus incompatibilist accounts of divine action in general (without restricting the scope of the distinction to OSDA), then we generate a whole series of confusions. That is because divine action can take several forms, and there are good reasons to make a differentiated judgment that some of them are, and others are not, compatible with a deterministic natural order. Note the three forms of divine action that I identified in section 1 above: (1) God acts directly as the creative cause of the being of all finite things; (2) God acts indirectly through the operation of created causes and agents as they bring about their effects in the world; and (3) God acts directly in the world's history to bring about events that created causes alone are insufficient to produce (OSDA). Typically, those who argue that OSDA is incompatible with closure among finite causes also *affirm* that the first two forms of divine action are perfectly *compatible* with determinism. This is obscured when we classify views of divine action as globally incompatibilist or compatibilist. Furthermore, thinkers who typically are identified as compatibilists often agree that OSDA is incompatible with causal closure (that is, they are in fact *incompatibilists* in the narrower sense). So how do "compatibilists," in this global sense, differ from "incompatibilists"? In most cases the "compatibilists" do not want to be committed to denying deterministic causal closure, and therefore they reject OSDA, affirming only those forms of divine action that are compatible with causal closure (viz., the first two), and speaking of special divine action only in the subjective and material senses. By contrast, "incompatibilists" affirm OSDA, and therefore reject determinism. The key point to note is that this disagreement is *not* about the incompatibility of OSDA and determinism, on which they *agree*. It is about whether or not to affirm OSDA, given that this affirmation puts our theology of divine action at risk of being refuted if a deterministic interpretation of the sciences were somehow to gain the upper hand. "Compatibilism" and "incompatibilism" are the wrong names for these positions, though the terms have come to be used this way. One example of someone who *does* appear to affirm the compatibility of OSDA and determinism is Arthur Peacocke, though I argue below that his intriguing proposal finally is unsuccessful.

[13] There are theological alternatives to this idea of particular responsive divine action within the course of historical events. For example, if divine omniscience includes knowledge of what any free being would decide to do in any possible circumstance of choice, then God would be in an epistemic position to respond to the free actions of creatures in the design of creation "before" (explanatorily, rather than temporally) the universe is called into being. Roughly this was the position of Molinists in the sixteenth century, and their views on divine foreknowledge and human freedom have been debated to this day. See, for example, Luis de Molina, *On Divine*

Do these familiar alternatives exhaust the possibilities? It might be objected that neither account of freedom can properly be applied to the relation between God and human agents precisely because they both remain within the grip of a contrastive understanding of this relationship. On this account, the relation of the divine agent to free human acts is unique precisely because God is not part of the causal nexus to which the finite agent belongs, but rather is its transcendent ground. God's agency, therefore, does not constitute a causal factor that is in danger of displacing the creature's free self-determination. Rather, God's activity operates at another level altogether, "beneath" or "above" or "before" (choose the prepositional metaphor you prefer) the contrast between causal determination and underdetermination of the agent's act. God brings about the very existence of the finite agent at every moment of that agent's activity, whether that activity is correctly described by compatibilist or libertarian views of freedom, both of which have to do with the relation of the created agent to other created causes. God's transcendent, creative agency posits the finite agent in her free act. There is no need to limit this divine action in order to "leave room" for the creature's freedom; it is precisely this divine action that brings it about that there is a creature acting freely. On this view, the notion that there is any tradeoff between divine action and human freedom involves a fundamental misunderstanding of the nature of God's creative agency.[14] Divine action in response to the free actions of creatures will not, therefore, present any *special* puzzles (that is, over and above the inherent incomprehensibility of God's utterly unique activity as the giver of being *ex nihilo*). God acts creatively to constitute both the free human agent-in-action and the events in the world that are God's response to that action.

This third way of thinking about divine action and human freedom is right to insist that God's activity as creator cannot be a determining cause *on the same level* as the secondary causes typically considered in debates over freedom of the will. In bringing about the existence of an agent of intentional actions, God does not merely influence or affect the finite agent, determining the action in the way a confluence of finite causes (e.g., a deterministic development of beliefs and desires) might do so. Rather, God constitutes, or posits, the agent performing this act. But we can nonetheless ask what powers of action God calls into being in the finite agent. Does God create a finite agent with some capacity to determine its own action, or does God's creative will fully specify the content of the finite agent's choice (viz., that there shall be an agent-so-acting)? If we say the latter, then it is hard to see any basis for denying that the action is determined by God. We can acknowledge that God's act of "causing being" is different in kind than the creature's act of "causing

Foreknowledge: Part IV of the Concordia, Alfred J. Freddoso, ed. (Ithaca, N.Y.: Cornell University Press, 1988); Thomas Flint, *Divine Providence: The Molinist Account* (Ithaca, N.Y.: Cornell University Press, 1998); Robert Adams, "An Anti-Molinist Argument," *Philosophical Perspectives* 5 (1991): 343-53.

[14] For positions of this sort see Placher, *The Domestication of Transcendence*, chap. 9; David Burrell, C. S. C., *Aquinas: God and Action* (Notre Dame, Ind.: University of Notre Dame Press, 1979); idem, *Freedom and Creation in Three Traditions* (Notre Dame, Ind.: University of Notre Dame Press, 1993); and Tanner, *God and Creation in Christian Theology*, chaps. 3-4.

change," so that the statement that God "makes" the agent-so-acting (i.e., makes this agent to be) must not be conflated with the claim that God "makes" the agent perform this act (i.e., that God acts upon the agent to compel the act). What follows from this, however, is not that determinism is averted, but rather that God's determination of the creature's action is fundamentally different than determination by finite efficient causes. If the agent's choice is built into God's creative will, and God's creative will is unobstructable, then we are left with determinism after all, albeit a distinctively theological one.

In sum, it appears that we cannot altogether avoid contrasting divine and created agencies in certain limited cases. The limits here are centrally important. We need not adopt in general a zero-sum conception of God's relation to the actions of creatures or treat God as merely one agent among others. If we understand God as creator *ex nihilo*, then the powers of action possessed by finite things come from the hand of God and depend absolutely upon God's sustaining activity. In this way God acts in, with, and through every activity of creatures in a uniquely intimate and internal way. We also suggested that God may act among creatures, engaging the world to turn events in a direction they would not have gone otherwise. One theological motive for defending the possibility of particular providence in this third, and strongest, sense is that this would allow us to speak of God acting in response to the actions of free creatures. The idea of responsive divine action can be developed in more than one way, depending in part on one's understanding of the nature of human freedom. Given a strong (incompatibilist) conception of human freedom, a viable concept of non-miraculous particular divine action may be especially helpful. This mode of divine action, however, would be just one among a variety of ways in which God acts, and it is not the fundamental one. If we think of theological views as taking shape within a dynamic network of (sometimes contending) considerations, then keeping the idea of non-miraculous special divine action in the mix will expand the range of options in interpreting central theological topics (e.g., in the theology of nature and of history, in theological anthropology, in soteriology, and so on).

2.2 Special Divine Action without Appeal to Causal Openness: Peacocke on Whole-Part Constraint

I want to consider a second challenge to the general theological program that we are exploring, a challenge that arises in the thought of Arthur Peacocke.[15] This response, like the first, contends that we do not need to find openings in the causal structures of nature in order to speak of divine action in the world. Furthermore, it shares the conviction that we ought not to represent God as acting directly among or upon natural causes and agents; divine action of this sort strikes Peacocke as "interventionist," whether it involves a miraculous interruption of ordinary causal sequences or an unobservable divine act of

---

[15]See especially, Arthur Peacocke, *Theology for a Scientific Age*, 2nd ed. (London: SCM Press, 1993), chap. 11; idem, "God's Interaction with the World," in *CC*, 263-88; idem, "The Sound of sheer Silence: How Does God Communicate with Humanity," in *NP*, 215-48; idem, *Paths from Science Towards God: The End of All our Exploring* (Oxford: One World Publications, 2001), chap. 5.

determining what nature leaves underdetermined.[16] Unlike the first objection, however, it does not interpret providence exclusively in terms of God's universal creative activity, but rather holds that God acts to affect the ongoing course of events in the world. In other words, Peacocke's position seeks to sustain all three of the following claims: (1) God does not act directly among or upon natural causes, and therefore (2) theology does not need to find causal incompleteness in nature (though he affirms the likelihood that nature at various levels is in fact indeterministic), but nonetheless (3) God's activity brings about particular effects in the world that would not have occurred had God not so acted.

At the center of Peacocke's development of this position is the idea that we might model God's relation to the world on the relation of a complex whole to the parts that are united within it. This proposal has its scientific underpinnings in the familiar practice of describing the natural world as a nested hierarchy of increasingly complex structures. Higher levels in the hierarchy are built up upon lower levels, and they subsume those lower levels as parts within functionally integrated wholes that set conditions for the operation of the parts. Systems of this sort display properties that are a function of their organization, and therefore are not expressible simply in terms of a description of the properties of their parts. Distinctive descriptive and explanatory concepts are needed to understand the dynamics of new levels of functional organization (hence we have diverse sciences). Peacocke notes the crucial explanatory role of the organizational principles of such systems, and he then cautiously takes a further step, suggesting that wholes act as causes in relation to their parts. "When the nonreducibility of properties, concepts, and explanations applicable to higher levels of complexity is well established, their employment in scientific discourse can often, *but not in all cases*, lead to a putative and then to an increasingly confident attribution of a *causal efficacy* to the complex wholes which does not apply to the separated, constituent parts..."[17] Peacocke explains the causal efficacy of wholes by noting that they constrain or influence their parts; the parts operate as they do because they are organized in a particular way within the overall structure, and for this reason we find it useful to explain the operation of a part by reference to its functional role within the whole. In this connection, Peacocke calls upon a helpful distinction between "structuring" and "triggering" causes.[18] The structure of the system sets boundary (or limiting) conditions for the activity of its parts; that is, it consists of a network of relationships among those parts that

---

[16] The term "intervention," therefore, is used in two crucially different ways by the authors in this volume. Peacocke applies it to *any* direct divine action among secondary causes, including non-miraculous action at points of causal openness in nature. The proponents of *non-interventionist* objective divine action obviously are using the term to describe only those divine actions that interrupt otherwise complete secondary causal chains (or as it is often put, that "violate the laws of nature").

[17] Peacocke, "The Sound of Sheer Silence," 219.

[18] Fred Dretske introduces this distinction in "Mental Events as Structuring Causes of Behavior," in *Mental Causation*, John Heil and Alfred Mele, eds. (New York: Oxford University Press, 1993), 121-8.

delimits their behavior. In this sense the system as a whole "causes" the parts to behave as they do in response various triggering inputs.

Peacocke combines this scientific picture of the layered structure of nature with a panentheistic conception of God as immanent within, yet not simply identified with, the universe. God is the highest level whole that embraces within the transcendent divine life the entire hierarchy of structure in the universe, the system-of-all-systems. This makes it possible to think of God as acting, not by directly affecting any of the myriad subsystems that make up the cosmos, but rather by setting the conditions (the structural constraints) under which they carry out the causal programs proper to them.

> By analogy with the operation of whole-part influence in real systems, God could affect holistically the state of the world (the whole in this context). Thence, mediated by the whole-part influences of the world-as-a-whole (as a *System*-of-systems) on its constituents, God could cause particular events and patterns of events to occur which express God's intentions. These latter would not otherwise have happened had God not so intended...Moreover, this action of God on the world may be distinguished from God's universal creative action in that particular intentions of God for particular patters of events to occur are effected thereby—and the patterns could be intended by God in response *inter alia* to human actions or prayers.[19]

God affects the system at the highest level, and this ramifies in the operation of lower levels. The divine action does not disrupt the causal structures governing the lower level systems; rather, it simply sets the conditions under which they operate, so that they conform to the causal laws appropriate to each system at its own level, but they do so in accordance with a higher level constraint. On this view, God acts as a "structuring cause" rather than as a "triggering cause"; God does not make particular inputs to the system, but rather acts as the boundary condition on the universe as a whole. In this way, Peacocke contends, God can affect the ongoing course of the world's history and bring about effects that would not occur without this divine action, and yet God never acts on the same level as created causes or produces particular effects directly rather than through the ordinary course of nature.

This proposal is carefully conceived, theologically appealing, and conceptually artful. As with any model of divine action, however, it faces various difficulties; if we press for further detail on how God's activity engages the world, these problems come into focus. How does God, acting at the level of the system-as-a-whole, bring about particular *changes* in the activity of creatures, so that events in the world unfold differently than they would have had God not so acted? Recall that a high level system operates as a structuring cause by constraining the behavior of the subsystems that make it up. How these subsystems function (the particular effects they produce) will depend on (a) their location within the network of relationships that constitutes the system, and (b) the inputs into the system (i.e., triggering causes). As long as the structure of the system remains constant, the only source of change in its operation will be through triggering causes. If God does not act upon parts of the system as a triggering cause, then in order to affect its operation God will

[19] Peacocke, "The Sound of Sheer Silence," 236.

need to modify its structure. But since the system *consists of* the network of relationships between its parts, modifying the structure of the system will require acting upon the parts to change their patterns of interaction.[20] This involves affecting the parts in a different way than simply by constraining their operation; it is to modify the constraining relationships in which they stand. Peacocke's position appears to require that God act not only as a structuring cause, but also as the efficient cause of changes in the world's structure, adjusting the relationships of its parts so that they will produce the particular effects that God intends. It is true, of course, that natural systems, understood as functional wholes that relate multiple subsystems, develop over time, and may be self-organizing. On a panentheistic view, these changes take place within the life of God. The difficulty for Peacocke's position, however, lies not in the fact of structural change, but in how it takes place. If God acts to modify systems, even at the highest level, then there will be causal discontinuities in the history of the system, namely, transitions in the development of its structural relationships that cannot be given a sufficient causal explanation in terms of the prior history of the system at any lower level. It appears that ontological openness is required after all if there is to be particular divine action that changes the course of events in the world.[21]

2.3 The Need for Metaphysical Mediation: Process Theology

Let me mention a third, and significantly different, line of objection to the strategy of exploring the contemporary sciences in search of structures that display a theologically interesting causal openness. It might be argued that this approach is problematic because it links theological claims to the interpretation of particular scientific theories without employing a comprehensive metaphysical scheme that can mediate the relation between science, religion, and other areas of human experience and knowledge. This objection contends that unless a general ontology is brought to the interpretation of natural science, current scientific theories will be treated as though they were philosophical proposals about the fundamental structures, entities, and processes that constitute the world. The danger is that science will be prematurely turned into metaphysics, without giving enough attention to problems of interpretation and to the provisional and open-ended character of scientific theory building. The corresponding theological hazard is that important faith claims may be tied very closely to physical theories which are almost certain to change. The critic might claim, therefore, that it is better to

---

[20] A change in the organization of the system *is* a change in the relations of the parts. In some systems it may be possible to change (e.g., substitute) parts without changing the operation of the whole, but it appears that it will not be possible to change the system without changing the relations of the parts.

[21] See Nancey Murphy, "Emergence, Downward Causation, and Divine Action," in this volume, for a comparable line of criticism. Also see Robert Russell's careful analysis of Peacocke's proposal from a scientific perspective, and in particular his critique of the idea of "the world as a whole" given that the topology of the universe in Big Bang cosmology does not allow for a boundary from which God could act, in *Cosmology: From Alpha to Omega* (Minneapolis, MN: Fortress Press, 2007), chap. 4, section 4B.

begin by adopting a comprehensive conceptual scheme that can embrace both science and theology, a scheme that is independent of all particular scientific theories and so holds the promise of being able to guide our interpretation of the sciences as they change over time.

There is merit to the underlying concerns expressed in this objection, and these issues are often cited as one of the reasons to adopt some form of process theology. It will be helpful to say a bit more about the relation of process theologies to the theological program we are considering, since there are some substantial areas of overlap as well as illuminating contrasts. Process theologies characteristically affirm that the natural world is causally open to God's activity. For the reasons just noted, however, they typically do not (at least initially) look to scientific descriptions of the world for evidence of open structures in natural processes. Instead, they build this openness into a metaphysical scheme that describes the fundamental processes by which anything at all comes to be. This is a difference of approach in pursuing a shared insight: namely, that particular divine action within the ongoing course of events requires that the structures of the world be open to that action.

For example, process theologies employing Whitehead's system make use of a set of concepts that are drawn from human experiences of interrelatedness and creative self-formation, and they extend these by analogy to provide general metaphysical categories describing the basic units of existence. Every discrete individual takes shape in a process of becoming that unites in a distinctive way the influence conveyed through its relations to other individuals, including God. This process is in part determined by the past state of the universe, but it is also a free creative achievement. The determinate properties of the other entities to which any individual is related provide the material for that individual's synthesis. But the contributions made by these other entities provide only necessary, not sufficient, conditions for what the new individual becomes. The coming-to-be of every entity includes a self-determining moment; this creativity may reduce asymptotically toward the vanishing point in the most primitive entities, but in can never be entirely absent. God's activity can then be interpreted in terms of this scheme. God contributes to each individual the crucial initial aim of its own creative development. God's influence, like that of all others, does not determine the individual's final identity, but it does shape the individual's own process of becoming more profoundly than any other relationship, and it does so in light of God's universal sympathetic appreciation of the experience of every existing thing.

This is just the briefest sketch of one version of a process metaphysic, but it illustrates the general strategy of approach to divine action. That strategy is to construct a system of metaphysical categories that guarantees, as a primitive feature of its ontology, openness to divine action within the world. We can then make use of this conceptual scheme in interpreting both natural science and theology. Since the metaphysic abstracts from particular scientific descriptions of the world and can be used to interpret them all, it holds out the promise that we can construct an account of God's activity without looking to the natural sciences themselves to provide the causal openness required for ongoing divine influence on the course of events. It is easy to see the appeal of this strategy, but there are at least two important problems with it.

First, the metaphysical system needs to provide an illuminating interpretation of what the sciences are telling us about the world, and if it fails to do so, it is in danger of becoming an isolated artifact of the philosophical imagination. Here the abstractness that helps make the metaphysical system theologically useful becomes a potential problem. The system presents a large-scale theoretical construct, an alternative world-structure, whose relation to the world described by the sciences is not easy to see. We then face the task of applying this metaphysical system to the descriptions of the world put forward by the natural sciences, and this can pose problems at least as daunting as those involved in offering a theological interpretation of current science that is not mediated through a large scale metaphysical system.[22] Note, for example, that if the processes of nature described by the sciences were relentlessly deterministic, then the causal openness built into the metaphysical system would have no correlate in nature as we actually find it to operate. Nothing is gained by appealing to creative freedom or the divine lure within the systematic metaphysic unless the causal openness it postulates is somehow reflected in the structures of nature described by the sciences, and that requires a plausible indeterministic interpretation of the science itself.[23] This suggests that we would do as well to proceed straight to the interpretation of the science.

Second, in adopting a set of universal metaphysical categories, we are setting the terms for theological interpretation and construction; our theological affirmations are filtered through this system and must be reexpressed in its categories. A number of familiar objections to process theology reflect discontent with this procedure. Whitehead's metaphysical scheme, for example, imposes a variety of constraints on what can be said about God's nature and activity: e.g., that God exemplifies, but does not establish or ground, the basic categories of creative becoming, and therefore cannot be creator *ex nihilo*; that there is no life with God beyond individual perishing, and that while God's experience incorporates what each individual has achieved, there is no redemptive overcoming of evil for those whose good is sacrificed along the way. The theological pros and cons of this conception of God have been energetically debated, of course. The point here is simply that one cannot have the interpretive power of Whitehead's particular metaphysical system without these theological results (though there are, of course, variants on Whitehead that have different theological implications).

In order to diminish this hermeneutical predetermination of the theological outcome, we may prefer to give up the advantages of a comprehensive metaphysical scheme. The alternative is to begin with careful attention, on the one hand, to the theological ideas and traditions that need to be explicated and, on the other hand, to the scientific theories and the interpretations of those theories that may present interesting connections with theology. This does *not*

---

[22] For a recent example of careful attention to these issues by process philosophers see *Process Studies*, 23, 3 & 4 (1997).

[23] Robert Russell, *Cosmology: From Alpha to Omega*, (Minneapolis, MN: Fortress Press, 2008), 137-40, and "Barbour's Assessment of the Philosophical and Theological Implications of Physics and Cosmology," in *Fifty Years in Science and Religion*, Russell, ed. (Burlington, VT: Ashgate, 2004), 148-50.

avoid all metaphysical assumptions and commitments; rather, it practices a candidly *ad hoc* and piecemeal metaphysics, introducing provisional ontological claims in response to the interpretive opportunities presented by the scientific and theological subject matter. This approach is reflected, for example, in adopting an attitude of "critical" (i.e., cautious and qualified) realism about at least some scientific theories. Note that we need not endorse any single version of critical realism in general, but rather we can fine-tune our conclusions about the representational value of different scientific theories. The result of proceeding in this way will be that we generate provisional and exploratory theological proposals, proposals that are offered in the expectation that they will be revised in light of new developments in the sciences and in response to ongoing critical assessment of their theological adequacy. Theological work of this kind carries on its reflection about God's relation to the world in light of the current state of science and in full recognition that science will change. It works at the boundaries of theological interpretation, and therefore does not aim at being definitive, at producing a settled position that remains fixed over time. It recognizes that the process of interpreting its own sources in light of current understandings of nature is perpetually unfinished.

3 Proposals: Can Theologically Relevant Causal Openness Be Found within the Sciences?

How might this theological program be carried out in particular cases? In the more than ten year history of the VO/CTNS research project a variety of specific proposals have been made that fit the general pattern I have described. A simple way to categorize them is in terms of the scientific theories to which they appeal in seeking a theologically interesting openness in the structures of nature. I will briefly consider three prominent proposals, which make use respectively of chaos theory, quantum mechanics, and developments in neuroscience and philosophy of mind.

3.1 Chaos Theory

Chaos theory has brought to light the extreme sensitivity to initial conditions displayed by some non-linear dissipative systems. Vanishingly small differences in the initial conditions of such systems can result in massively different outcomes after a relatively small number of iterations. This sensitivity severely limits our ability to predict future states of the system; we simply cannot specify the initial conditions with sufficient precision to provide predictions for more than a brief projection of the system into the future. This behavior can be elegantly modeled by relatively simple equations, such as the logistic equation discussed by Wesley Wildman and Robert Russell.[24] Though the equation defines a perfectly deterministic development over time, the values it generates quickly become unpredictable in principle. Russell and Wildman point out that this unpredictability entails, ironically, that we cannot

[24]Wesley J. Wildman and Robert John Russell, "Chaos: A Mathematical Introduction," in *CC*, 49-90.

determine whether or not our deterministic model of the system is correct. Chaos theory, then, vividly illustrates the insurmountable epistemic limits of any finite intelligence.

John Polkinghorne appeals to this phenomenon of sensitive dependence and unpredictability as the starting point for his metaphysical reflections on openness in the structures of nature. One cannot, of course, move directly from noting limits on our ability to predict the future (an epistemic point) to the conclusion that the future is causally open (an ontological claim). Although it has been alleged that this is what Polkinghorne is doing, he steadfastly (if perhaps with growing weariness) insists that it is not.[25] Instead, he is offering an interpretive extrapolation beyond chaos theory; his proposal represents a metaphysical conjecture motivated by the surprising emergence of ordered unpredictability within the bounds of a deterministic science. He argues that in this respect his proposal parallels the interpretive decisions made by those who side with the indeterministic interpretation of quantum mechanics offered by Niels Bohr and Werner Heisenberg rather than with the determinism of David Bohm.[26] Scientific theory underdetermines our ontology; we can give various accounts of the structures in the world that are captured by the theory, and our metaphysical claims cannot appeal simply to science for their justification. Polkinghorne proposes that unpredictability in chaos theory should be regarded as a deep feature of nature, deeper than the deterministic equations that are used in modeling chaos. These equations, he suggests, can be regarded as "downward emergent approximations to a more subtle and supple physical reality."[27] On this account, deterministic chaos represents a special case generated by theoretical simplification and abstraction from open textured and holistic natural structures that we are not yet able to describe.

This metaphysical hypothesis opens up some intriguing possibilities for understanding the action of God in the world. It is tempting to suggest that God acts by making minuscule adjustments in the initial conditions of chaotic systems. Polkinghorne resists this approach, and contends instead that what we know of chaotic systems suggests that they are extensively linked in interrelated wholes that are open to "top-down" or "whole-part" influence. God may be understood to act upon these systems in roughly the way Peacocke described, namely, as a structuring cause that affects the context in which systems operate. God's influence takes the form of an "input of active information"[28] that selects between alternative paths of development ("this way, not that way")[29] without any input of energy. This is allowed when the phase space describing possible behaviors of a chaotic system constitutes a

---

[25] See for example, John Polkinghorne, *Faith, Science and Understanding* (New Haven, Conn.: Yale University Press, 2000), 99-101.

[26] John Polkinghorne, "Physical Process, Quantum Events, and Divine Agency," in *QM*, 181-90.

[27] John Polkinghorne, "The Metaphysics of Divine Action," in *CC*, 153.

[28] Ibid., 153-4.

[29] John Polkinghorne, *Belief in God in an Age of Science* (New Haven, Conn.: Yale University Press, 1998), 62.

"strange attractor" that approaches the limiting case of infinite density, so that there is no energy difference between vanishingly close paths of development.

Carefully framed in this way, Polkinghorne's account of divine action through chaotic processes offers an interpretive hypothesis that goes beyond chaos theory, and so it cannot be rebutted simply by noting that the theory itself is deterministic. Nonetheless, this objection contains the germ of an important criticism. Let us grant that chaos theory can be interpreted in the way Polkinghorne suggests, namely, as revealing an underlying openness in nature. The question we need to ask is whether there is any justification, *in chaos theory itself*, for this metaphysical conjecture. It is of course possible to generate a general philosophical view of the structures of nature according to which those structures are inherently flexible and open to divine influence. We just considered one such proposal, *viz.*, process metaphysics. A comprehensive metaphysical view of this sort can be brought to bear on each of the sciences, and it would be justified in part by its power to provide an integrated and illuminating interpretation of them. But note that it would not appeal to particular scientific theories directly for its justification, precisely because it is a second-order view operating at a higher level of abstraction, the *meta*-physical level.

Polkinghorne's proposal, however, begins with chaos theory as its inspiration and (at least partial) warrant, in accordance with the principle that "epistemology models ontology." The effect of this principle is to affirm that "what we can know is a reliable guide to what is the case."[30] It is this cautious realism about scientific theories that presumably powers the move from chaotic unpredictability to the interpretation of nature as flexible and open. The problem, however, is that it is doubtful whether this principle provides support for this interpretation of chaos theory. That is because in chaos theory, unpredictability arises *strictly as a function of* deterministic equations. It is not as if we have a well-established science of unpredictable systems, and then ask how we are to model these systems, with the result that deterministic non-linear equations provide one alternative among others. On the contrary, the particular sorts of unpredictable behavior discussed by chaos theory emerge within and are artifacts of a fully deterministic mathematics. There is no warrant for affirming the unpredictability of these systems while denying their deterministic character. Indeed, within the terms of chaos theory itself this cannot be done coherently; unpredictability of *this* sort does not arise apart from particular patterns of deterministic development. The same point can be made about the idea of divine action through non-energetic inputs of information; this possibility emerges only when dealing with a deterministic mathematics that generates a strange attractor in which there is (at the limit) no difference in energy between "nearby" trajectories in the phase space.[31] The

---

[30] Polkinghorne, "The Metaphysics of Divine Action," 148.

[31] Nicholas Saunders discusses this point in detail; see *Divine Action and Modern Science* (Cambridge: Cambridge University Press, 2002), chap. 7: "Active information input relies again on the determinism of mathematical chaos to produce the required fractal structure in attractors, the required infinite limit of that structure, and the corresponding region in which energy differences between alternative possible trajectories tend to zero *in the infinite limit*" (194). Saunders also argues that the

principle that epistemology models ontology, therefore, would warrant the conclusion that the world depicted by this theory is deterministic.

Of course, everything Polkinghorne says about the openness and flexibility of nature might be true, and deterministic chaos might in fact simply reflect the limits of the models used to explore these systems in nature. But chaos theory, at least in its current form, does not give us a reason to adopt this view. The question, then, is why we should accept this interpretation of chaos theory, attractive though it is. The motive for doing so, I think, has to do with the appeal of a metaphysics that sees all or most deterministic causal theories as approximations (under simplifying assumptions about, e.g., the separability of physical systems in nature) to more flexible and intimately interrelated underlying structures. This is a bold claim of wide scope, and it would need a philosophical and scientific defense with a much broader base than can be provided simply by an appeal to chaos theory.

3.2 Quantum Mechanics

An alternative approach looks to quantum mechanics as a resource for theological construction. This will require that we offer an interpretation of quantum theory that meets the conditions we initially noted in introducing the idea of non-miraculous special divine action. First, the domain represented by the theory must include events that do not have causally sufficient antecedent conditions. Second, at least some of these underdetermined events must have the potential to make the right sort of difference in the subsequent course of events. Third, the underdetermined events that make this difference must be an integral part of regular structures in nature, and not merely a random interruption of those structures. Only if these three conditions are met will an interpretation of quantum mechanics provide a useful resource for the theological program we have been considering.

3.2.1 Issues of Interpretation

The interpretation of quantum theory, of course, is a deeply puzzling enterprise characterized by strikingly diverse alternative proposals. It is worthwhile to note a contrast between the predicament of the interpreter of quantum mechanics and that of the interpreter of chaos theory. Chaos theory, we just observed, does not currently have a non-deterministic scientific formulation, or even a formulation that is neutral with regard to the determinism/ indeterminism alternatives; mathematical chaos is a function of particular deterministic equations under specific regimes. That is why chaos theory, coupled with a critical realist principle about scientific theories, does not provide a promising jumping off point for an indeterministic metaphysic. Quantum mechanics, on the other hand, presents us with a well-established formalism that can be interpreted in multiple ways, and the formalism remains

limiting case of infinite complexity, at which there will be zero energy difference between paths through the attractor, is unlikely to occur in actual systems in nature. He writes, "Any model which assumes that there exits an infinite fractal intricacy to attractors may not represent the physical world" (199).

intact as the scientific core that various interpretations seek to preserve. Significantly (and in contrast to chaos theory), the formalism of quantum mechanics permits (indeed, it invites) an indeterministic interpretation. That is because the formalism itself does not provide a complete, closed causal account, but rather is stubbornly probabilistic at important points. David Bohm has shown that it is possible to fill in a supplementary story about quantum entities/systems according to which their properties are determinate (position being basic) and the probabilistic uncertainty in the formalism is merely epistemic.[32] Further, some "many worlds" interpretations are also deterministic. One might hold, for example, that all of the possibilities contained in the state description of a quantum system are actual in discrete worlds. Since the wavefunction representing those states develops deterministically according to the Schrödinger equation, the probabilistic character of quantum theory simply reflects our inability to predict which of these branching actualities will be the one in which we find ourselves when a measurement is made. The important point for our purposes, however, is that *the quantum formalism does not commit us to any particular interpretive view.* Deterministic interpretations of quantum mechanics must compete for our metaphysical affections alongside indeterministic interpretations, and both can count as forms of (highly) qualified realism about quantum theory. In the current state of science, no single interpretation has been able to establish itself as *the* correct account of the structures in the world whose behavior is captured by the theory.

It looks, therefore, as though quantum theory satisfies the first condition for being enlisted in our theological program; it is open to indeterministic interpretation. Because any interpretation is one among others in an ongoing discussion, and because quantum mechanics provides perplexing challenges for even the most chastened scientific realism, we will do well carefully to qualify our claims about what quantum mechanics tells us about the world. But it is not possible to think theologically about quantum mechanics without adopting some interpretation or other, and it is perfectly legitimate to explore viable (and in fact widely held, even dominant) indeterministic options, since these hold particular theological interest.

The question, then, is whether such an interpretation of quantum mechanics meets the other two conditions of theological usefulness. It is here that some important, though not fatal, difficulties arise. I have previously discussed two of these difficulties: the measurement problem and questions about the amplification of quantum effects.[33] Rather than repeat that discussion here, let me briefly restate my conclusions and comment on some related issues that have recently been presented as objections to divine action at the quantum level.

---

[32] This determinism does not rescue classical physics, however, since the Bohmian scheme is strongly nonlocal, that is, it involves instantaneous links between spatially distant regions. See David Bohm, "A Suggested Interpretation of Quantum Theory in Terms of Hidden Variables, I & II," *Physical Review* 85 (1952), 166-93; idem, *Wholeness and the Implicate Order* (London: Routledge & Kegan Paul, 1980).

[33] See "Creation, Providence, and Quantum Chance," sec. 3, in *QM*. On these issues also see the papers by Russell, Ellis, and Polkinghorne in *QM*.

3.2.2 Measurement

The measurement problem is a central puzzle for current indeterministic interpretations of quantum mechanics. The family of views broadly known as the Copenhagen interpretation typically takes the position that those properties of a quantum system which must be described probabilistically (e.g., the spin orientation but not the spin magnitude of an electron) do not, in fact, have determinate values unless "measured." The stuff of the universe at this level is in certain respects indefinite, characterized by irreducibly probabilistic structures of potentiality that have no counterpart in classical mechanics. The development in time of this quantum state of superposed potentials is described by the Schrödinger equation, which evolves in a smoothly deterministic way until an interaction takes place that results in the irreversible collapse of the wavefunction to a determinate value for the measured property. It is here that we find chance in quantum systems; this discontinuous reduction of potentiality to actuality has necessary but not sufficient causal conditions in the prior history of the quantum system and its environment. In these transitions, the principle of sufficient reason is not satisfied within the order of natural causes.

If we see these underdetermined quantum events as providing occasions for non-interventionist objective divine action, then our account of this mode of action will be inextricably tied to a tangle of difficult scientific and interpretive issues. When and why does the wave function collapse? How does the strange realm of quantum superposition and entanglement give rise to the familiar macroscopic world of objects with determinate properties and deterministic causal relations? Shouldn't quantum uncertainty "spread" to include the apparatus of measurement itself?[34] These puzzles lie at the heart of indeterministic interpretations of quantum mechanics, and we should qualify our theological uses of the theory in light of them. If we say that God acts in quantum events, then we must acknowledge the profound incompleteness of our understanding of the place of these events in nature.

In addition to these unresolved issues in the interpretation of quantum theory itself, two specific objections often are lodged against the idea of divine action at the quantum level. John Polkinghorne and others have contended that because chance quantum transitions occur only when measurement takes place, God's action would be too discontinuous and limited in scope to have much theological significance.[35] In response, we can make both a theological and a scientific point. The theological response is that we need not and should not understand God to act only through quantum transitions; God acts

---

[34] These problems have been given their best known expression in the thought experiment involving Schrödinger's unfortunate cat, which must apparently exist in a superposition of being dead and alive until we open the apparatus and observe its condition.

[35] See, for example, Polkinghorne, "The Metaphysics of Divine Action"; Nicholas Saunders, "Does God Cheat at Dice?: Divine Action and Quantum Possibilities," *Zygon* 35 (2000): 517-44.

pervasively as the creator who gives every finite thing its powers of operation and who sustains each of them as they exercise those powers in an interactive web of natural relationships. Objectively special divine action at the quantum level presupposes these more fundamental forms of divine action. The scientific response is that state reduction takes place ubiquitously throughout nature on many levels, as Russell has argued in detail.[36] The language of "measurement" here is misleading, since it is tied to the activity of recording an event in a laboratory and obscures the point that these transitions occur wherever there is an interaction that results in the irreversible specification of an actual (determinate) value for one of the potential (indeterminate) states of the system.

Another familiar objection is that quantum divine action, not withstanding its proponents' claims to the contrary, involves intervention in the order of nature after all. This claim has weaker and stronger forms. The weaker form simply amounts to a disagreement over what sorts of divine action should be described as "intervention." I noted above that Peacocke applies this term to any divine action within the nexus of created causes, whether or not that action interrupts a natural causal chain. Defenders of non-interventionist objective divine action, on the other hand, reserve the term for events in which God overrules natural law. The weaker version of the objection is readily dealt with, therefore, by making the needed distinction between these two senses of this ambiguous term; the debate over intervention in Peacocke's broad sense can then be recognized as a form of the general theological dispute we noted above (in section 2.1) about whether it is ever appropriate to say not only that God acts as primary cause in sustaining the existence and operation of secondary causes, but also that God acts directly among or upon them to affect the course of events, although never as a secondary cause.

The stronger form of the objection attempts to present a more direct challenge. Nicholas Saunders has argued that each of the possible ways of understanding God's action within a quantum system involves precisely the sort of disruption of lawful processes that the advocates of quantum divine action set out to avoid. Saunders identifies four possible "routes" for this form of divine action: by changing the wave function, by performing a measurement, by altering the probabilities of measurement outcomes, and by selecting a particular outcome of measurement.[37] I think he is right that the first three of these alternatives (depending upon how they are construed) involve violations of natural law. But defenders of quantum divine action have seen this, and have never seriously entertained these three possibilities, focusing instead on the fourth option: namely, that God acts in the measurement event by selecting one value from within the range of possibilities provided for by the probabilistic laws that describe the system. There certainly appears to be no violation of natural law here; indeed, it is not easy to say just how God's action in any single quantum transition *could*

---

[36] Robert J. Russell, "Special Providence and Genetic Mutation: A New Defense of Theistic Evolution," in *EMB*, sec. 2.3.2; idem, "Divine Action and Quantum Mechanics: A Fresh Assessment," in *QM*, sec. 4.

[37] Nicholas Saunders, *Divine Action and Modern Science*, 149-57.

violate a probabilistic law (as long as that action does not produce a result with zero prior probability).

It is a bit tendentious, therefore, for Saunders to describe this fourth option as proposing that "quantum SDA occurs by means of God ''ignoring' or intervening against the measurement probabilities 'predicted' by the orthodox theory."[38] As long as individual quantum events fall within the range of possibilities allowed by the wave function, nothing in nature is being ignored or overridden by God's action. In order to defend the claim that this form of divine action "intervenes against" the requirements of the relevant probabilistic laws, we would need to adopt a highly idiosyncratic understanding of such laws. Saunders contends that defenders of non-interventionist special divine action are committed to a "necessitarian" conception of natural law that is incompatible with their proposal about God's action in quantum events. Wesley Wildman further unpacks this claim, suggesting that Saunders assumes a "strong ontological" conception of probabilistic law according to which these laws refer to "principles or deep structures of nature that statistically govern each individual event within an ensemble of events."[39] It is not clear what it means to say that a law "statistically governs each individual event within an ensemble." If this simply says that each event must fall within the range of probabilities spelled out in the law, then as we have already seen, there is no problem here for quantum divine action. In order for this view to have any critical bite against its intended target, it must assert that a probabilistic law somehow uniquely prescribes each event within its domain. Perhaps the idea here is that a probabilistic law operates rather like a device that keeps track of the relevant past events and calculates what value the next event should have in order to maintain at each moment "the right" total probability distribution. The probabilistic law, on this account, would precisely specify each outcome; that is, it would function deterministically. It is not easy to see how one might formulate and defend such a position, but we do not need to pursue that question, since *no* defender of non-interventionist quantum divine action has adopted it, and no reason has been given to think they are implicitly committed to it. It appears that Saunders's critique, as Wildman elaborates it, depends on attributing to his opponents a strikingly peculiar view that they do not in fact hold.

---

[38] Ibid., 155.

[39] Wesley Wildman, "The Divine Action Project, 1988-2003," *Theology and Science*, Vol. 2, No. 1 (2004): 31-76, and in this volume. I think that Wildman's development of this idea produces an argument rather different than Saunders' own. Saunders claims that the concept of a "violation" of natural law presupposes a "necessitarian" conception of those laws. He then argues that the hypothesis of quantum divine action commits those who hold it to a "regularitarian," or neo-Humean, conception of causal laws. This makes it impossible to say that any event contradicts a natural law, and therefore it destroys the very distinction between interventionist and non-interventionist divine action. So a defender of quantum divine action starts out by assuming, and concludes by denying, a necessitarian view of natural law. See Saunders, *Divine Action and Modern Science,* 155. I comment briefly on the shortcomings of this argument in a review of Saunders in *Notre Dame Philosophical Reviews*, October 2003, published on-line at http://ndpr.nd.edu/review.cfm?id=1319.

It is surprising, therefore, to find Wildman making use of this view of natural law in his critique of attempts to construct a non-interventionist account of objectively special divine action.[40] He presents four claims that he says provide "the most demanding criterion for an adequate theory of SDA" and that express the "constraints that theories of SDA use to generate traction (as consistency) between science and theology."[41]

1. Objectivity: God undertakes objectively special action in the world.
2. Incompatibilism: non-interventionist objective special divine action is incompatible with causal determinism.
3. Noninterventionism: God may undertake objectively special divine actions without intervening (i.e., without violating any natural law).
4. The strong-ontological view of the laws of nature.

Since the strong-ontological view entails that every event is uniquely specified by the laws of nature, propositions (1)-(3) jointly entail the negation of (4), and this set of claims as a whole is straightforwardly self-contradictory. These four propositions, therefore, cannot possibly define "the criterion for success" of proposals regarding non-interventionist special divine action. Nor does the failure to affirm all four represent a "weakening" of such proposals;[42] it hardly needs to be said that a theory is not compromised by its inability to accomplish a logically impossible task. This tetralemma has force only against a position that unwittingly commits itself to all four of these propositions. But the proponents of non-interventionist objective special divine action do not make this flatfooted mistake. On the contrary, a recognition of the incoherence of these claims, and specifically the rejection of (4) and its deterministic kin, is a precondition for the project of exploring how we might conceive of objective special divine action.

Wildman assumes that it would be advantageous for a theory of divine action to adopt the strong-ontological interpretation of the laws of nature because this would "maximize traction" (that is, engagement and consistency) with the sciences.[43] There are at least two compelling reasons *not* to say this. First, this treats the strong-ontological view as though it has some special standing in science (e.g., either as a presupposition of scientific rationality or as a result of scientific research). But since this view involves a deeply problematic account of probabilistic law, it is not at all plausible as a founding principle that we abandon at the risk cutting our intellectual ties to natural

---

[40] Parts of the response I offer here to Wildman's argument also appear in an earlier reply to him. See my "Scientific Perspectives on Divine Action? Mapping the Options," *Theology and Science*, Vol. 2, No. 2 (2004): 196-201.

[41] Wildman, "The Divine Action Project," 57, 59.

[42] Ibid., 57.

[43] Ibid., 59. The metaphor of traction between theology and science was introduced by Philip Clayton into our discussions at the Capstone Conference. See Clayton's chapter, "Toward a Theory of Divine Action That Has Traction," in this volume.

science. Wildman strengthens the philosophical pedigree of the strong-ontological view by suggesting that it is an expression of Kant's causal principle. I think the two are in fact quite different, but suppose that we drop the strong-ontological view from the tetralemma and replace it with Kant's principle that every event must have sufficient natural causal conditions. This simply makes it explicit that defenders of non-interventionist objective special divine action must reject Kant's transcendental determinism. This comes as no surprise; the first move in opening up conceptual space for a proposal of this type is to challenge determinism. The *Critique of Pure Reason* was not handed down from heaven on tablets of stone (an exceedingly heavy book!); part of what is so fascinating about the interpretive puzzles presented by quantum mechanics is that they have led a significant proportion of the scientific community to reject the claim that causal determinism is necessarily presupposed in our efforts to understand the world. As Wildman himself notes, "Kant's transcendental philosophy has not survived the transition away from deterministic Newtonian physics as well as Kant might have wanted."[44] Once this is acknowledged, the tetralemma loses whatever force it might have had.

Second, if we refuse to privilege a deterministic interpretation of the sciences, then it becomes clear that "traction" can be maintained, and even increased, in a carefully formulated proposal about non-interventionist objective divine action at the quantum level. If we are interested in consistency with science, and if a leading interpretation of quantum mechanics is indeterministic, then an appropriately cautious use of this interpretation will in fact constitute a full-fledged engagement with science. Any theological proposal of this kind makes itself vulnerable to changes in physical theory and to reassessments of the relative validity of competing interpretations. As we noted above, if Bohm's determinism proved in the end more convincing than Heisenberg's indeterminism, then that would be a serious blow to non-interventionist approaches to divine action at the quantum level. Here we have a particularly vivid instance of traction; indeed, the riskiness of establishing links to a developing scientific discussion provides one of the reasons why many theologians draw back from such proposals and may retreat altogether from talk of particular divine action in the world. When theological speculation ties itself so intimately to the interpretation of a scientific theory, it is safest to offer the result as a tentative and exploratory thought experiment. Given the hazards of the enterprise, there is certainly no basis for the claim that this thought experiment backs away from full engagement with natural science.[45]

---

[44] Ibid., 58.

[45] Philip Clayton repeats this claim in his response to Wildman in "Wildman's Kantian Skepticism: A Rubicon for the Divine Action Debate," *Theology and Science*, Vol. 2, No. 2 (2004): 186-90. Clayton contends that Wildman presents a fundamental challenge to those of us who explore the possibilities for non-interventionist objectively special divine action. "You can't play the traction game half way. If you seek to develop a theology that is consistent with natural science, you cannot suddenly in the middle of your work make an appeal to faith to solve an argumentative problem or avoid a potential criticism" (189). This is right, of course, but it bears no relation to what is going on in proposals about God's action via underdetermination in nature. The only reason to think that such proposals give up consistency with science is if you

3.2.3 Amplification

We noted above that one of the necessary conditions for the viability of the idea of divine action through quantum chance is at least some of these transitions make the right sorts of differences in the development of events at the macroscopic level. If underdetermined transitions at the quantum level simply accumulate to form higher level regularities, then they will be part of God's general providence in establishing the lawful structures of nature, but they will not be of help in understanding special divine action that redirects the course of events in the world. On the other hand, if relatively small numbers of such events sometimes have significant macroscopic effects, then they may provide one of the routes by which God affects the course of events. The question here is a scientific one: are there naturally occurring structures (i.e., other than in our technologies) that amplify quantum events so that their effects register at the macroscopic level? There clearly are at least some significant structures of this sort, as Robert Russell, George Ellis, and others have pointed out. For example, the role of quantum events in genetic mutation, coupled with the amplification of these effects through natural selection, suggests that quantum events play an important role in the evolution of life.[46]

It appears, then, that quantum mechanics can reasonably be construed to meet all three conditions for enlistment in the theological project that seeks openings for non-miraculous special divine action in the world. This conclusion obviously hinges upon a number of unsettled questions, and it is important to offer any such proposal with a diffidence that reflects both the open-endedness of the science and the pluralistic uncertainty of the interpretation of that science. But if our best current physical theory (on at least one widely accepted interpretation) presents us with a description of the natural world as a causally open structure, then this needs to be taken into account, not only by those of us interested in particular divine action, but any theologian committed to reflection upon God's relation to the natural world. It is important to note that divine action at the quantum level will not be the most pervasive and profound way in which God acts, certainly not if we affirm that God is the creator *ex nihilo* who acts at the foundation of every event. Nor will this form of divine action provide an adequate account of some of the most important classical claims about particular acts of God; traditional doctrines of the incarnation and resurrection require something more robust and "miracle-like" than the orchestration of quantum events.[47] Nonetheless, divine action by

suppose that science must necessarily be deterministic. But Clayton himself rejects precisely that view in his response to Wildman, so there is no Rubicon to be crossed here.

[46] Russell, "Special Providence and Genetic Mutation," *EMB*; Alastair Rae, *Quantum Physics: Illusion or Reality?* (Cambridge: Cambridge University Press, 1986), 61. On other instances of amplification see Russell, "Divine Action and Quantum Mechanics," and George Ellis, "Quantum Theory and the Macroscopic World," both in *QM*.

[47] If we think of miracles in the Humean way as "violations of natural law," then the resurrection is far more than a miracle; it is rather the inauguration of a new creation that transforms and completes the old order. Robert Russell suggests that the

means of quantum chance provides one way in which God could affect the course of events without disruptive interventions in the structures of nature. The possibility of conceiving of special divine action in this way will depend on developments in science, but how seriously we take this possibility will be determined primarily by theological considerations of the sort I mentioned earlier (for example, by whether we judge it theologically problematic to involve God so deeply in physical events in the world).

3.3 Neuroscience and Cognitive Science

Our inquiry into causal openness in the structures of nature need not focus exclusively on the lowest levels in the hierarchy of organization; there may be flexibility at higher levels as well. The obvious place to look is in the lives of rational moral agents. The idea of responsive and personal relationship with God is deeply rooted in the theistic faiths, and this suggests that the mental (i.e., emotional, rational/reflective, moral, spiritual) life of persons will be one of the loci of divine action. For a theologian interested in non-miraculous divine action the question is whether we might understand God to affect mental life without disrupting the lawful structures of the natural world.[48] This is not simply a return to the familiar assumption in modern theology that we can side-step questions about divine action in the physical world by turning to the human self as the location of God's activity.[49] If we seriously grapple with our growing understanding of the intimate dependence of mental life on the body, then this path will be less appealing.[50] It appears that if God affects human persons, then God affects the human body.

The philosophical project of thinking about the place of rational and moral agency within the lawful structures of nature has been around for a long time. Contemporary neurobiology and cognitive science, coupled with analytic philosophy of mind and action (which has been inspired in part by these sciences), bring new data and conceptual precision to the discussion. Nonetheless, the ancient and essential puzzles remain about how phenomenal consciousness (the experiencing subject's point of view) arises from the physical organization of the body, and about how freedom of action is to be understood. I want to comment briefly on just one aspect of this discussion that is particularly relevant to the theological program we have been considering. If

---

resurrection might be understood as the "first instantiation of a new law of nature," in *Cosmology: From Alpha to Omega*, chapter 10, 309.

[48] Nancy S. Wiens has explored questions about spiritual discernment and divine action in, *Discernment and Nature: Exploring their Relationship through Christian Spirituality and the Natural Sciences* (unpublished Ph.D. dissertation, Graduate Theological Union, 2007).

[49] Rudolph Bultmann, with his neo-Kantian conception of a self that cannot be captured in "objectifying" scientific discourse, provides a notable example of this turn to the transcendental subject. See, for example, *Jesus Christ and Mythology* (New York: Scribner, 1958).

[50] It is important to note, however, that various forms of mind/body dualism continue to have skillful defenders. See Richard Swinburne, *The Evolution of the Soul*, rev. ed. (Oxford: Clarendon Press, 1997); Charles Taliaferro, *Consciousness and the Mind of God* (Cambridge: Cambridge University Press, 1994).

a case could be made that human thought and action is not strictly determined by its natural conditions, then this would provide a point at which we might understand God to act without disrupting the causal structures of nature. One family of views about human freedom, as we have already seen (in section 2.1.2), makes just such a claim; defenders of incompatibilist freedom hold that a necessary condition of free human action is that it not be causally determined by events prior to the agent's choice.[51]

Now it may be that quantum chance, appropriately located and amplified within neurological structures, provides the indeterministic physical correlate (and necessary condition) of free human deliberation and/or decision.[52] If so, then God's action in the mental lives of persons could be encompassed within the discussion of divine action at the quantum level. Such an account, however, would remain strictly "bottom-up," locating non-interventionist objective divine action solely at nature's lowest organizational levels. If God affects mental life by acting at higher levels, then we need to argue that new forms of causal openness (other than quantum chance) appear as more complex structures are built upon this base. We might seek to show, for example, that the lock step of causal determinism, which appears to be reestablished among classical objects, is interrupted in rational free agency. This would require that we formulate an incompatibilist account of freedom that is not dependent upon quantum indeterminacies, but that finds causal gaps at higher levels such as neurobiology.

This is a challenging project that faces significant initial skepticism from many scientists and philosophers. One way to begin to address these matters would be to consider the difficult question of the causal efficacy of mental states. We typically explain human behavior by reference to our own and other's beliefs, desires, preferences, values, plans, and so on. Do these mental states actually play any role in bringing about our actions? If mental states are realized in (or "supervene" upon) physical states, and if the underlying physical states constitute a closed causal series, it appears that the content of the mental states plays no role in the production of the action. Could it be convincingly argued that mental causation requires a denial of causal completeness of physical events?

A great deal of recent philosophical effort has gone into arguing that this is not the case. Two very clear and careful discussions of mental causation in *NP* (*viz.*, by Theo Meyering and Nancey Murphy) argue that we can uphold the causal relevance of mental states (considered in terms of their content as mental states, and not just their token identity with physical states) and also grant that the underlying (subvenient) physical states may form a closed,

---

[51] Part of the reason incompatibilist freedom continues to be so vigorously disputed is that it is difficult to give a satisfying account of the *sufficient* conditions of free action, namely, agent self-determination.

[52] Robert Kane makes an interesting proposal of this kind in *Free Will and Values* (Albany, N.Y.: SUNY Press, 1985). For a quite different approach see Roger Penrose, *The Emperor's New Mind: Concerning Computers, Minds, and the Laws of Physics* (New York: Penguin Books, 1989); Sir John Eccles, *Evolution of the Brain: Creation of the Self* (New York: Routledge, 1989); idem, *How the Self Controls its Brain* (New York: Springer-Verlag, 1994).

deterministic series. For Meyering, the affirmation of mental causation amounts to the claim that psychological (mentalistic) explanations identify causally relevant structures in the world that are not captured by explanations at any other level. He introduces a number of arguments for this failure of reduction, two of which we can note briefly. The first is the familiar argument from "multiple realizability." It appears possible in principle for mental states to be realized in a great a variety of different physical states (e.g., the belief that "some mushrooms are edible" might be realized in somewhat different neuronal pathways in different human brains, in the brains of members of other species, or perhaps in the electrical states of a sufficiently complex computer). As a result, the mental predicate picks out features of the world that are not captured by any single physical state, including the one that happens in this instance to realize the mental predicate. It would require a potentially massive and messy disjunction to express all and only the physical states that are realizations of this mental predicate. Second, Meyering makes an argument from "multiple supervenience." Let us suppose for the moment that a mental state is a functional property of a physical system mapping inputs into outputs. Not only can a single functional property be realized in various physical systems, a physical system can realize multiple functional properties (each activated by different inputs into the system). As a result, a description of the physical system alone cannot do the work performed by the functional explanation, since such a description will not pick out the explanatorily relevant causal pathway. In turning from explanation at the higher (mentalistic) level to one offered at the lower (micro-physical) level, we lose crucial information about the causal structure of the world. Yet, as Meyering points out, this explanatory nonreducibility of the mental is compatible with complete causal closure at the physical level.[53]

Murphy offers a related and especially helpful argument for the causal role of mental states.[54] She notes that standard accounts of the supervenience of the mental on the physical ignore the context, or circumstances, of the supervenience relation. Whether a particular mental state is realized by a particular physical state will typically depend upon the causal history through which that physical state arises and comes to bear its functional role. Lower-level physical processes may be subject to "downward causation," that is, to various constraints or selective pressures from the environment and from higher levels in the system. Using this insight, Murphy generates an illuminating model for the formation of brain structures that are the physical realization of rational relationships. Let us hypothesize, as some neuroscientists do, that learning involves a selective development of some neurological pathways in preference to others. The feedback loops involved in, say, learning the multiplication table will select for those neurological processes that constitute giving correct answers. When we offer an explanation of those brain processes, therefore, it will not be sufficient just to describe causal relations at the micro-physical level. We must also cite the role of rational norms, in this case the rules of arithmetic, in the processes by which those physical pathways were structured and "edited." As a result, "causal

[53] Meyering, "Mind Matters."

[54] Murphy, "Supervenience and the Downward Efficacy of the Mental."

closure at the neurobiological level does not conflict with giving an account at the mental level in terms of reasons."[55]

Both of these accounts cogently make the case that mental causation is consistent with physical determinism, and they thereby undercut the claim that rational agency requires a new indeterminism at the level of mental life. Can a case be made for this stronger claim? As I read Philip Clayton, he is moving in this direction as he develops a position he calls "emergentist monism." Murphy indicates that this as the equivalent of her own "nonreductive physicalism," but I think that Clayton's view is in fact significantly different from hers.[56] Physicalism, as Meyering puts it, is "roughly the view that the stuff that physics talks about is all the stuff there is."[57] Clayton says something similar. "Monism asserts that only one kind of thing exists. There are not two substances in the world with essentially different natures."[58] Both positions, then, affirm that physics describes the basic entities and relationships upon which all higher level structures depend. Further, both nonreductive physicalism and emergentist monism insist on the irreducibility of mental explanations and of the causal patterns those explanations identify. Clayton makes the additional claim, however, that at the mental level new causal powers emerge that bring about effects which do not have an adequate physical explanation. That is, once mind emerges from its physical basis, it can initiate events that have necessary but not sufficient conditions at the physical level. "A study of the emergent person is genuinely holistic, however, only if it retains a place for speaking of one higher-order event (for example, a thought or *quale*) causing another without insisting that the whole story can be told in terms of neuronal firings."[59] This appears to affirm mental causation in a stronger sense than that discussed by Meyering and Murphy; let us call their view MC1 and Clayton's MC2. MC1 can be expressed within a nonreductive but (macro)deterministic physicalism, and MC2 cannot, since it requires a distinctive indeterminism as a correlate of at least some thought and action.

MC2 is a metaphysically adventurous position, and its exact contours are not entirely clear in Clayton's *NP* essay. He commits himself to defending the "insufficiency thesis," *viz*., that "neuroscience will *not* be sufficient to explain all we come to know about the human person...There are parts of what it is to be a person that lie in principle beyond their reach."[60] This can be read as asserting either MC1 or MC2, and a number of the arguments he offers on behalf of the insufficiency thesis advance only the weaker claim of MC1 about the irreducibility of mental cause explanations. It seems clear, however, that he is aiming at more than this. He comments that nonreductive physicalism "appears to leave no room for genuine mental causes; all the determination of outcomes seems to flow from the bottom (the physical substratum), leaving no

---

[55] Ibid., 160.

[56] Ibid., 147, no. 1.

[57] Meyering, "Mind Matters," 165.

[58] Philip Clayton, "Neuroscience, the Person, and God," in *NP*, 209.

[59] Ibid., 194.

[60] Ibid., 188.

"room for play" for the mental actually to do anything."[61] For this reason he favors a position he calls weak supervenience, which acknowledges the dependence of the mental on the physical but which also allows that "there are genuine mental causes that are not themselves the product of physical causes. The causal history of the mental cannot be told in physical terms, and the outcome of mental events is not determined by phenomena at the physical level alone."[62] This holds out for a robust concept of mental causation as a power of action that arises from the net of physical causation but which can act back upon it. The insufficiency thesis boldly predicts that one of the things neuroscience will discover is that it cannot give a fully deterministic account of the physical processes that underlie human thought and action.

We have, then, two significantly different accounts of mental causation. There is much yet to be done in articulating and defending these two positions.[63] But for our purposes in considering non-interventionist objective divine action, the key point is that only the second view offers a new form of underdetermination in nature. Because the first approach seeks to formulate a theory of mental causation that is compatible with physical determinism, it yields the familiar alternatives for divine action. God (1) sustains the finite moral agent in being, and (2) brings about each development in the agent's life by means of the deterministic operation of natural causes. If God also (3) acts to affect the agent's thoughts or feelings or desires once the world's history is underway, then this will involve intervening in a causal order that would otherwise be closed at the levels that are relevant to human mental life. Unless this account is supplemented in some way, it entails that objectively special divine action in our "inner" lives will involve miracles, though they may be hidden in the untraceable complexity of neurophysiology.[64] The second view, on the other hand, claims that a new indeterminism emerges at the level of human thought and action, and this provides a distinctive locus for non-interventionist objective divine action. Views of this kind, however, present deep puzzles about how mental causes arise from and come to possess the power to act back upon their physical underpinnings. The difficulties here are substantial, and skepticism about this position runs deep in contemporary neuroscience and philosophy of mind. Without this stronger version of mental

---

[61] Ibid., 200.

[62] Ibid.

[63] Some of the recent philosophical literature on free agency promises to be helpful here. See, for example, Timothy O'Connor, ed., *Agents, Causes, and Events: Essays on Indeterminism and Free Will* (Oxford: Oxford University Press, 1999); William Hasker, *The Emergent Self* (Ithaca, N.Y.: Cornell University Press, 1999). Hasker defends a position he calls "emergent dualism."

[64] While this account of mental causation aims at compatibility with neurophysiological determinism, it need not affirm the *truth* of such determinism. Perhaps it will turn out that mental states are realized in physical processes that at certain points are underdetermined by natural causes (e.g., because quantum transitions are involved). MC1 is committed only to claiming that causal openness of this sort is *not required* in order to make sense of mental causation, e.g., of rational deliberation and intentional action. If underdetermined events play some role in mental life, then they would provide an opportunity for non-interventionist divine action.

causation, however, there appears to be no basis for claiming that a distinctive indeterminism makes its appearance in rational free agency.

4 Where Do We Go from Here?

This discussion has generated a complex agenda of issues that resists any simple summary. Let me briefly list, under three headings, just a few of these issues for continuing research. First, there are, of course, scientific issues that need attention. Insofar as our theological proposals are tied to particular sciences, new theoretical developments in those sciences will be of crucial importance. Each of the approaches to special divine action that I have discussed faces specific pressure points at which it is particularly vulnerable. By identifying these points of critical dependence on physical theory, we can set an agenda for continuing work along the boundaries of science and religion. The case of theological appeals to quantum mechanics is particularly vivid. As I have assessed the options, quantum mechanics holds out the greatest promise of being helpful to the theological project of constructing an account of how God might act within the course of events in the world without disrupting the structures of nature. But the possibility of conceiving of special divine action in this way depends upon a number of unsettled but fundamental questions in quantum theory and its interpretation. This theological proposal will need to pay continuing attention, for example, to new approaches to the measurement problem, to efforts to understand the relation between quantum mechanics and chaos theory, and to the ongoing give and take between alternative interpretative strategies.[65] Similarly, claims about part-whole or top-down divine action can be enriched by deepening scientific understanding of the operation of complex hierarchical and self-organizing systems.

Second, there are some crucial issues in metaphysics on the agenda. The discussion has raised questions about what might be called metaphysical method (e.g., systematic versus "*ad hoc*" metaphysics) as well as about which metaphysical categories are most useful in the conversation with the sciences (e.g., some version of the Thomistic scheme or of process metaphysics or of contemporary physicalism). There are, in addition, specific topics in metaphysics that need sustained attention: for example, (1) questions about causality generally and ideas of "downward causation" and "part-whole causation" in particular; (2) questions about the causal efficacy of the mental; (3) questions about the nature of free will and the place of agent self-determination within the causal structures of the world; (4) questions about concepts of reduction, supervenience, and emergence. These are among the

---

[65] Robert Russell included a helpful appendix, "Directions for Future Research," in "Divine Action and Quantum Mechanics," *QM*, 324-8. Russell sketches an "architecture of philosophical problems" in the interpretation of quantum mechanics, identifying issues (e.g., superposition and nonlocality) that arise for all current interpretations, both deterministic and indeterministic. There are aspects of quantum mechanics that may be theologically relevant but which await fuller exploration; for example, in his doctoral dissertation Kirk Wegter-McNelly has worked on the theological significance of nonlocality and entanglement. See *Created Wholeness: The Significance of Quantum Nonlocality for the Christian Doctrine of Creation*, Ph.D. dissertation, Graduate Theological Union, Berkeley, California.

most long-standing and thorny issues in philosophy, and neat solutions are not to be expected, but additional focused attention to them will help to clarify and advance our theological projects.

Third, there are underlying theological issues that need continuing discussion. The project of looking for openings in nature for special divine action faces objections from a historically well-established and conceptually rich tradition in theology that denies that such openings are needed. The disagreements here turn on classical questions about God's attributes and about God's relation to creatures as their creator. Attention to this debate helps to identify the particular theological concerns (e.g., about responsive divine action) that motivate the search for openness in the structures of nature. Further, in giving an account of special divine action within the structures of nature, our differing approaches reflect underlying judgments on precisely these classical questions about how to speak appropriately of God (e.g., as an agent, yet not just one agent among others), about God's powers of action, about the scope of God's knowledge (especially of future contingent events and the actions of free creatures), and about God's relation to time. Finally, the problem of evil arises forcefully in this discussion, since to the extent we make progress in giving an account of God's activity within the world's unfolding history, we may find that we deepen our perplexity about the place within that history of suffering, death, and susceptibility to moral failure. The proposals we make are shaped as deeply by these ancient theological topics as they are by the natural sciences. In fact, theological concerns help to motivate the positions that participants in these discussions take on disputed questions in the interpretation of science, with the result that scientific debates sometimes mask underlying theological disagreements. It is important, therefore, to bring the theological issues into the foreground and address them explicitly.

Acknowledgement: I want to thank the conference participants for their helpful responses to this essay. A special word of thanks is due Robert Russell for a very careful and perceptive reading that led to significant improvements.

DIVINE ACTION IN AN EMERGENT COSMOS

Keith Ward

The topic of divine action is one of the most important issues for theology in its relation to science. How can we conceive of purposive divine action in a universe that seems to many scientists to be governed entirely by universal, mathematically expressible laws of nature?

This is not just a problem about God. In a culture deeply influenced by the success of the natural sciences we also find it difficult to account for the possibility of free and morally responsible human acts. We have not yet recovered from Immanuel Kant's desperate division of reality into a deterministic phenomenal world and a free noumenal world—a division that makes every human act both perfectly free and wholly determined at the same time.

This proposal does not seem acceptable to many, but it is very difficult to see how free action, human or divine, can be rendered consistent with a deterministic account of physical laws. One possibility is to give up such a deterministic account. In a non-deterministic cosmos, which is dynamic and emergent, divine causality can be seen as integrating in a plausible way with a creative drive towards realizing personal values.

God's acts will be guiding and influencing factors, normally operating within the probabilistic limits of physical law, but occasionally transcending the regular powers of objects for reasons of religious significance. They will not be interferences in a mechanical system, but causally operative factors in the Universe's emergent generation of value.

In what follows, I shall explore this possibility, though within this series of conferences the question of how far scientific laws are deterministic, and in what sense, remained disputed, and certainly more work remains to be done in this area.

1 Free Action and Indeterminism

In English law there are three rules, known as the McNaughten rules, for determining when an agent is responsible. The agent must know the difference between right and wrong, must know that what he or she is doing is right or wrong, and must be able to do otherwise.

The sting is in the final rule, and there is much philosophical debate about what 'being able to do otherwise' amounts to. One interpretation is that in the same causal conditions, all prior causes and laws of nature being the same, at least two different actions could ensue. The agent knows some of the alternatives available, and simply intends to realize one of them, when the agent could have intended otherwise.

The intention is a new causal factor that is not itself sufficiently caused, though of course it has many influencing causes. As I shall use the terms, *sufficient cause* is such that, if it exists, then there is one and only one effect that will necessarily follow. A *necessary cause* is such that, if an effect exists, then the cause necessarily existed. An *influencing cause* is neither necessary nor sufficient, but together with other causal factors it in fact brings about an effect, though it need not have, and the effect could have been brought about in other ways.

I assume that every event has a great number of influencing causes. Every event probably has a number of necessary causes, though it is difficult strictly to establish their necessity. Whether events have any sufficient causes is disputable. But it seems reasonable to think that many events or processes, like the movements of the planets, do have sufficient causes, at least in a weak sense of 'necessity,' meaning something like, 'necessary given the laws of nature and the absence of other possible causal factors.'

On an indeterministic view of freedom, intentional acts are not events that have no cause; they are events that have no sufficient cause. That is, we intend something because of what we think, what we tend to desire in general, and how we are influenced by others—these are all influencing causes. But none of those factors, taken singly or together, entails our intention, which is simply a matter of free decision. An instance of a morally significant choice would be a situation in which we can intend either to seek our own satisfaction or to bring about a state of affairs because we believe it is good in itself. We are morally responsible if it is in our power to do either of these things.

If this indeterministic view of freedom is correct, then at least on some occasions (not on all occasions) the laws of nature must permit such free choice. So it cannot be the case that the laws plus an initial state always entail some specific outcome. They may entail a limited number of possible outcomes (I can raise my arm, but not fly out of the window). A free action is one where an agent intends some state to come about, and where the laws of nature limit what may come about, but do not entail one specific outcome.

I think this is the commonsense view of human action. I intend to do something, and my intention brings about states in the physical world that would not otherwise have existed. Physical laws on their own therefore do not give a complete account of all that happens, for intentional action introduces causal factors that modify the regular operation of physical laws. If this is so in the case of humans, it is also likely to be so with God.

Despite being a natural commonsense view, this is philosophically controversial. It is contested especially by those who would identify intentions with brain states, and subsume changes of brain state under deterministic physical laws.

I think that most participants in this series of discussions would favor some form of identity between conscious states and brain states. As Philip Clayton puts it, the "dualist, individualist, cognitivist view of the person needs revision." Conscious states are something like brain states as experienced by the embodied brain itself. On the other hand, reductionist explanations were eschewed by most participants, because there are emergent properties of complex systems that cannot be understood by knowledge only of the simpler elements of which such systems are composed. Intentional acts therefore do not modify the operation of physical laws. It is rather that physical laws get much more complex when they deal with complex and structured systems. If this is so, physical laws do not need to be indeterministic, leaving room for a specific sort of mental causality. Yet new sorts of physical laws will come into play when complex physical systems exist. Free human acts, on this view, do not entail either immaterial causality or physical indeterminism—though a major problem remains about stating what sort of causality is involved in such emergent phenomena. For my own part, I believe this problem, and the problem of specifying the sort of identity that exists between mental and brain states, is worse than the problems involved in positing causally effective non-material minds. But mine is a minority report!

2 Physical Laws and Indeterminism

Whatever may be the case with human acts, I think that theists are committed to the ontological and causal priority of immaterial consciousness, since they usually believe there is an immaterial consciousness, God, who is the cause of the whole material universe. God knows what states could be brought about, has the power to bring them about, and brings some states about for the best possible reason, namely their intrinsic goodness or desirability (though possibly this will entail also bringing about states that are necessary or at least non-preventable conditions or consequences of such good states).

It is obvious that God does not create the cosmos in accordance with any physical law, since God creates all physical laws. And it is clear that in this case intentions do bring about physical states, and explain why such states exist (though humans may not have access to such explanations, since they do not have access to the mind of God). So theists are committed to at least one case of immaterial and indeterministic action, and to affirming the incompleteness of any explanation of why things occur purely in terms of physical laws.

Just because there is at least one divine act (the act of creation) that introduces an immaterial cause and is non-deterministic, it does not follow that there are particular instances of immaterial causality within the processes of the universe, or that physical laws are in principle non-deterministic. It does follow, however, that these are logical possibilities, and whether or not they are actualized is a matter of fact, to be decided by observation or argument.

The evidence of the natural sciences on this question is hardly decisive. Suppose I take the deterministic view that every event is sufficiently caused by prior events plus the laws of nature. That commits me to saying that no event could be other than it is, that no other outcome than the one that occurs is possible, and that every state (where this is specified accurately and comprehensively enough) entails one and only one succeeding state.

But how could I ever come to know such things? Even if I successfully predict everything that happens in a series of events, I cannot rule out the possibility that something else could have happened. As David Hume pointed out, there is no logical entailment between physical events, so in what sense can one event entail another? We might say that the existence of a law of physics forces one event to succeed another. But in what sense do laws of physics exist, that they might 'force' any event to exist at all? Even if they do exist, might they not easily change?[1]

---

[1] Cosmologists like Lee Smolin claim that they have changed. Cf. Smolin, *The Life of the Cosmos* (London: Weidenfeld and Nicolson, 1997), esp. chap. 6.

William Stoeger, in the first set of conference papers, suggests that laws are primarily descriptive, and are "idealized models" of how things regularly happen, when they are left alone. Or as John Polkinghorne puts it, "laws of physics are an asymptotic approximation" to physical processes. Quantum physics strengthens these conceptual points by claiming that different outcomes (different measurements) can follow from the same initial quantum state (wave function). Whether or not this rules out determinism is highly disputed, and David Bohm's deterministic view had strong defenders in this symposium. But a conceptual scheme of wave functions, superposition and nonlocality is very different from one of mechanistic atomism, and it is at least agreed that determinism is not a necessary condition of scientific methodology any longer. That is to say, if the Copenhagen interpretation of quantum theory is true, and there are objective indeterminacies in nature, scientific theory can still proceed without any difficulty, using probabilistic calculus.

A probabilistic representation of what quantum theorist Bernard d'Espagnat calls a 'veiled reality' throws doubt on what some have called the reality criterion—that every physical element must have a theoretical counterpart, so that physics maps the world in a one-one correspondence.[2] But it does not undermine the amazing predictive accuracy of physical theory. Our mathematical representations do mirror nature in important ways, but, as d'Espagnat states, they may not give complete access to the objective causal powers of nature.

Wesley Wildman and Robert Russell, at the second conference, argued that practical limits to control of the environment, inability to obtain totally precise initial conditions and inability to cope with the number and speed of the required calculations, ensure that we cannot establish the truth of determinism, or guarantee that there is a closed causal web which rules out divine action as one causal influence among others.

The exasperated physicist may say that events simply do obey general laws, and there is no point considering merely abstract possibilities. However, there is a point, and it is this: if a physicist claims that "science shows" that every event is sufficiently determined, or that indeterministic free-will cannot exist, since there is no room for it, we need to ask in what way such a thing has really been, or could ever be, shown by experiment or observation.

3 The Case for Indeterminism

Present scientific knowledge does not, and perhaps cannot, rule out indeterminism. This decision remains an open question, on present scientific evidence, and so may properly be based on other considerations. In my view, there are two main considerations favoring indeterminism—and let me restate that I do not mean quantum indeterminacy, but the thesis that initial physical states and general physical laws do not always sufficiently determine subsequent physical states.

One consideration rests on observation of one's own continuity and agency as an agent capable of choosing possible futures on many occasions.

---

[2] Bernard d'Espagnat, *Reality and the Physicist* (New York: Cambridge University Press, 1990), 175.

This sense of free choice between alternatives is a primary datum of observation, and should be accepted as such in the absence of decisive arguments against doing so. The other is that the attribution of praise, blame and moral responsibility presupposes that moral agents could have done otherwise, in a strong and realistic sense. In the absence of overwhelming counter-evidence from the natural sciences, these considerations are strong enough to induce me to adopt scientific indeterminism.

One of the features of the conference on quantum physics is the huge range of divergences between quantum physicists about the ontological implications of their work. Views range from some form of idealism (Eugene Wigner), for which observers play a constitutive part in constructing reality, to forms of objective determinism (David Bohm), for which there exists in principle a formalism which will correspond in detail to real objective features of reality. Where there is such wide disagreement among competent practitioners about what objective physical states our theories are describing, we are clearly very far from being able to say whether it is even possible to give a complete and correct description of an objective physical state.

What the physical sciences actually provide is a very high reliability of predictability in controlled or closed causal systems, together with a strongly warranted hypothesis that such closed systems are deterministic, and a much more speculative generalization that the more uncontrolled and open physical systems in the natural world are likely to be deterministic too.

However, an equally plausible hypothesis is that while stateable physical laws completely govern the behavior of physical particles and states in controlled or isolated systems, when those systems interact with other systems or with wider environments of types that cannot be exhaustively specified in advance, the operation of the known set of physical laws alone does not completely govern physical behavior. There can be causal influences that affect physical behavior that arise from outside any particular specified system (for example, how does dark matter, until quite recently unsuspected, causally affect the observable universe?).

Among those causal influences could be the intentional acts of agents, including God. If God acts, God brings about a physical state in order to achieve some divine intention. It is scarcely credible that God's intentions should always coincide, by a sort of pre-established harmony, with what a set of natural laws alone entails.

It is possible that at least sometimes God's acts should bring about non-physically-determined states of nature, that God should exercise a causal influence on how things go, an influence which cannot be expressed by any purely physical law. At first it looks as though we could conduct experiments—to see, for example, if supernatural causality exists in the universe. But on closer inspection the possibility of such experiments largely disappears. We would have to measure with absolute precision a given physical state, ensure that it was not interacting with any other physical state, and detect every possible sort of causal interaction that could cause a physical change. Both the exact measurement of all physical quantities and forces (can we be sure we have exhaustively specified them?) and the rigorous exclusion of extra-systematic factors are impossible for a great many physical systems, including the human brain, which is in continual interaction with a much wider and experimentally uncontrollable environment in ordinary life situations. So what the natural sciences give is a widely generalized hypothesis from the success of predictions in controlled and closed conditions to the uncontrolled and open conditions of ordinary life. This generalization is not experimentally verified and perhaps not even realistically testable. It cannot therefore be said to be well established.

The first, second and fifth conferences all raised the question of whether the physical world is governed by a deterministic set of inviolable laws, which provides a sufficient explanation for all that happens. There was no agreed answer to this question, but modern physics at least presents a real possibility that natural laws are asymptotic, idealized models, often probabilistic, and not capable of capturing exhaustively the physical features of the cosmos. Since the rise of experimental science, the idea of law has replaced that of direct intentional agency (by God, angels or demons), for all normal physical explanations. But it remains an open question whether there are forms of causal influence not captured by the law-like descriptions of the physical sciences.

4 The God of the Gaps

Does a non-deterministic view of natural laws return us to a view of God who only acts in the gaps in the structure of nature? Not if God is the creator of the laws and the structure of the universe as a whole. The most basic action of God is that of causing the whole universe to exist, with the laws that it has. So in a sense God is causally active, in the most important way possible, at every time. Yet does God not also act in particular ways within the structure of the universe? And is God not then confined to acting in the gaps that non-sufficient causal laws offer? The term 'gap' belongs to the old Newtonian mechanistic world view.[3] It is as though there could be a wholly deterministic explanation, but it does not quite exist, and the deficiency must be made up from elsewhere. If one gives up the hankering for deterministic explanations, one might see the universe as open to contingent novelty at many points. These are not gaps in explanation, but contingent possibilities inherent in an open and emergent universe.

Such possibilities can be actualised either by chance or by human or divine intention. What is needed is not a God who can fill the gaps in a closed deterministic universe, but a God who can influence the actualization of the future in an open emergent universe. This is not a 'God of the gaps,' but a God of creative possibility, a God who exerts a continuing influence on an emergent future.

Ilya Prigogene spoke of the "re-enchantment of nature," in the final chapter of the book, *Order Out of Chaos*, which he wrote with Isabelle Stengers.[4] This is based on Prigogene's work on unstable dynamical systems far from equilibrium, which, he says, suggests that time is not, as classical

---

[3] The phrase 'God of the gaps' was first used by the Cambridge mathematician C. A. Coulson in *Science and Christian Belief* (London: Fontana, 1958), 41.

[4] I. Prigogene and I. Stengers, *Order Out of Chaos* (Toronto, Canada: Bantam, 1988).

physicists sometimes thought, an irrelevant factor in the operation of physical laws. Rather, as Prigogene and Stengers put it, "Randomness and irreversibility play an ever-increasing role. Science is rediscovering time."[5] In nature there is genuine becoming and room for free creativity and the sort of directionality that may well speak of purpose. There is a sensitivity to context and a seemingly non-deterministic, and certainly non-predictable, character to basic physical processes that suggests they have a more holistic and open character than classical physics allowed. Thus nature may be less like a deterministic machine than it is like an emergent and dynamic process, with many possibilities of development and creative novelty.

Prigogene's claim that his discoveries vindicate indeterminism remains controversial, and in particular the view that chaos theory somehow suggests indeterminism is not widely accepted. Yet he shows that good science can work with an indeterministic view, and that if the flow of time, the phenomena of emergence, and the possibility of creative novelty in nature are taken seriously, it is possible and plausible to see God as playing a causal role in the emergent development of a partly open future.

The postulation of such a God does not introduce some nonphysical agent as part of a scientific explanation. The original expression, "God of the gaps," referred to Newton's postulate that God would have to adjust the orbits of the planets very occasionally, to stop them falling into the sun. On his view, the laws would not account for the stable orbits of the planets, but needed some non-physical force to supplement them. It was this hypothesis of which LaPlace's more exact mathematical formulation had no need.

Divine action does not, however, depend upon there being any gaps in the regular processes of nature. It provides a teleological explanation of the structure of nature that, however plausible it may seem, may always be denied or ignored without contradiction.

If God creates the universe for a purpose, reference to divine intention will be an essential part of the explanation of why the universe exists. But it will not be part of a scientific explanation, and so it can be ignored by explanations in terms of physical laws. After all, a non-embodied agent cannot be seen to be acting, so one may always deny that any agent is involved, who would be able to make states exist for the sake of the values they express or make possible. Such values can be interpreted simply as natural consequences of non-purposive physical laws, or as chance occurrences that may appear to violate the laws of nature, though they do not really do so. Even their value may be seen as a purely subjective evaluation of such states, made by human beings. So it seems that teleological explanation, in terms of God's action in the world, can always be denied.

However, if one has a long and complex process that culminates in states of great value, it is also possible that it has been brought about intentionally. The main relevant considerations in deciding whether this is so are: whether the process is efficiently or elegantly ordered, whether it does result in states of great value, and whether the process could, so far as we can see, easily have been otherwise. Perhaps the process could have been much less efficient, or it could have been an inevitable result of the operation of physical laws alone, or

[5] Ibid., xxviii.

the consequences could have been of much less value. If one has a contingent process that efficiently issues in states of great value, there is quite a high probability that it is the result of an intention, that it comprises an action or series of actions.

It is not easy to decide whether or not the universe is a divine action, on these criteria. There is evidence of an amazing mathematical elegance and integration in the structure of the basic laws of nature, yet there also seems to be a great deal of wastefulness and redundancy in natural processes that one might imagine being eliminated by a wise creator. It is not clear whether the course of evolution is necessary, or whether it could easily have been very different. And while the universe does manifest very great value, not least in sentient experiences of beauty, understanding, love and creativity, yet there is also a great deal of suffering and pain that is difficult to account for, if the process is purposive.

The picture of the universe given by contemporary science is thus rather ambiguous, with regard to divine action. It may seem to suggest the existence of a being of great wisdom and power, who is concerned for the realization of certain sorts of value, but who seems to be relatively indifferent to the suffering of individuals. It is not surprising that physicists and mathematicians are often sympathetic to the idea of an intelligible source of the universe that has immense wisdom and power, but tend to be unmoved by religious pictures of a heavenly Father who loves all his children.

5 Humanity and Cosmic Purpose

The plausibility of theism largely turns on its ability to account for the seeming wastefulness and suffering involved in the emergent processes of reality, to produce some account of a divine purpose that would make such processes intelligible.

One way of doing this is to see how much of the seeming wastefulness of evolution is in fact an inevitable outcome of the probabilistic nature of its basic laws, which is in turn a necessary condition of the sort of relative autonomy that creatures possess. Probability may be a necessary condition of the emergence of indeterminist freedom, and if so it must be seen as desirable. If God sets up a probabilistic system of laws, God cannot at the same time always guide events to their best possible outcome, since that would eliminate probability. In the course of evolution, the existence of probabilistic laws can ensure that a desired outcome results eventually. But many undesirable or apparently wasteful states will also result.

What is the desirable outcome at which a creator God might aim? Traditionally, the main purpose of God has been seen as leading humans to knowledge and love of the Creator. But can we any longer see human persons as central to the existence of the cosmos?

The third and fourth conferences focus on this question. In the third conference, Camilo Cela-Conde says, "no straight line can be drawn from our ancestors to the modern human species."[6] This may suggest that humans cannot be seen as the intended consequence of the evolutionary process. They

---

[6] Camilo J. Cela-Conde, "The Hominid Evolutionary Journey," in *EMB*, 59.

are the products of random mutation, of (from their point of view) fortuitous accidents. On the other hand, evolution is not wholly random. It is governed by physical laws, which may make the emergence of life virtually inevitable, as Simon Conway Morris argues.[7] Yet there are undeniably many false trails and much waste in the process. So William Paley's benevolent and designer God, concerned always that all things in nature should be for the best, does seem implausible. Arthur Peacocke and John Haught suggest that the evolutionary process is one instance of the emergence of complexity, "characterized by propensities towards increase in complexity, information—processing and storage—consciousness, sensitivity to pain, and even self-consciousness."[8] It is as if there is an insistent drive to novelty and complexity, but no divine providential direction of every specific event.

Haught suggests that this implies a vulnerable, loving, participatory God, a God who is self-giving, persuasive to emergence, protective of autonomy, oriented to the future. From an evolutionary perspective, God may be seen not so much as an all-determining sovereign, but as the ideal that invites, the energy that drives, the mathematical wisdom that orders the cosmos. If the universe creatively advances into novelty, conflict and failure will be ineliminable from the process. For such a view, where nature is given a much greater self-organizing autonomy, God is not the wholly free all-determiner, but a rational orderer and persuader, the co-operative shaper of the cosmos rather than its benevolent dictator.

The human species may still be the reason for the creation of such a cosmos, since even on a probabilistic view of physical laws, the existence of humans is probably predictable by any omniscient being. And the conditions for life are so very improbable that life may exist only once even in an incredibly large universe. In that case, humans could move back to a central position in the cosmos, however large it all is.

If, on the other hand, humans are a contingent result of the general drive to emergent complexity, they are unlikely to be a unique instantiation of divine purpose. In that case the cosmos could still be oriented to the existence of personal life forms, though humans would be only one of those forms.

From a cosmic viewpoint, the purpose of the universe could thus be to bring into existence forms of personal life that are capable of contemplating beauty and truth, and of relating to one another and to their creator in knowledge and love. These created persons are evolved from simple material elements, and are points at which the spiritual potential of the physical order begins to be consciously realized. It is within their power to continue the cosmic process of spiritual transformation by attending to and realizing the good. But they also have the power to obstruct that process by giving way to self-regarding desires and ignoring the good of others and of their world.

---

[7] Simon Conway Morris, *Life's Solution* (New York: Cambridge University Press, 2003).

[8] Arthur Peacocke, "Biological Evolution—A Positive Theological Appraisal," in *EMB*, 367.

6 Divine Creation and Divine Intention

Suppose God the creator intends to create finite beings who have this responsibility for shaping their part of the cosmos, who have the power either to choose selfish or altruistic goods, either to love creation and its creator or to ignore them. If God intends to give them that power, then God might want them to be altruistic, but God has now put it beyond even the divine power to intend that. God can wish or command it, or forbid self-regarding choice, but even God cannot intend to do what it is not in the divine power to do, because God has decided to give that power to another. So if human beings choose self-regard, God must accept that as an unintended consequence of having intended to create beings with the power of free choice. That is how there can be things in the universe that God does not intend, and even forbids.

It follows that the free acts of creatures are not, properly speaking, acts of God. God has delegated power to them, and God's act can consist in creating them, keeping them in existence, and maintaining their power of choice. God can command them to do right, but cannot compel them to do so, if God creates them to be free. So it is that many things in our world are contrary to God's will, and are not intended by God, since they are due to the selfish acts of creatures.

Another way in which God may create events that God does not intend, but the possibility of which is entailed by what God does intend, can be brought out by analogy.

Suppose I intend to climb Mount Everest. I may do so for the sake of the values involved, which will include my courage and endurance, my capacity to accomplish something difficult, my feelings of success in overcoming obstacles, and the sheer thrill of climbing dangerous cliffs. None of these values would be possible if there were no hardships, danger and difficulties. But if there are dangers, then there is the possibility that I may fail in my attempt. Only if I could fail will my success be worthwhile to me. Suppose, then, that I set out but do fail. It could not be said that I intended to fail, with all the disappointment that involves. Yet failure was always a possibility entailed by what I did intend. So if God intends to create values that involve risk, courage, conflict and endurance, there will be some events that God does not intend, but that God cannot logically prevent.

This is also going to be the case if God creates a world that is largely governed by intelligible laws of nature. This may be to enable creatures to predict things that normally happen, to be able to exercise responsible choice in a relatively neutral environment, or to manifest the supreme elegance and beauty of an intelligibly ordered universe. In this case, God may intend there to be general laws, without intending every specific event that falls under such laws.

For example, God intends there to be a law of gravity. Under this law, if someone falls from a high place, she may be injured or killed. It must be the case that the value of having such a law outweighs the evil of that death (perhaps given the addition of some religious axioms, like the possibility of eternal life for that person), or God could not set up such a system. Nevertheless it would be quite untrue to say that God intends that person to die. Perhaps God did not want them to be in that place, or perhaps God foresaw that such things would happen, did not want them to happen, but could not prevent them without changing the laws, which would contradict the divine intention in creating a law-like world.

In a law-governed universe, God cannot be said to intend every specific event that happens, even though God intends that the universe should be as it is, in general. Biblical writers sometimes tended to interpret every event as a directly intended act of God, though they usually saw that God's intentions apply to the general structure of nature, and not to every individual occurrence. The laws must be such that most occurrences that fall under them must be of distinctive value. But it does not follow that all of them will be.

7 Particular Acts of God

Both a law-like universe and creaturely freedom thus limit the scope of those things that can properly be called acts of God. But neither entails that God never acts through direct intention, and it would be very odd if a God of *chesed,* of loving kindness and a concern for moral conduct never did so. There must be a place in the universe for specific, directly intended divine acts. What can be said about such particular acts of God?

On the view I am taking of responsible human freedom, there must be regularities of nature. These may be construed as normal exercises of the powers of objects, causally interacting with one another. That such regular interactions continue is itself an expression of the providence of God, who alone can establish that necessity of connection that Hume rightly failed to find in the natural powers of objects themselves. If responsible human freedom is indeterministic, there must also be probabilistic features of natural laws, such that there will, under appropriate circumstances, be alternative futures open to causal influence. In particular, it must be possible for human thought and intention so to affect the structure of the brain that physical interaction-patterns are influenced in some directions rather than in others.

Such influences are not arbitrary, and they are not 'interferences' in an otherwise inviolable natural order. They are parts of the natural order, and proceed in accordance with intelligible principles, though of a non-physically determined, sort. Human minds are part of the natural order, and they correlate with the complex physical structure of the central nervous system.

If there are open systems within the physical universe, making possible creative shaping of the future by humans and perhaps by other personal beings, then it is plausible to see the universe as a whole as an open system in relation to God, its spiritual origin, foundation and goal. Why should God not creatively shape the future, making new creative decisions, rather in the way an artist may decide to shape her work in new ways as it progresses?

The Biblical writers sometimes thought in this way, seeing God as a potter shaping pots on the wheel. This metaphor, however, makes the pots rather too passive, and it is clear that the prophets usually regarded God as responding to human prayers and actions, not just as determining them completely. God's action is that of creatively composing the universe, but of doing so in relation to the creative activity of sentient beings within the universe. It is their acts that partly shape the future, to which God responds in judgment, forgiveness, healing or liberation.

God will not act in ways that destroy the structure of law and freedom that characterize the universe. God will act only to advance the divine purpose. Suppose a main part of the divine purpose is that creatures shall come to realize the source of their being in God, and come to know and love God. How can love exist without any responsive actions between lover and beloved? If there is to be anything like a personal relationship with God, there must be some form of responsiveness between God and creatures. So God will act in ways that manifest such a responsive personal relation. God will make the divine presence known to creatures, and that will involve some change in the brain states of such embodied persons. If knowledge of God ever occurs; if, as knowledge, it is partly caused by the presence of God; and if knowledge is a mental state that has, as its concomitant, some brain-state; then it follows that God is part of the cause of some specific physical states in the cosmos.

God's particular actions will occur within a more general process of exercising a continuing influence on the emergence of the future. How may this be conceived? An analogy is perhaps the way in which a human being may form a long-term intention to write a novel, many years in the future. In the meantime, she may accumulate experience, practice writing skills and write short stories. The writing of the novel may come only after years, and take place over many years. It is not just one discrete action, like raising an arm. She may never consciously think of it for months at a time, yet still be directing her actions towards that aim, having previously decided to do so. Her long-term intention influences all her particular actions, shaping them into a pattern often imperceptibly. In the end, of course, she will have to write the novel, doing specific things that realize her intention.

So one may think of God as having the universe-long intention to bring conscious beings into a community of freely chosen loving relationships. This intention will shape the initial laws of the universe and the emergence of more complex possibilities within it.

There is wide agreement among symposiasts that if there is divine causal influence, it must integrate with the law-like nature of physical reality in a rational way, and not be an arbitrary interference with an otherwise smoothly running system. Tentative suggestions range from some form of holistic or 'top-down' causation, communication of information, determination of sub-atomic indeterminacies, or non-episodic "tuning" or "resonance." The drive is to seek a form of divine causality that in some sense complements or completes natural laws, rather than interferes with or violates them. Sympathetic to this endeavor, my own way of picturing divine causality is to think of God as the spiritual basis or macro-environment of the cosmos, ordering its regularities, co-operatively leading it towards an ideal, and expressing the divine being, in all its complexity, in the cosmic evolutionary process. The goals of creation are like blueprints (intentions) in the mind of God that are realized, not at some first moment of creation, but progressively throughout the history of the cosmos. They become more specific in response to the relatively autonomous character of that history. There will be a divine influence for good that preserves the relative autonomy of nature and its probabilistic laws, and the freedom of creatures to accept or reject the invitation to respond to the divine presence. As creatures come to sense possibilities of conscious relationship to the divine, there will be specific divine acts that initiate such relationship, acts that make the presence and character of God apparent. These will be revelatory acts, and many of them, though not all, may be termed miracles.

8 Miracles and Natural Laws

Miracles are not violations of immutable laws of nature—David Hume's picture, intended to make one think they are both immoral (instances of law-breaking) and irrational (since the laws should have been better designed in the first place). They are law-transcending events, extraordinary events manifesting divine causality that modifies the normal regularities of nature with the purpose of manifesting the basis and goal of the physical world in a wider spiritual realm. They show the power of Spirit to relate matter to Spirit so as to transcend normal material patterns of interaction, and establish a new conscious interaction of knowledge and love between the material and the spiritual.

For Christians, the key example of a miracle is the resurrection of Jesus (for Jews and Muslims, it would be the dictation by God of the Torah or Qur'an). The resurrection should not be seen as an event that violated a law that dead people do not appear in bodily form. Its role is to disclose a higher principle, that matter can be united to God to become a sacrament of divine life. When it is so united, it is immune from decay and dissolution. It is transfigured into a perfectly receptive vehicle of the divine will. Thus the body of Jesus was transfigured into a spiritual form, showing the goal of the whole physical process to be the transformation of the physical into an incorruptible vehicle of divine life. The resurrection is not an arbitrary breaking down of the physical process. It is the foreshadowing, at one point in human history, of the goal of material evolution, and thus a proleptic fulfillment of the physical process.

In miracles, God does not 'interfere' in a closed physical process. God perfects the physical process, showing what the ultimate divine purpose is.

Every authentic miracle has such a disclosive function. In the Johannine phrase, it is a *semeion,* a sign of God's purpose. It occurs only when that purpose needs to be declared or guided in a new way. Thus miracles will not upset the order of nature, which is not undermined by occasional modification of its laws. Miracles have the function of showing what the basis and purpose of the order of nature is. They will occur rarely, in contexts in which great spiritual teachers have prepared the way for a new declaration of the divine will, and in ways that members of their cultures are ready to understand.

There are thus three main forms of divine action, on the sort of account I am proposing. First, there is the general action of creating and holding in being a universe for the sake of the values it is able to generate out of itself. This action selects the basic parameters and constants of physical law. Second, there are actions shaped by the controlling intention to encourage a community of sentient and moral agents to come into existence. These acts will be within the probabilistic limits of physical law, and will normally be influencing, but not sufficient, causes of processes tending to the emergence of values. The physical sciences can ignore such divine actions, since it is possible, even if often improbable, that they happen simply by chance. Third, there are specific acts of revelation or response in which the normal physical powers of objects are transcended. God is always acting to sustain the universe in general and to guide its emergence to new forms of value in co-operation or sometimes, regretfully, in conflict with the acts of creatures. But God will only act occasionally in miraculous ways, since the probabilistic processes of nature and the autonomy of creaturely choice cannot be overthrown without destroying the general divine purpose.

This account of divine action is my own, and while it has been strongly influenced by the Castel Gandolfo conferences, it does not express an agreed position among the group's members. One would not expect agreement among all the members of such symposia.

But there is a sort of metaphysical view that has arisen out of this series of discussions between scientists and theologians. It puts in place a background world-view for a restatement of religious doctrines that is probably as important as Aristotle was for the Christian church in the thirteenth century.

The scientific understanding of a universe of intelligible law and emergent creativity changes the perspective within which one sees divine action. God will not be seen as an interfering designer correcting a partly incomplete mechanism. God will be more like a universe-environing field of Spirit, setting the parameters of nature, guiding its emergent development, and ensuring the eventual fulfillment of the divine intention for its existence. If one is to speak of divine action in such a universe, it will be in the context of a cosmic process of emergent value, within which humans can begin consciously to realize the spiritual potential of the physical. This vision of the nature of reality provides a new context for a reformulation of religious ideas, but it does not render obsolete the idea of revelation.

Religious revelation can be seen as a process in which the Supreme Spirit discloses its final purpose to beings who are parts of the natural order, and who have developed consciousness and rational agency so that they can pursue that purpose in conscious relation to the creator and goal of all things. Science finds in the natural order a general disclosure of the wisdom of God. Religion finds in the lives and teachings of its saints and prophets particular disclosures of the final purpose of God. Only when we are able to hold both together can we truly be said, at least in some small way, to know the mind of God.

DOES GOD NEED ROOM TO ACT? THEO-PHYSICAL IN/COMPATIBILISM IN NONINTERVENTIONIST THEORIES OF OBJECTIVELY SPECIAL DIVINE ACTION

Kirk Wegter-McNelly

1 Introduction

The Center for Theology and the Natural Sciences (CTNS) and the Vatican Observatory (VO) series of research conferences constitutes the most sustained and rigorous effort to date within the growing interdisciplinary field of religion-and-science to examine the question of divine action in light of the sciences.[1] The commitment of those involved to theological scholarship directly and critically engaged with the sciences has made the resulting volumes a prominent touchstone in the field. My reflections on these conferences and publications are divided into two sections. The first includes a variety of general observations, some overtly theological and others more sociological, which are intended to bring the series' past achievements as well as its future prospects into focus for critical assessment. The second section brings to light an important issue that has been at play—though more implicitly than explicitly—in the various proposals put forward regarding what arguably became the central agenda of the series: the development of a noninterventionist account of objectively special divine action (henceforth, NIODA). Briefly, the issue I discuss is whether or not NIODA is conceivable *only* in a world that unfolds indeterministically. Some of the series' contributors adopt this viewpoint, which amounts to saying that objectively special divine acts and physical processes must relate to one another in a zero-sum fashion: the operation of one necessarily precludes "room" for the other. For those of this mind, the world's processes cannot be fully determined by natural causes if God is to act at particular times and places—some sort of indeterminism must be present as an essential component of NIODA. I label this perspective "theophysical incompatibilism" and view it as a distinctively "modern" perspective, for it takes its cue from ways of thinking about God's relation to the world that have predominated since the Enlightenment. Other contributors adopt the opposing perspective, namely "theo-physical compatibilism," aligning themselves with pre-modern thinkers such as Thomas Aquinas for whom God's objectively special activity is neither interventionist nor incompatible with physical determinism. I develop this compatibilist–incompatibilist distinction at length below because I believe that it reveals a significant but previously unnoticed rift among the theologians contributing to the series. This rift needs to be acknowledged because it has influenced and in some cases obscured the debate over the relative merits of different NIODA proposals. After assessing each of the series' main NIODA proposals in this

[1] I refer to all *QC* citations following the pagination of the second edition, designated *QC2*, which differs slightly from the first edition.

light, I conclude by arguing that future work on the problem of special divine action should adopt a theo-physical compatibilist framework.

2 General Reflections

One of the great strengths of the VO/CTNS series was the active participation of leading scientists and philosophers of science in the conference discussions. The essays they contributed, which will no doubt stand for years to come as clear and authoritative guides to the relevant scientific disciplines, reflect the organizers' insistence that theologians must engage science not just in summary but in detail and with full awareness of the debates surrounding current scientific research. Such attention to the current state of the sciences has had an impact well beyond the conference participants by fostering an expectation among those working in the religion-and-science field that serious scholarship can only come from being well versed in both areas of thought.

In spite of this important role played by non-theologians in the series' development, the potential for collaborative work between scientists and theologians was unfortunately not fully realized. Only three of the 112 essays were jointly authored by scientists and theologians with the express purpose of intertwining scientific and theological themes.[2] Only one essay contained anything resembling new scientific research motivated specifically by the scientists' or others' theological interests.[3] Although scientists were not chosen to participate primarily on the basis of their interest in writing collaboratively with theologians or developing new research proposals but because of their ability to report accurately on the content of their fields, this lack of collaborative work is nonetheless disappointing—especially in light of the initial desire expressed by Robert Russell for the conferences to be a venue for "two-way interaction" between scientific and theological research programs.[4]

One important development that occurred in later conferences was an increased willingness of the theologians to explore the implications of the edges of scientific research. Many in the field of religion-and-science would echo the caution expressed by Arthur Peacocke in the first edition of his *Theology for a Scientific Age* (1990) that a theologian who marries the science of today risks becoming the widow of tomorrow.[5] In the early conferences

---

[2] The authors of these essays are C. J. Isham and J. C. Polkinghorne, "The Debate over the Block Universe," in *QC*, 139-48; Wesley J. Wildman and Robert John Russell, "Chaos: A Mathematical Introduction with Philosophical Reflections," in *CC*, 49-90; and Wesley J. Wildman and Leslie A. Brothers, "Intimations of Transcendence: Relations of the Mind and God," in *NP*, 449-74.

[3] Camilo J. Cela-Conde and Gisele Marty, "Beyond Biological Evolution: Mind, Morals, and Culture," in *EMB*, 445-62.

[4] Robert John Russell, "Introduction," in *QC2*, 3. Computer programming communities that write open-source programs (such as Linux) may provide an instructive model for future collaborative writing. These communities successfully maintain the integrity and internal consistency of large and complexly structured computer programs while at the same time allowing all participants the opportunity to edit any portion of the code.

[5] Arthur Peacocke, *Theology for a Scientific Age: Being and Becoming—Natural and Divine* (Oxford: Basil Blackwell, 1990), 28.

some participants felt that theologians should take on board science only in the most general way. Michael Heller, for example, explicitly advised against engaging particular scientific theories.[6] Gradually, though, and at the encouragement of others such as Russell and Philip Clayton, many of the participants came to adopt a more hypothetical, fallibilist mode of theological discourse.[7] Several, including Heller, later wrote essays exploring the implications of particular scientific theories as well as the debates surrounding them.[8]

This way of relating theology to science is indeed risky, but its benefits are also apparent. Theology done in this manner becomes more readily, though still indirectly, falsifiable. Theologians need to establish connections between their own theological theories and current theories in science if they want to talk with credibility about *this* world and about the God of *this* world. Also, exploring areas of consonance and dissonance between their own theological perspectives and those of science encourages theologians to wrestle more deeply with the relationship between the particularities of their own traditions and those of others, a step on the road to resolving longstanding confessional disputes. Finally, as Russell has pointed out, theologians working in this manner will benefit from a clearer sense of what is at stake theologically in various scientific research programs being pursued and thus will be able to take a more active role in promoting particular programs. This is not to say that they are qualified to judge the adequacy of scientific theories, but rather that theologians, like all members of society, have a responsibility to encourage research in directions relevant to their communities.[9]

Having briefly identified some of the general strengths and weaknesses of the series, I want to take a moment to compare the volumes with regard to the issue of theological convergence. What kind of theological unity or cohesion emerged within each of these volumes? I will not include *PPT* in this discussion because it was the product of a much more open agenda than the

---

[6] Michael Heller, "On Theological Interpretations of Physical Creation Theories," in *QC,* 2, 99, 102–3.

[7] Russell advocated this approach already in "Finite Creation without a Beginning: The Doctrine of Creation in Relation to Big Bang and Quantum Cosmologies," in *QC2*, 322, reiterating it in "Special Providence and Genetic Mutation: A New Defense of Theistic Evolution," in *EMB*, 217–8. Philip Clayton has championed this approach in many of his writings on the relation between scientific and theological methodology. He affirms his continued commitment to this approach in "Tracing the Lines: Constraint and Freedom in the Movement from Quantum Physics to Theology," in *QM*, 211–2.

[8] See, for example, Michael Heller, "Generalizations: From Quantum Mechanics to God," in *QM*, 191-210.

[9] See, for example, Robert John Russell, "Bodily Resurrection, Eschatology, and Scientific Cosmology," in *Resurrection: Theological and Scientific Assessments*, Ted Peters, Robert John Russell, and Michael Welker, eds. (Grand Rapids, Mich.: Eerdmans, 2002), 3-30, diagram and discussion on pp. 10-17; idem, "Eschatology and Physical Cosmology: A Preliminary Reflection," in *The Far Future Universe: Eschatology from a Cosmic Perspective*, George F.R. Ellis, ed. (Philadelphia: Templeton Foundation Press, 2002), 266-315, diagram and discussion on pp. 275-9, 284-8.

later volumes. In my judgment, *EMB* shows the highest degree of theological convergence of any of the volumes.

The essays in *EMB* present a collectively powerful argument for a coherent and compelling understanding of divine action in and through biological evolution, and an equally powerful argument against construing evolution as a denial of God's existence. Perhaps this convergence resulted from a shared sense among the participants of the cultural importance of developing constructive relations between evolutionary theory and religious belief. The Americans had the creationists to think about (e.g., Gish, Morris, and now, in a different key, the proponents of Intelligent Design) and the Europeans had the militantly atheistic evolutionists to contend with (e.g., Dawkins, Atkins, and Monod). Each of the participants clearly had some "local" stake in the success of this volume. It is also interesting to note that *EMB* is the only volume to include a related statement by Pope John Paul II or a substantial discussion of ethical issues, a distinction one might have expected of *NP*. Besides the important contribution of this volume to the series' exploration of divine action in light of the sciences, it provides a potent example of the growing strength and breadth of Christian theological interpretation of biological evolution.

The essays comprising *QC* are somewhat less convergent but the contributors do agree on the importance of working from a clear understanding of the status of the laws of nature. The most significant shortcoming of *QC* is the lack of any theological consideration of space. Essays in two of the five sections deal with issues related to time but none deal with issues related to space. This is a curious lacuna given the famously equal treatment accorded them by Einsteinian relativity. Perhaps this omission stems from the absence of any sustained tradition of reflection on the relation between created space and divine omnipresence in the manner of the classic discussions of time and eternity inaugurated by Augustine and Boethius.[10] The Christian tradition has not struggled with the limitations imposed on human life by spatial extension in the way it has with those imposed by temporal flow. Spatial extension, after all, plays a mostly positive and fundamental role in life by allowing the simultaneous existence of multiple, distinct creatures, whereas time plays a mostly negative role by separating us from our own experiences and from those we love. Is only human temporality in need of redemption, then, or does spatiality also in some way need to be redeemed? A more integrated theological analysis of these two dimensions of created existence is still needed.

Each of the three remaining volumes, *CC*, *NP*, and *QM*, has various strengths, but all lack the convergence of thought present in *EMB* and *QC*. In *CC* this is due primarily to a disagreement over whether or not chaos theory reflects genuine openness in physical processes. Most of the contributors judge that it does not, and therefore conclude that it will not be helpful in furthering the NIODA agenda. In *NP* the lack of convergence owes as much to the current absence on the scientific side of a comprehensive and widely-accepted

---

[10] Note, however, Isaac Newton's relatively more recent description of space as God's "boundless uniform sensorium" in his *Optičks*; see, for example, the second edition (London: 1717), Query 31, 379.

theoretical framework that explains the emergence of mind as it does from theological disagreement. Nancy Murphy notes in her introduction that there *was* a general consensus among those at the conference on the nature of the *challenge*: getting beyond the dualist and individualist view of the human person bequeathed by Western philosophy and theology.[11] But there was no agreement on how to formulate a solution to this challenge, however partial, from a theological, philosophical, or scientific perspective. It is also interesting to note that the only essays in *NP* to address the question of divine action at any length are those of Peacocke and George Ellis. Fraser Watts mentions divine action as the basis of religious experience but opts for a two-language approach instead of attempting to describe how God might act in the world to bring about religious experience. Wesley Wildman and Leslie Brothers also sidestep this issue and speak instead of "experiences of ultimacy." There is some irony in the fact that this volume—the only one to focus specifically on religious experience—deals least of any of the volumes with the notion of divine action. In *QM* the lack of convergence is mainly a result of the serious and unavoidable difficulties surrounding the interpretation of quantum theory. This volume includes many strong essays, but the theological conclusions drawn are varied and often at odds with one another. It is a testament to the scholarly integrity of the conference and editorial deliberations that disagreements were allowed to stand, and even highlighted, in the final content and structure of these three volumes.

3 Theo-physical In/Compatibilism in NIODA

In this section I wish to make several comments regarding the NIODA agenda advanced in different ways by a number of the essays in the VO/CTNS series and clearly identified as a central goal of the series since 1995. Russell provides an historical introduction to NIODA in *CC*[12] and, in updated form, in *EMB*.[13] I begin with a brief summary of this history.

NIODA attempts to resolve a key dispute between liberal and conservative theologians, the origins of which can be traced to the rise of Newtonian science. This account of the physical world painted a closed, deterministic, clock-like picture of natural processes and appeared to undercut the idea of God's presence and ongoing activity in the world. Nineteenth-century liberal theologians took Newtonian determinism to constrain God's activity in creation and responded by developing a theology divested of the notions of objectively special providence and miracle. God's only objective act, they agreed, is the enactment of history itself. According to this view, one might *perceive* God acting specially at a particular time and place in the natural world, but this would only be a matter of one's subjective perception (i.e., such an event would be a *subjectively special* divine act but not an *objectively special* divine act). God cannot act objectively in particular natural events, liberals assumed, because natural events are completely determined by the antecedent arrangement and events of the physical world. To guard human

[11] Nancey Murphy, "Introduction," in *NP*, xv.

[12] Russell, "Introduction," in *CC*, 2-6.

[13] Russell, "Special Providence and Genetic Mutation," 197-200.

freedom against Newtonian determinism, liberals adopted the strategy of philosophers such as Immanuel Kant who insulated humanity from the methods and descriptions of the natural sciences. Humans, for Kant, were actors in the realm of "history" and "freedom," not objects in the realm of "nature" and "law."

This understandable but ultimately stifling solution eventually led to such developments as the "biblical theology" movement of the mid-twentieth century, which had been inspired by Barth's discovery of "the strange new world within the Bible" but which had also inherited the closed causal view of the liberal theological tradition.[14] As Langdon Gilkey famously noted, biblical theologians wanted to proclaim the Bible's mighty acts of God in the world but were unable to say what, if anything, God had actually done.[15] Conservatives, on the other hand, held fast to the idea of objectively special divine action but instead portrayed God as breaking the laws of nature to achieve God's special purposes (e.g., the miracles of the Bible). To liberals, this made God appear to act in contradictory or at least capricious ways—here upholding nature's regularities, there breaking them, to suit God's purposes.

In response to these developments the theologians of the VO/CTNS series have developed their various NIODA theories. In agreement with the liberal tradition they argue that God must be understood to work with the grain of natural processes, not against it. However, they reject the liberal's purely subjective account of special divine action and instead attempt to show that God can act *objectively* in the world. In doing so, they affirm the basic thrust of the conservative point of view while rejecting its interventionism in favor of a God who acts without abrogating or intervening in the laws of nature. They do not aim to establish that this or that event is in fact the result of a special divine act but only that the notion of objectively special divine action is compatible with a robust scientific account of natural processes. The strategies developed by Russell, Ellis, Thomas Tracy, Murphy (quantum theory), John Polkinghorne (chaos theory), and Peacocke (top-down and whole-part influence)—in the VO/CTNS series and in other writings—can all be considered attempts to provide a noninterventionist account of the possibility of objectively special divine action in light of science.

From this brief account of the history leading up to NIODA I want to highlight one central theme, namely, the liberal assumption that objectively special divine acts are incompatible with deterministic natural processes. To accept the legitimacy of physical determinism as a constraint upon theological accounts of God's special activity in the world suggests a zero-sum understanding of the relation between special divine acts and natural processes. From a zero-sum perspective, if natural processes are completely determined by the immediately prior state of affairs in the world then there simply is no

---

[14] See Barth's essay by this title in, *The Word of God and the Word of Man*, Douglas Horton, trans. (New York: Harper Torchbooks, 1957), 28-50.

[15] Langdon B. Gilkey, "Cosmology, Ontology, and the Travail of Biblical Language," *The Journal of Religion* 41 (1961): 194–205, reprinted in *God's Activity in the World: The Contemporary Problem*, Owen Thomas, ed., Studies in Religion Series/American Academy of Religion, No. 31 (Chico, Calif.: Scholars Press, 1983), 29–44.

"room" for God to act in the world. William Placher has convincingly argued that this modern view "domesticates" the pre-modern understanding of divine transcendence.[16] Pre-modern theologians such as Aquinas and John Calvin did not understand divine activity and natural processes to be related in this manner, but rather took God's action to be the source of all of creation's processes, including free creaturely activity. They had no difficulty in seeing special acts of God in a world which had been given its own causal integrity, its own capacity to unfold and develop in time. One sees this most clearly in their discussions of the relation between human and divine activity, where humans are understood to be fully responsible for their actions even though their freedom is ultimately subject to God's freedom.[17] Placher's argument leads me to ask whether or not a similar kind of domestication is at work in some or all of the NIODA strategies of the VO/CTNS series.

To clarify the issue, I have adapted and expanded the idea of "in/compatibilism" from the familiar philosophical debate over the relation between human freedom and physical determinism.[18] I apologize for the cumbersomeness of the terms that follow, but I hope the distinctions they express will prove worthwhile. Here are the three varieties of in/compatibilism I wish to distinguish:

A. *Anthropo-physical in/compatibilism:* human freedom is in/compatible with physical determinism;
B. *Anthropo-theological in/compatibilism:* human freedom is in/compatible with divine determinism; and
C. *Theo-physical in/compatibilism:* objectively special divine action is in/compatible with physical determinism.

The distinction between these three varieties is relatively straightforward: variety A, as I just mentioned, corresponds to the debate among philosophers regarding the question of human agency *vis-à-vis* natural processes; variety B is the subject of ongoing discussion in the area of theological anthropology (and has been the subject of discussion long before the rise of modern science in the guise of the grace/free will debate); and variety C has to do with God's special acts within creation. It is the last of these, variety C, which pertains most directly to the debate over the manner in which scientific accounts of physical processes are understood to constrain (or not) theological accounts of special divine action.

It is important to be clear about the scope of variety C as I develop it here. Notice that the concept, as defined above, is limited to the issue of *special* divine action. One also could, of course, analyze theories of general divine action in terms of this idea. Almost all of the NIODA perspectives surveyed below adopt a theo-physical compatibilist perspective with regard to general

[16] William C. Placher, *The Domestication of Transcendence: How Modern Thinking About God Went Wrong* (Louisville, Ky.: Westminster/John Knox Press, 1996).

[17] Russell, "Introduction," in *CC*, 4.

[18] See Clayton, "Tracing the Lines," 221, for a clear summary of the philosophical discussion.

divine action. In other words, they accept the view that the particular character of natural processes has no bearing on God's ability to create *ex nihilo* or to sustain creation in its own being. (Process thought, with its rejection of the classic doctrine of divine omnipotence and thus of *creatio ex nihilo*, is an important exception.) My discussion here is focused exclusively on the issue of special divine action. Thus it should be understood that whenever I use the term "theo-physical in/compatibilism" I refer only to this limited context. Beyond this, whether or not it makes sense to adopt different perspectives with regard to general and special divine action is a question that goes beyond the scope of this paper. Suffice it to say that some NIODA proponents do appear to mix perspectives. (Russell, for example, is a theo-physical compatibilist with regard to general divine action but a theo-physical incompatibilist with regard to special divine action.)

One virtue of distinguishing these three varieties of in/compatibilism in this way is that it allows us to separate various aspects of the wider discussion and to inquire into the relations among them. One interesting question to ask is whether a particular stance on variety A entails a particular stance on variety C. For example, does the notion that human freedom is incompatible with physical determinism thereby commit one to the notion that special divine acts are also incompatible with physical determinism? The answer to this question will depend in part upon one's understanding of the relation between human and divine agency, i.e., upon one's attitude toward variety B. If human and divine agency are judged similar enough to warrant an analogy between the two with regard to the issue of physical constraints on particular acts—a judgment that points in the direction of anthropo-theological incompatibilism (B)—then being committed to anthropo-physical incompatibilism (A) will commit one to theo-physical incompatibilism (C) as well. However, if human and divine agency are judged more dissimilar than similar—a judgment that points in the direction of anthropo-theological compatibilism (B)—then being committed to anthropo-physical incompatibilism (A) will not necessarily commit one to theo-physical incompatibilism (C). This logic can be summarized quite compactly: whether or not a particular stance on A entails a particular stance on C depends upon one's particular stance on B. What this discussion serves to point out, perhaps unsurprisingly, is one way in which theories of special divine action inevitably depend upon prior judgments regarding scientific, philosophical, and theological theories of human action.

I am persuaded, along with those contributors to the VO/CTNS series who adopt a physicalist understanding of human nature (see *NP* for an extended discussion of this issue), by the perspective of anthropo-physical incompatibilism (A). As thoroughly physical creatures, our freedom, if such exists, must inevitably be constrained by physical processes. Any successful defense of the existence of genuine (counterfactual) human freedom *vis-à-vis* physical processes must show that these processes allow for its presence (either through, say, quantum indeterminism or some strong form of emergence). However, following Placher (as well as others such as Karl Barth, Wolfhart Pannenberg, and Ted Peters), I am also persuaded that anthropo-theological compatibilism is a more theologically sound view than anthropo-theological incompatibilism (B). That is, I think human agency and special divine agency can work together without competing against one another because they are radically different things. Special divine acts ought not to be construed as closely analogous to and thus in competition with human acts, but rather as something that gives rise to and sustains them. One point in favor of this view comes from noting that the incarnation might reasonably be construed as the *sine qua non* of a special divine act. In whatever way divine and natural agency come together in "ordinary" special divine acts, this conjoining of agencies must be especially true in the case of the incarnation, which has traditionally been understood to be the most intimate possible union of human and divine agency—a union in which neither agency overpowers or destroys the other. This kind of agential unity can only be articulated from an anthropo-theological compatibilist perspective. An incompatibilist account inevitably leads down the blind alley of parsing out which acts are "human" and which are "divine." Because I am suspicious of a zero-sum understanding of the relation of special divine acts to creation's own powers generally—that is, not only with respect to human action but with respect to all physical events and processes—I am increasingly drawn not only to the perspective of anthropo-theological compatibilism (B) but also to the perspective of theo-physical compatibilism (C). But is this option even possible in light of my commitment to anthropo-physical incompatibilism (A)? If I agree that physical determinism undermines human action, why would I also agree that it undermines special divine action? Because my view of B blocks this implication.[19] If special divine action is not analogous to human action, it need not be similarly undermined by determinism.

Let me turn now to the various NIODA strategies developed in the VO/CTNS series and sort them according to the issue of theo-physical in/compatibilism (C). We can easily incorporate this distinction into the VO/CTNS divine action typology.[20] This typology originally distinguished between noninterventionist views that accepted or rejected the claim that objectively special divine action would be apparent without religious presuppositions. Over the course of the conferences, however, none of the participants argued for the possibility of perceiving objectively special divine acts without religious presuppositions, i.e., no one argued for a strict "natural theology" approach. Therefore, I suggest replacing what proved to be an unimportant distinction with the one I have developed and referred to as theo-physical in/compatibilism. In addition to making this substitution, I have expanded the table by including two additional NIODA strategies (5 and 6 below) that did not appear in the original table. Only the relevant portion of the table is reproduced here.

[19] Clayton has wagered, contrary to my own point of view, that there *is* a sufficient analogy between human and special divine acts to warrant the conclusion that both are similarly constrained by the physical world (Ibid., 214).

[20] This table was first presented in Russell, "Introduction," in *CC*, 11, and subsequently reprinted in Russell, "Introduction," in *QM*, iv.

VO/CTNS **Divine Action Typology** — MODIFIED PORTION ONLY —	Objectively Special Divine Action	
	Noninterventionist	
	Theo-physical Incompatibilist	Theo-physical Compatibilist
(1) Top-down or Whole-part *Peacocke*		X
(2) Bottom-up (e.g., quantum indeterminacy) *Russell, Ellis, Tracy, Murphy*	X	
(3) Lateral Amplification (e.g., chaos) *Polkinghorne*	X	
(4) Primary/Secondary *Happel, Stoeger*		X
(5) Process *Barbour, Birch, Haught*	X	
(6) Functionalism Discussed by *Tracy* in *QM*		X

Table 1: VO / CTNS Divine Action Typology

This modified portion of the typology reveals that the four NIODA strategies originally included divide equally between theo-physical compatibilism and incompatibilism. Russell (strategy 2) is clearly motivated by a commitment to theo-physical incompatibilism. He implicitly endorsed a theo-physical incompatibilist perspective in *CC*, writing that Owen Thomas had correctly identified "overdetermination" or "double agency" as a genuine problem for relating divine and creaturely activity.[21] And his concern that no adequate noninterventionist account could be developed within a deterministic framework has led him to build his own account of NIODA around the concept of quantum indeterminism.[22] Interestingly, though, Russell recently expressed a desire to rethink the importance of quantum theory for NIODA from a theo-physical compatibilist position.[23]

---

[21] Russell, "Introduction," in *CC*, 9. Stoeger mentions this issue as well, but he does not consider it to be a serious problem when understood in terms of the Thomistic (and theo-physical compatibilist) distinction between primary and secondary causes (William R. Stoeger, S. J., "Describing God's Action in the World in Light of Scientific Knowledge of Reality," in *CC*, 254). More on Stoeger's views below.

[22] See his contributions throughout the series, and especially in Russell, "Special Providence and Genetic Mutation," 215, and idem, "Divine Action and Quantum Mechanics: A Fresh Assessment," in *QM*, 317–8.

[23] *Russell,* "Divine Action and Quantum Mechanics," 318, no. 87.

Polkinghorne (strategy 3) is not as clear as Russell on whether indeterminism is necessary for his own account of NIODA using chaos theory. To my knowledge, Polkinghorne never appeals in writing to the need for any kind of "indeterminism," but he does see a need for "ontological openness" and "genuine novelty" in nature.[24] His distinction between passive and active information (the latter allegedly requiring no energy exchange) suggests that he is searching for a kind of openness that does not rely upon indeterministic physical processes.[25] It is difficult, however, to imagine how such openness would work.[26] In spite of Polkinghorne's reluctance to use the language of indeterminism to describe the world's gappiness, his acceptance of the importance of gaps for providing a robust account of special divine action puts him clearly in the theo-physical incompatibilist camp. In this regard it is interesting to note that he sees in both human and divine agency the same relation to the world's intrinsic gaps. "We are unashamedly 'people of the gaps,'" he writes, "and there is nothing unfitting in a 'God of the gaps'...either."[27] Despite the ongoing dispute between Russell and Polkinghorne over whether quantum theory or chaos theory is the better choice for advancing the NIODA agenda, both come across as committed theo-physical incompatibilists.

Moving now to the theo-physical compatibilist column, let us turn first to the advocates of the Thomistic distinction between primary and secondary causation (strategy 4). Neither Stephen Happel nor William Stoeger, who are among those contributors most deeply aligned with the Thomistic tradition, thinks that indeterminism offers any advantage for constructing a theory of divine action. Stoeger, interestingly, thinks that we *should* link the idea of God acting in the physical world to our understanding of the regularities of physical processes as formulated in the laws of physics. But he takes care to distinguish between how God "appears" to operate to us in these laws and how God "actually" operates "behind the veil."[28] This distinction is crucial for Stoeger because he is convinced that there is little similarity between how God acts in creation and how we act.[29] God's uniquely transcendent relation to the world as its creator allows the divine presence in each entity to constitute a "direct, immediate, relationship of that entity with God, and therefore [to constitute] the channel of divine influence in secondary causes."[30] Here is an example of

---

[24] John Polkinghorne, "The Laws of Nature and the Laws of Physics," in *QC2*, 433.

[25] But see John Polkinghorne, "Physical Process, Quantum Events, and Divine Agency," in *QM*, 189.

[26] Russell's theo-physical incompatibilism is again evident in his critique of Polkinghorne's approach, namely, that the subtle nature of chaotic systems has no bearing on the question of divine action unless chaos theory can some day be given an indeterministic interpretation (Russell, "Introduction" in *QM*, v).

[27] Polkinghorne, "The Laws of Nature and the Laws of Physics," 438.

[28] William R. Stoeger, S. J., "Epistemological and Ontological Issues Arising from Quantum Theory," in *QM*, 96.

[29] Stoeger, "Describing God's Action in the World in Light of Scientific Knowledge of Reality," 252ff.

[30] Ibid., 257.

an approach to divine action that takes both noninterventionism and the dialogue with science (including quantum theory) seriously—but from a theo-physical compatibilist perspective.

This leaves us with Peacocke's approach (strategy 1). It might at first glance seem a mistake to identify Peacocke's strategy with theo-physical compatibilism. After all, he construes the indeterminacy of quantum theory as pointing to the "open-endedness" and "flexibility" of creation, which he takes to limit God's activity.[31] He also has God "influencing" the course of physical events through his top-down and whole-part models. Although this sounds like the language of a theo-physical incompatibilist, let me explain why I am convinced that Peacocke is at heart a theo-physical compatibilist. For Peacocke, the flexibility of physical processes places a limitation on divine foreknowledge—God cannot compute the future from the present. This flexibility does not provide, *à la* Polkinghorne and Russell, a locus of physical underdetermination that would give God "room" to act.[32] Would Peacocke need to reconfigure or abandon his whole-part model of divine action in order to square it with, say, Bohm's deterministic version of quantum theory? I suspect not. In *NP* Peacocke adds a telling clarification to an early passage taken from his *CC* essay:

> If God interacts with the "world" at a supervenient level of totality, then God, by affecting the state of the world-as-a-whole, could, on the model of whole-part constraint relationships in complex systems, be envisaged as able to exercise constraints upon events in the myriad sub-levels of existence that constitute that "world" without abrogating the laws and regularities that specifically pertain to them—and this without "intervening" within the unpredictabilities we have noted [I had in mind here the in-principle, inherent kinds, i.e., quantum events, though the remarks would also apply to the practical unpredictabilities of chaotic systems].[33]

Elsewhere in his *CC* essay Peacocke argues that his whole-part model allows God to act at particular moments "without ever abrogating at any point any of the natural relationships and inbuilt flexibilities and freedoms operating at all of the lower levels..."[34] Peacocke does not want to appeal to any sort of physical indeterminism (i.e., nature's "inbuilt flexibilities and freedoms") as a locus for divine action because this would amount to saying that God sometimes ignores or overrides the divinely-bestowed integrity and freedom of these processes. He does not wish to have God acting *in* the indeterminacies of the world but rather *on* and *with* them through whole-part influence so that God's ongoing activity in creation can be affirmed without destroying the integrity of natural processes. For God to act *in* the indeterminacies of the world would be, from Peacocke's theo-physical compatibilist perspective, a kind of needless and disrespectful intervention. Of course, theo-physical

---

[31] Ibid., 279, 281.

[32] Ibid., 280.

[33] Arthur Peacocke, "The Sound of Sheer Silence: How Does God Communicate with Humanity?," in *NP*, 235, no. 73.

[34] *CC*, 286.

incompatibilists such as Russell do not accept that such acts are appropriately described as interventions, since no law is actually broken. Whether or not a God who makes a difference by affecting particular events but whose actions remain hidden (from the gaze of science, anyway) within the statistical regularities of quantum processes should be labeled "interventionist" or "noninterventionist" is a difficult question that I will not attempt to answer here. At any rate, these considerations suggest that Peacocke is in fact a theo-physical compatibilist at heart, notwithstanding a certain aura of incompatibilism that surrounds his writing.

A process account of special divine action was not included in the original typology, though this perspective is discussed and espoused to varying degrees by several VO/CTNS authors such as Ian Barbour, Charles Birch, and John Haught. Process theologians, known for their rejection of divine omnipotence and their characterization of God as a "divine lure," opt for an essentially theo-physical incompatibilist account of divine action. But in this case the strategy is not one of finding room in creation for God to act but one of imagining a mode of divine action in which God refrains from directing the course of events so that creation itself has room to act. God participates in this process by luring creation forward, not by dictating its future. An example of this kind of approach can be found in John Haught, who develops his views in terms of the idea of divine kenosis. He argues that a kenotic God who intimately embraces creation actually allows the world to be differentiated from God by allowing creation the freedom to bring about its own future.[35] The logic of this perspective is reversed: God does not find room to act in an underdetermined world but rather the world can act because God does not act, at least not in a traditional controlling manner. Nonetheless process thought fits within the theo-physical incompatibilist perspective because it, like the other views in this perspective, assumes that God and natural processes cannot both be fully responsible for bringing about the same natural event.

In addition to the process perspective, I have also added to the typology a theo-physical compatibilist strategy discussed (though not directly espoused) by Tracy in *QM*. In the early volumes of the series, Tracy committed significant energy to the theo-physical incompatibilist strategy of locating objectively special divine action within quantum indeterminacy. In *CC*, for example, he explored the option that special divine acts occur in some but not all quantum events. There he also rejected Brian Hebblethwaite's account of divine action because it did not explicitly appeal to gaps in physical processes. According to Tracy, "any account of the relevant events that is couched strictly in terms of finite agencies must contain some gaps."[36] An account such as Hebblethwaite's could not do justice to the idea that objectively special divine acts ought to "make a difference" in the way the world unfolds. In *QM*, however, one finds Tracy pondering whether there might, after all, be a way of affirming the possibility of noninterventionist objectively special divine acts from within a deterministic framework. In a section of his *QM* essay entitled "Special Divine Action in a Deterministic World" Tracy writes that even

---

[35] *EMB*, 407.

[36] *CC*, 306.

though God might be acting uniformly in all events, some event could nonetheless

> play a particularly important role in advancing God's purposes, and this will be a fact about their *function* within the causal series and not just about our perception of them. History may have turning points, and the special causal significance of these events is in no way diminished if they arise smoothly within the causal structures of the world.[37]

I refer to this account of NIODA as "functionalism" because in it objectively special divine acts are identified with natural events that occur in the context of strict physical determinism but which by virtue of their role in the unfolding of history *function* to advance the world in a special way. God's special activity is not, according to this view, a result of God's acting differently in some events than others or of "making a difference" in the world by causing it to unfold differently than it would have under its own divinely-given power. Instead, God's special activity is located in the (fully determined) events upon which history turns. The fact that this approach embraces determinism means that it is best understood as a theo-physical compatibilist version of NIODA.

Tracy is understandably at a loss as to where to locate this NIODA strategy in the original divine action typology.[38] Functionalism draws no sharp distinction between uniform divine action and objectively special divine action, and so it would appear that the typology divides the issue in a way that functionalism will not allow. However, I believe that Tracy wrongly understands the relation of functionalism to the other approaches. Yes, functionalism allows natural events to be expressions of God's uniform action and objectively special action at the same time. But in the original typology the heading "Objectively Special Divine Action" is not meant to identify those options that reject God's uniform divine action; rather, it designates those options that accept both uniform and special divine action. Each of the original "theology and science" strategies (1–4) affirms both. None takes God to cease acting uniformly—i.e., to withdraw from upholding an event in its existence—in an event where God chooses to act specially as well. In this regard functionalism is no different from the other NIODA strategies. The real problem with locating functionalism in the original typology is that there is no way to signal its acceptance of a certain kind of theo-physical compatibilism (i.e., one that embraces determinism) with regard to the issue of objectively special divine action.

An important question for the functionalist account is whether it can sufficiently express God's presence in those events that function in objectively special ways to turn history in new directions. If not, then this strategy would quickly collapse into a form of deism. Tracy seems to think that functionalism *can* meet this challenge, as long as one remembers that God's role as Creator—sustaining all of creation in its being at every moment—guarantees

[37] Thomas F. Tracy, "Creation, Providence, and Quantum Chance," in *QM*, 242, emphasis mine.

[38] See Ibid., no. 14. In fact, it was Tracy's unsuccessful attempt to fit this account within the typology that originally led me to develop the modified portion presented above.

that God will be directly and intimately present to each event as it contributes to the unfolding of creation. I would add that although God might not play a special role in such deterministic events with regard to the means by which they come about, God could still be present to them with regard to the divine intention for the special role they play in the world's development. One might perhaps speak from this perspective of God's "intentional presence" in such events.

The proponents of the four NIODA perspectives included in the original typology have on occasion acknowledged the importance of combining their strategies to create a more complete picture of God's activity in the world. However, if my analysis has merit, then a significant theological rift lies between those who think that special divine acts and physical processes exist in a zero-sum relationship and those who do not—a rift which stems from differing theological assessments of the capacity of scientific description to constrain theological assertions about divine activity. It is important that the divine action typology developed in the VO/CTNS series reflect not only the differences between the "uniform" and the "uniform + special" approaches, or the differences between the interventionist and noninterventionist approaches, but also the differences that exist among the noninterventionist approaches themselves, such as the difference regarding theo-physical in/compatibilism.

For those who are interested in the theological import of quantum indeterminacies as well as the unpredictabilities of chaos theory but who also see the value of theo-physical compatibilism, it will be important to reexamine in detail the theo-physical compatibilist character of strategies such as Stoeger's and Peacocke's, as well as functionalism. My interest in the last should not be taken as a rejection of the theological importance of physical indeterminism revealed by modern science. I find functionalism interesting not primarily because it embraces determinism but because it provides an interesting additional example of a theo-physical compatibilist approach. Future work on special divine action must find ways of discussing indeterminism from a theo-physical compatibilist perspective, i.e., from the desire to explore indeterminism as a reflection of the character of God's activity from our human, creaturely perspective, rather than from the desire to find room for God to act. Stoeger puts it well when he says that Newtonian determinism posed a problem for theology not because it would have prevented God from acting in the world if it were true, but because it portrayed a world devoid of potentiality, flexibility, and scope for newness.[39] The Jewish and Christian theological traditions look to a God who does "new things" (Isa. 43: 19) in the world. They have a sense not only of history's linearity but of its novelty as well. Determinism might be compatible with the God of theism (per my discussion of *functionalism* above), but I believe it is deeply incompatible with the Christian God.

The vexing problem of divine action dealt with so comprehensively in the VO/CTNS series finds its roots in the rise of modern science and the Enlightenment, especially in the attitude inculcated during this period which refused to treat God any differently from other objects of human experience.

---

[39] Stoeger, "Describing God's Action in the World in Light of Scientific Knowledge of Reality," 244.

Theo-physical compatibilism holds promise as a framework for advancing the NIODA agenda because it does not share this refusal but instead honors the distinction between divine and creaturely activity. Exploring the meaning and significance of physical indeterminism from this perspective will not be an easy task. Theo-physical compatibilism renders connections between divine action and science much more difficult to discern and develop because, unlike theo-physical incompatibilism, it does not provide any clear way of linking scientific and theological theories. God can act in the world, according to the theo-physical compatibilist perspective, regardless of the exact nature of the world's physical structures. But given this, there is no reason to suppose further that the nature of the world's physical structures will not in some way reflect the character of God's actions. The existence and character of the world's openness and novelty still ought to matter deeply for how we talk about God's presence in the world.

Acknowledgments: I wish to thank the participants in the Capstone Conference for their insight and comments, and especially Robert Russell for his help with the revision process. I am also indebted to Fred Sanders for his unfailing knowledge of Karl Barth's writings.

REFLECTIONS ON THE VO/CTNS SERIES ON DIVINE ACTION FROM THE PERSPECTIVE OF THEOLOGY AND SCIENCE

Mark Worthing

The decision to choose 'Divine Action' as the theme of the inaugural Vatican Observatory/Center for Theology and the Natural Sciences (VO/CTNS) conference and publication series has, in my view, proven itself justified. The topic has given the project both focus and broad scope. Indeed, it would be difficult to think of any branch of the natural sciences that do not touch upon the question of how we are able to understand divine action in the world today. The real challenge of the series was never going to be finding sufficient topics, but achieving some common understanding of what is and is not meant by divine action. The other challenge was going to be the coordination of a sustained theology-science dialogue with a level of focus and depth not previously carried out. From the perspective of theology and science as both an emerging discipline and as an on-going dialogue the VO/CTNS series broke new ground. I am not aware of any other project approaching the scope of the decade long series involving leading scientific, theological and philosophical voices and with a focus on a single theme. It has provided an effective model for such dialogue—though other models exist.[1] The series has advanced both the field and the dialogue and the organizers and participants deserve recognition for this accomplishment.

1 Divine Action, Intervention, and the God of the Gaps

Clearly, one of the most difficult and frequently encountered problems in the science/theology dialogue is that of the nature of the ongoing relationship between God and the physical universe. How do we understand God's relationship to an evolving universe? Does or can God act in any real and 'special' sense within the processes of the physical universe? On the one hand, we are faced with the model of a God constantly intervening or tinkering with the creation, after the manner of Newton's cosmic plumber. What kind of God

[1] In the Australian context the Australian Theological Forum (ATF) runs a number of conferences, including theology and science, in which, like the VO/CTNS series, a limited number of participants are invited to present papers and these are revised in light of the discussions. In the ATF model two significant variations are that the conference itself is open to a restricted number of non-presenters, and each paper has a formal response. The responses are not in all cases published, but when they are both the paper and response go through a series of revisions. In ecumenical dialogues two distinctive positions are identified (along denominational lines) and each tradition works together to present their position as clearly as possible and to find common ground with the other point of view. This model forces differences within each 'camp' to be at least provisionally resolved for the purpose of addressing the major points of difference. While this approach has produced some significant ecumenical advances I am not aware of this model being applied to the science and theology dialogue. If it were, the two camps would not be 'science' and 'theology' but would be defined according to distinctive interpretative and theological approaches.

would this be? And how could legitimate science ever take place under the shadow of an omnipotent being constantly changing or breaking the rules? On the other hand, if God does not or can not act in special instances at all, where does that leave the truth claims of Christian theism? Either way we face a dilemma. If we admit to special divine action, then it might appear that we undermine science. If we deny special divine action, then it appears that we erode the very core of Christian faith. This, at least, is how one might traditionally pose the question.

One of the real benefits of the VO/CTNS series has been what would appear to be the dual consensus that (1) God does act in the world, and (2) this action is best understood in noninterventionist terms.[2] Although no agreed model was arrived at as to just how this should be viewed, Bob Russell has suggested that four such models did emerge from the discussions. The four distinct ways of understanding divine causation within a noninterventionist framework, according to Russell, are: (1) Top-down or whole-part causation, (2) bottom-up causation, (3) lateral causality, and (4) primary/secondary causality.[3] But this apparent consensus itself raises questions. Can a case still be made for an 'interventionist' view of divine action? Although this view, when understood in terms of God only being 'active' in the world when 'intervening,' carries with it a number of problems, it is still common among such groups as evangelical Protestants. Further, if divine action can be perceived in events regarded as 'special,' in what way is this clearly distinct from more nuanced versions of interventionist thinking? My suspicion is that there is a fair degree of overlap between more nuanced interventionist models, and what many contributors mean by special divine action. Finally, if some events are regarded as potentially constituting recognizable special divine action, this would seem to link God directly to secondary causes while shifting focus away from divine action in the individual processes of the physical universe?

In the end, the series did not, nor could it, resolve the big questions about divine action and the physical world. It did, however, make some significant strides in clarifying what some of these questions are and in exploring several significant models for understanding divine action.

One question that arises concerns the significance of open processes for the question of divine action. While never far from the surface, this issue comes to the fore particularly in *CC*. In the past, misunderstandings about the deterministic nature of reality and the fixed status of physical law has produced statements that would seem to rule out any possibility of divine action, particularly special divine action. It is important to point out that quantum theory, chaotic systems, and more fluid understandings of physical laws all make such statements now more difficult, if not impossible. At the same time we must be cautious in exploring the implications of open systems that we do not give the impression that it is precisely (or only) in these open systems that we find room for divine action.

---

[2] This apparent consensus, however, should not be taken to suggest that all participants would agree on these points. I am not sure in what sense either Frank Tipler or Paul Davies, for instance, valuable as their contributions have been, could affirm that God acts in the world.

[3] Robert John Russell, "Introduction," in *QM*, v.

One intriguing approach made possible by taking open systems into account is that of Thomas Tracy in his article, "Particular Providence and the God of the Gaps." Working from the assumption that "explanatory gaps in principle occur within some of the most powerful contemporary physical theories," Tracy draws particular attention to quantum theory and chaos theory.[4] Tracy goes on to argue that "contemporary theologians ought not summarily...dismiss the appeal to gaps in scientific accounts of the world" and that "theology may not be able to get along altogether without gaps."[5] Yet Tracy also argues that the theologian who takes such explanatory gaps into account "will not simply be returning to the earlier 'God of the gaps' theology which sought to exploit explanatory gaps...and claimed that these represented causal gaps. Rather, the theologian will be working with explanatory gaps...which represent fundamental limitations upon what we may know."[6] I do not think this distinction between causal gaps and gaps of what Tracy calls 'the second type' that represents fundamental limits on what we may know alleviates the problems traditionally associated with the older style of God of the gaps approach.[7]

Michael Heller, in the same volume, argues also that a distinction is to be made between types of gap. For Heller the distinction is between spurious gaps and genuine gaps. "Spurious gaps," he explains, "are temporary holes in our knowledge usually referring to an incomplete scientific theory or hypothesis." This kind of gap, which would seem to include what Tracy calls explanatory gaps, Heller does not believe exist. The only genuine gaps, he suggests, are the ontological, epistemological and perhaps axiological gaps.[8] The gaps that exist concern questions of being and meaning. Why is there something instead of nothing, and why is the world comprehensible? The natural sciences cannot answer these questions—but this is a long way from finding gaps in our knowledge of the processes of the physical universe.

Most contributors go out of their way to reject this perspective or to distance themselves from it. Yet as soon as unpredictability/chaos and indeterminism become focal points of discussion for the possibility of divine action there is a real challenge to distinguish what we are saying from such an approach. For this reason the contribution by Willem Drees, "Gaps for God?"—which is at the other end of the spectrum to the article by Tracy—raises valid concerns.

Drees illustrates the difficulty with viewing open processes as representing apparent 'gaps' in our understanding of nature when he writes:

> One can never exclude particular divine action hidden in the unpredictability [of chaotic processes]. However, as I see it, if there is no indication of or need for such an assumption of openness and divine action, the assumption is not justified. Quantum uncertainty, such as in the decay of a nucleus, may be of a different kind.

---

[4] Thomas Tracy, "Particular Providence and the God of the Gaps," in *CC*, 291f.

[5] Ibid., 292.

[6] Ibid.

[7] When Tracy returns to the question of providence in "Creation, Providence, and Quantum Chance," in *QM*, 235-258, he seems to have modified somewhat his approach.

[8] Michael Heller, "Chaos, Probability, and Comprehensibility," in *CC*, 120f.

> Here we have good grounds to exclude an ordinary cause or 'hidden variable,' and thus an explanation of the limited predictability as a consequences [sic] of an unobserved but real physical process. However, even with quantum physics we need to be cautious, as quantum physics will be modified or replaced, and is open to various interpretations.[9]

I remain convinced, however, that open systems are significant for understanding divine action at some level. While Drees's concerns should not be easily dismissed, most theologians (myself included) would not be comfortable with his own naturalistic approach.

While the problem of divine action and physical law will always elude easy answers, the problem can at least begin to be addressed by clarification of terms. If we suggest that at heart the difficulty lies with the apparent incongruity of special divine action and natural law, then we must begin by clarifying what we mean by both concepts. The difficulty here is highlighted in Bob Russell's introduction to *QC* in which he identifies at least six distinct understandings of divine action: (1) Neo-Thomist, (2) Process, (3) literal divine action, (4) embodiment models, (5) non-embodiment models, and (6) interaction models. Russell draws our attention to Owen Thomas's conclusion that all approaches ultimately move toward the metaphysical assumptions of either the Neo-Thomist approach of primary and secondary causation, or the process philosophy approach.[10] He concurs that these seem to be the most likely options at present, by default. But rather than endorsing one or the other, Russell suggests that the existence of only two strong metaphysical models serves to highlight the need for an alternative approach. In any event, it is a matter of some significance that the two models identified by Thomas as offering 'full-blown metaphysical theories' both see divine action in all processes—though understood differently. My suspicion is that other possible models, while perhaps not having developed a 'full-blown metaphysic,' do not at the end of the day fall into line under either a Neo-Thomist or Process approach.

2 Finding a Common Language

The language of divine action within the context of a scientific worldview is inherently problematic. For many, and I include myself in this number (though with increasing reservations) the term 'intervention' is still occasionally employed, especially in theological reflections outside the context of the science and theology dialogue. But this term is loaded with so much baggage it is probably best avoided in most cases—unless one wants to provoke a discussion. I have sensed a tendency among some to replace 'intervention' with 'special divine action' without further reflection about what significance, if any, this terminological shift actually has.

Both scientists and theologians have complained at times that terminology from their respective disciplines has been misunderstood or not

[9] Willem Drees, "Gaps for God?," in *CC*, 231.

[10] Cf. Robert John Russell, "Introduction," in *QC*, 7ff.

properly nuanced.[11] This requires effort on the part of both theologians and scientists. While many of the contributors to the series have been at pains to define key terms, not all have done this. While the glossary at the end of the last volume in the series, *QM*, is useful, it is also interesting for what is not there. The terminology is scientific. Theological terms, traditionally more difficult to define concisely, do not appear. Would it have been possible to include terms like divine action, special divine action, intervention, non-intervention, providence, special providence, God-of-the-gaps and *creatio continua* in the glossary? I suspect this would have been very difficult. The contributors themselves suggest that in many instances their understandings may not be in consonance with one another. Of course, only through sustained dialogue do these differences begin to come to light and it is probably only at the conclusion of the series that serious attempts at common definition (including recognition of areas in which this does not exist) could have been made.

The problem might be illustrated with a concept that virtually all contributors appear to agree on—at least by the conclusion of the series—namely that 'intervention' is not an appropriate or helpful word for speaking about divine action. But what exactly is meant by 'intervention?' In theological discussions about divine 'intervention' (and these still occur!) one usually gets the sense of at least two different uses of the word. In the stronger or more radical sense an act of divine intervention is one in which God interrupts, sets aside or violates natural laws or processes in order to bring about a divinely willed outcome. If we insist on this sense of divine intervention there would appear little room for genuine theology and science dialogue. There is, however, a more integrated use of the concept that views divine intervention as God working with and through natural processes and laws—in ways we cannot fully comprehend—in order to bring about an outcome that would not have occurred without some 'special action' on God's part. In this later view intervention need not and must not be seen in opposition to the laws of nature that God established when God called the material universe into being. It could be argued that this represents a noninterventionist understanding of 'intervention'—and perhaps this is what the contributors mean by non-interventionist divine action. In that case, the series may even provide alternative and less confusing terminology to that traditionally used in theological circles.

Bob Russell gives a very useful clarification of some of his own terminology in "Divine Action and Quantum Mechanics: A Fresh Assessment," in which he explains as noninterventionist the view that "God does not violate the laws of quantum physics but acts in accordance with them. In essence, God creates the universe such that quantum events occur without

[11] Jeffery Wicken, for instance, taking to task a theologian of no less stature than Wolfhart Pannenberg, illustrates our point well. He wrote: "If we want to use the word *energy* or *field* in science-theology discourse, let us do so in some way commensurate with their understandings in physics. Talking about 'spirit' as 'energy' and granting it by implication the status of physical law runs dangerously close to usurping the hard-won denotative language of science for the physicalzing of theology. This serves neither enterprise." "Theology and Science in the Evolving Cosmos: A Need for Dialogue," *Zygon* 23:1 (March 1988): 48.

sufficient natural causes and acts within these natural processes and together with natural causes to bring them about."[12] As long as God does not violate any of the laws of nature, even if working to bring about specific outcomes where events (i.e., quantum events) lack sufficient natural causes, this is not understood as intervention. Russell acknowledges the concern that God might appear to be reduced to a natural cause. Yet God, in this view (in contrast to Intelligent Design with its introduction of such concepts as agency and a designer into scientific theory) is not seen as "an explanation within science." By viewing God's action at the quantum level as "hidden from science" Russell intends to preserve the integrity of science while at the same time "allowing science to be integrated...into constructive theology where 'God' as an explanation of natural events is appropriately and fully developed." This understanding of intervention/nonintervention has advantages in being clear and able to make specific distinctions with reference to God's action vis-à-vis natural laws. I suspect, however, that not all would agree with limiting intervention to a violation of natural laws (though most in the science and theology dialogue at this stage probably would) or that there is still not some question as to whether God might appear to be implicitly identified with natural causes when direct divine action "at the quantum level gives rise to phenomena that cannot be explained by science."[13]

It seems that what some theologians still refer to as 'intervention' in this later sense is not far removed from what many of the VO/CTNS conference series participants speak of under the category of special divine action. For instance Bob Russell, to pick one example, argues in the final paper of the final volume of the series for a "noninterventionist objective special divine action."[14] Russell and others, I believe, want to be clear in their affirmation that God does act, and not just in some vague, subjective sense, but specially and objectively. That the series can conclude on such a note is both significant and welcome. I also understand the necessity, from the scientific standpoint, that this 'objective special divine action' be noninterventionist. This position, if it can be maintained, would be the best of both worlds for those committed to the theology and science dialogue. My concern is that many may find it difficult to understand at what point an objective special divine act moves from being noninterventionist to interventionist. For many, if God acts in such a way so as to alter the outcome that would otherwise have been expected, then this—by the ordinary way of thinking, is an intervention. My sense is that Russell and many others involved in the series would not dispute that God acts so as to alter the outcome, but rather that this action never intervenes (or contravenes) the regular patterns and processes of the natural world. Hence there is scope for confusion, especially on the part of theologians not familiar with the concerns of those working within the natural sciences. What we seem to have arrived at in the VO/CTNS series is not so much a theologically noninterventionist model (for this must be more specifically defined), but a scientifically non-interventionist model.

---

[12] Robert John Russell, "Divine Action and Quantum Mechanics: A Fresh Assessment," in *QM,* 295.

[13] Ibid., 296.

[14] Ibid.

3 Laws of Nature and Open Processes

Of course, tinkering with the definitions of 'intervention,' 'action,' and 'special divine action' and seeking subtler understandings of what these may or may not entail do not entirely overcome the problem. Nor does it address the other horn of our semantic dilemma. We are still left with the difficulty of what we understand by the concept of natural laws. These, too, taken in a rigid sense, can appear to leave little or no room for divine action—especially those models that are open to so-called 'special' acts. If the concept of divine action has been prone to misunderstanding, the concept of natural law has suffered no less from inflexible representations of what it may or may not allow.

A century ago the scientific community seemed agreed that the universe was static, without beginning or end, and governed by a set of unchanging, unbending physical laws. In such a context any sort of divine action smacked of the most crass variety of interventionism. Just what role a transcendent creator could play in such a universe remained to be seen. Yet we now view the universe much more fluidly. Like the life it contains, the universe, too, is in process, engaged in a dynamic dance of cosmic evolution. In the context of the current view of the universe the question of divine action must also be a much more fluid one. A survey of the current situation in the natural sciences, such as what occurs via the many scientific contributions to the series, would seem to indicate that the apparent necessity to exclude all forms of special divine action may not be so necessary as it first appears. Three areas of contemporary science come up in the series frequently in this regard, especially in *QC, CC,* and *QM.* These are the uncertainty principle of quantum mechanics, current understandings of the nature of physical law, and chaos theory.

The laws of nature, like the universe they govern, are more complex than they once appeared. Yet pointing to this complexity must not be seen as making room for God to work, as if God cannot work within the structure of understood, regular physical processes. The assumption that God can only be seen to act when and where natural laws are unclear or imperfectly understood would be a rather unfortunate return to a god-of-the-gaps mode of thinking, however we dress it up. God is not able to act simply because physical law is more complex and more 'flexible' than we are sometimes prone to think. Rather, a more realistic assessment of natural laws—like a more realistic assessment of divine action—demonstrates the difficulty of assuming that God's action in the physical cosmos is simply not possible. William Stoeger summed up the situation regarding the physical laws well in making the following two points:

> ...(i) the laws of nature as we actually formulate them in the sciences are but imperfect and incomplete descriptions of the regularities, structures, relationships and processes that obtain in reality itself (the laws of nature as they actually function); and...(ii) the laws of nature as we describe them are descriptive, not prescriptive.[15]

[15] William Stoeger, S. J., "Epistemological and Ontological Issues Arising from Quantum Theory," in *QM*, 95.

The background and significance of this assessment merits further consideration. Heisenberg, along with Erwin Schrödinger and Paul Dirac, developed the theory of quantum mechanics based on the uncertainty principle. Their theory predicted no specific observable events, but rather a range of possible results along with formulae for predicting statistically the chances of obtaining each possible result in any given instance. This does not leave an opening for God to work in the sense that God could not act if it were not for the uncertainty principle. Whether this provides a clue as to how God actually does work is another question altogether. At the very least, the complexity of natural laws and the existence of the uncertainty principle remind us that the nature of the physical world and processes are such that one cannot dogmatically exclude divine action—even 'special' actions of God.

Providence, or divine action, need not, however, be understood only in terms of 'special' acts, sometimes identified with the miraculous or even interventionist. It is entirely possible for a 'special' act of providence to occur without violating any laws of nature. Arthur Peacocke contends that particular events or clusters of events "can be intentionally and specifically brought about by the interaction of God with the world in a top-down causative way that does not abrogate the scientifically observed relationships operating at the level of events in question." Such a possibility, according to Peacocke, is of value in that it "renders the concept of God's special providential action intelligible and believable within the context of the perspective of the sciences."[16] It would be difficult as a theologian to disagree with Peacocke's position. The question is whether it says enough.

Are God and the processes of nature simply two separate ways of describing the same thing? In this compatibilist approach there can be no conflict ultimately between God and natural laws because everything God does is also described by the laws of nature. The incompatibilists among the contributors to the series, such as Bob Russell and Nancey Murphy, agree that much of what God does is described by the laws of nature. Their contention, however, is that who God is and what God does are ultimately more complex than this. Divine action and laws of nature are not necessarily always strictly compatible. While this view presents its own difficulties, it holds on to the important theological point that God is always more than we expect or understand God to be. Peacocke's view presents well and leaves no room for conflict between the affirmation of a God who acts and the laws of nature. But such a view of God, despite its attractiveness on some levels, may ultimately be too limited.

4 Chaos, Order and Divine Freedom

Into this whole mix now comes the question of order and chaos. In this sense, the appearance of *CC* as the second volume of the series is a logical progression. For centuries the belief in an orderly God went hand in hand with the belief in an orderly universe. The God of Newton, whether we are consciously aware of it or not, is the God that most of us have grown up with

---

[16] Arthur Peacocke, *Theology for a Scientific Age: Being and Becoming—Natural and Divine* (Oxford: Basil Blackwell, 1990), 182.

and most easily conceive. This God is a God of order. A God who has methodically created a methodically run universe—a God who is predictable and whose world is predictable. This God, we have been taught, is opposed to the 'demonic' influences of chaos and unpredictability.

In the last century these safe and comfortable perceptions began to come unraveled, at first almost imperceptibly. Slight tremors appeared on the landscape of physics that only a small few seemed to really understand, but which have grown in magnitude to the point that we now have something akin to two worldviews existing side by side in uncomfortable juxtaposition. The classical view is one of order and predictability accompanied by an orderly God knowingly and sovereignly presiding over it all. The more recent view is one of inherent unpredictability and chaos that seem to exist in paradoxical juxtaposition to the God whom most of us conceive. The question of interest from the perspective of Christian thought is (or should be) how we are to understand God's relationship to such a world. Indeed, James Cushing has stated that "the question of 'openness' (indeterminism/chance) versus a 'closedness' (determinism) in the world can arguably be seen as the fundamental issue in relation to possibilities for a particular way of God's action in what is taken to be her creation."[17] Yet the reality is that large sections of contemporary theology have yet to take any real cognizance of this situation, at least in the context of the discussions within the physical sciences. The long and complex history of theological discussion of predestination and human freedom present very similar problems. But the benefits of these substantial discussions within theology have rarely been applied effectively to the related question of determinism versus apparent indeterminism within contemporary science.

John Polkinghorne indicates something of the possibilities many have seen in chaotic systems for divine action when he writes:

> The exquisite sensitivity of chaotic systems certainly means that they are intrinsically unpredictable and unisolable in character… I propose that this should lead us to the metaphysical conjuncture that these epistemological properties signal that ontologically much of the physical world is open and integrated in character. By 'open' is meant that the causal principles that determine the exchange of energy among the constituent parts (bottom-up causality) are not be themselves exhaustively determinative of future behavior. There is scope for the activity of further causal principles. By 'integrated' is meant that these additional principles will have a holistic character (top-down causality).[18]

This approach, however, leads in some very interesting directions with regard to traditional theological views about divine (fore)knowledge. As Polkinghorne explains, "a world open to both bottom-up and top-down causality is a world released from the dead hand of physical determinism." It also is a shift away from the determinism of divine foreknowledge. It is also a world of becoming in which not even God knows the future exactly. Rather,

---

[17] James Cushing, "Determinism Versus Indeterminism in Quantum Mechanics: A 'Free' Choice," in *QM*, 99.

[18] John Polkinghorne, "The Metaphysics of Divine Action," in *CC*, 153.

"God knows all that can be known, but the future is still inherently unknowable."[19]

Of course it may be argued that there is a significant distinction between God knowing the future and God actually determining the outcome of future events. This also reflects a theological divide over how divine foreknowledge is to be understood. For my part, I believe there are strong philosophical and theological grounds for viewing God's knowledge as determinative. If God knows a future outcome, then this knowledge also fixes or determines that outcome. It's more than an infallible calculation based upon known data. There is clearly much scope here for some very significant theological reflection of the implications of this approach, especially in light of the traditional theological discussions on divine foreknowledge and human freedom.

We might even ask whether the hoped for unification of physics will bring with it a resolution of the tension between determinism and indeterminism? Or must either determinism or indeterminism ultimately prevail? The Oxford mathematician Roger Penrose, although himself skeptical that this will indeed prove to be the case, suggests that, "it is even possible that we may end up restoring determinism in quantum mechanics."[20] Of course, a number of physicists believe relativity theory may well have to be adapted to make room for some degree of indeterminism.[21]

The German physical and theoretical chemist O. E. Rössler has written "it could turn out...that a universe that is chaotic itself ceases to be chaotic as soon as it is observed by an observer who is chaotic himself."[22] 'Chaos,' in other words, may be simply a matter of perspective. A hypothetical chaotic observer would view a chaotic universe as entirely 'in order'—and because it is observed as such it would indeed become orderly. It is an assertion of the fundamental unpredictability of the quantum, and perhaps also the large-scale systems of the universe. Michael Heller makes a similar point when he says that the cosmic elements (which include predictability) and the chaotic elements (which include unpredictability) are equally as mathematical and explicable. There are no gaps in the sense of natural laws or scientific explanation (or what Heller calls spurious gaps). Hence, "*Cosmos* and *chaos* are not antagonistic forces but rather two components of the same *Logos* immanent in the structure of the Universe."[23]

The uncertainty principle of quantum physics has led, in many ways, to a re-examination of the nature of 'chaos,' or what in classical terminology would be called non-linear systems. Most significant among such re-examinations have been those that highlight a sort of dialectical tension between determinism and indeterminism by suggesting that chaos is not so chaotic as we might imagine. In other words, if chaotic systems are not radically nondeterministic, but actually strengthen the theory of metaphysical

---

[19] Ibid., 156.

[20] Roger Penrose, "Big Bangs, Black Holes and 'Time's Arrow,'" in *The Nature of Time*, R. Flood and M. Lockwood, eds. (Cambridge, Mass.: Basil Blackwell, 1965), 60.

[21] Paul Davies and J. R. Brown, *The Ghost in the Atom: A Discussion of the Mysteries of Quantum Physics* (New York: Cambridge University Press, 1986), 30.

[22] O. E. Rössler, "How Chaotic Is the Universe?," in *Chaos*, Arun V. Holden, ed. (Manchester: Manchester University Press, 1986), 317.

[23] Heller, "Chaos, Probability, and Comprehensibility," 121.

determinism, then we have moved beyond popular assumptions that deterministic systems are ultimately predictable and that chaotic systems are indeterministic and governed by chance.

A great deal of attention has been given recently to the idea of a deterministic indeterminism as a possible harmonization of the implications of classical and quantum physics. Cushing, taking both seriously, appears to move in a similar direction, albeit without using this terminology, by putting a 'deterministic gloss on physical phenomena, even at the quantum level."[24]

But how would contemporary views of a deterministic indeterminism differ from a Laplacian determinism in which the future state of the universe could potentially be calculated from its present state? Roger Penrose indicates the direction in which such discussions are leading when he writes that he believes that some "new procedure takes over at the quantum-classical borderline which interpolates between" deterministic and probabilistic 'quantum jump' parts of quantum mechanics, and that this new procedure "would contain an essentially non-algorithmic element," which would in turn imply "that the future would not be computable from the present, even though it might be determined by it."[25]

In distinguishing between computability and determinism Penrose would seem to represent a break from a strict Laplacian determinism that did not make such a distinction. The concept of a determined but not completely predictable universe also has been seen to contain possibilities for theological reflection on the providence of God. As a model of God's providential direction of the universe Penrose's tentative outline of a "correct quantum gravity" (CQG) theory would seem to demonstrate the possibility of conceptualizing divine providence and divine freedom as consistent aspects of God's providential sustenance of the physical cosmos. In fact, Penrose himself suggests that a CQG theory would also leave a role for human free will within an essentially deterministic universe.[26] Is this, then, a 'humanity of the gaps?'

Wesley Wildman and Bob Russell, in "Chaos: A Mathematical Introduction," also point to the problem posed for assumptions of free will by chaos theory. Chaos theory, they point out, "adds its considerable weight to the side of determinism. However...chaos theory highlights an epistemic limit in the macro-world of dynamical systems, tethering the deterministic hypothesis even as it advances it." They conclude that while scientists are likely to be affirmed in their metaphysical deterministic assumptions, even in light of seemingly random processes, theologians will find this more problematic. So long as we assume that "a stronger case for the deterministic hypothesis necessarily implies a weaker case for the possibility of free divine and human acts" then the deterministic nature of chaotic systems is going to be bad news for the theologian. They query, however, the assumption that a stronger case for metaphysical determinism necessarily means a weaker case for free will (human or divine). They suggest that chaos theory itself makes it impossible to rule out God as a freely acting causal factor. Chaos theory not only strengthens

---

[24] Cushing, "Determinism Versus Indeterminism in Quantum Mechanics," 109f.

[25] Roger Penrose, *The Emperor's New Mind: Concerning Computers, Minds, and the Laws of Physics* (Oxford: Oxford University Press, 1989), 431.

[26] Ibid., 353ff. and 431f.

the case for determinism, but it also sets limits that ensure the case for determinism cannot get any stronger and that the question of divine action can never be excluded.[27]

The debate about determinism and indeterminism in the sciences has intriguing parallels in the history of theological reflection on God's providential direction of the universe (and of human history). These discussions highlight a similar tension between the 'indeterminism' of human 'free will' and the 'determinism' of divine predestination. Even within the divine economy (excluding the problem of human free will) there exists a tension between the 'indeterminism' of divine freedom and divine predestination. These traditional theological discussions deserve continuing attention by anyone seriously interested in models of understanding the relationship between deterministic and indeterministic aspects of the physical world.

If God governs the creation in such a way that the universe evolves according to a strict, predetermined order, what place is left for divine freedom? On the other hand, if God is really free to act in the physical universe, what happens to the traditional doctrine of divine predestination? Theological systems have tended to ultimately opt for one or the other. In a sense, theologians proceed much like physicists in this regard. On the micro-level of miracles, prayer, human moral decision, etc. we proceed as if God and individual persons were able to actually alter the course of history or affect in some way the state of the physical cosmos (e.g., the human decision to pollute or not to pollute, to engage in nuclear war or not). On the macro-level of the teleological outcome of history or the consummation of creation the pre-determined program of God takes over. Christ will return, evil will be defeated, God will be all in all. These outcomes are seen as givens by traditional theology. Like physicists, theologians would very much like to know how these two levels fit together. Does one ultimately prevail over the other? While current discussions among physicists cannot possibly provide answers to these theological questions, they may very well provide useful paradigms for overcoming, or at least learning to live with the tension between indeterminism and determinism.

If physics can learn to live with so-called deterministic chaos perhaps theology can accept a providential freedom of God that accommodates the 'determinism' of divine predestination and the 'indeterminism' of human and divine freedom without denying the inevitable tension between the two or subsuming one under the other. The mathematician Ian Stewart put the matter well when he suggested: "Perhaps God can play dice, and create a universe of complete law and order, in the same breath."[28]

Again, the result of a closer look at the state of our current understanding of the nature of the physical universe is not that there exist holes or cracks of such nature as would allow some insipid god-of-the–gaps to tinker—but rather that the nature of physical reality and the processes that govern our evolving

---

[27] Wesley Wildman and Robert John Russell, "Chaos: A Mathematical Introduction," in *CC*, 83ff.

[28] Ian Stewart, *Does God Play Dice?: The New Mathematics of Chaos* (New York: Penguin Books, 1990), 2.

cosmos do not appear to prohibit the concurrent free action of a transcendent creator.

5 General Assessment of the VO/CTNS Series from the Perspective of Theology and Science

The very fact that an explicitly theological theme, that is, divine action, could serve to hold together such an in-depth and sustained series of conferences and publications says something about both the theological significance of the theme as well as the possibilities it presented from the standpoint of the sciences. This is to be commended. If the series were to be done again (and I am not suggesting that the theme of divine action be pursued further as a joint project) there are some things that I as a theologian would like to see. I think some important movement toward consensus on at least some key points was clearly visible. But a greater number of participants who took very different points of view from the very beginning could have strengthened the impact of these achievements. For instance, some reputable theologians hold moderate and nuanced interventionist views and it would have been good to have had that voice added to the discussion. Perhaps these folk are the heretics of the science theology dialogue but the history of Christian doctrine teaches us that it is often the presence of heretics within our discussions that assists the orthodox view in shining forth as it distinguishes itself from views that are either unacceptable or unworkable.

There is also a strong theological tradition, especially within Roman Catholic theology, that distinguishes between primary and secondary causation and does not look for divine action among secondary causes. The focus is on God acting as primary cause within the 'regular' processes of the natural world. Even what might be termed the miraculous can be understood in these terms. While this view appeared present especially in Bill Stoeger and Denis Edwards, it was probably underrepresented. From the standpoint of the sciences the number and quality of scientists involved was very good and many of these were veterans of the theology and science dialogue. Apart from the volume on neuroscience, however, the series stuck to established fields of dialogue such as physics (particularly quantum theory) and biology (particularly evolution). This allowed the series to build on an already well progressed discussion in most cases. Yet there is a continuing need to involve more fully scientific fields of enquiry that have not traditionally been involved in the dialogue. In part this may entail an examination of what we mean by 'science' and to what extent both stated and unstated hierarchies of sciences play a role in determining the field of dialogue.

For various reasons, the number of established theologians involved in the series was not great. But if the insights of these and similar efforts are to have an impact on the way Christians think about God and the world, then more theologians need to be involved in the discussions at some level—especially theologians unfamiliar with or wary of the natural sciences. Most of the theologians who participated in the series have had a significant history within the theology and science dialogue. The bulk of our most prominent theologians however, both Protestant and Roman Catholic, are still in need of conversion to the dialogue.

In my experience within the Australian context over the last decade most new people coming to theology/science conferences and other activities could be characterized as scientists with a personal interest in and/or commitment to the Christian faith. We occasionally have folk from non-Christian religious traditions getting involved. What we seldom have, however, are theologians from the many theological and Bible colleges that dot the landscape getting involved in the discussions. I have personally spent a great deal of time seeking to persuade some of my theological colleagues to involve themselves (occasionally with success) but for the most part I encounter a good deal of reluctance. Many come from a theological perspective (often influenced by Barth—who gets blamed for a lot!) that does not value the significance of this sort of dialogue. Others claim they are too busy. A few admit they are intimidated by science and scientists and find they are out of their depth in such discussions. I suspect that the situation is much the same in other parts of the world. The problem, of course, is that while very significant progress is being made in the science-theology dialogue, it is often not reflected in the theological lecture theatres.

I am saying nothing new in pointing out this situation. Within this context, however, an assessment of the VO/CTNS series from the standpoint of the theology-science dialogue gives opportunity to ask whether the series has helped to bring more theologians onboard. My suspicion is that it probably has not. Of course, this was never the goal of the series, so these comments must be viewed in that context. With the series subtitle: "Scientific Perspectives on Divine Action" we would expect that scientists would be more prominent than theologians in the conferences. Yet the involvement of more theologians like Moltmann, who have had some interest in the sciences but who have not been heavily involved in the dialogue, may have helped. I sense that for many of my theological colleagues, still struggling with the question of whether theology and the sciences really have anything to learn from one another, the discussion has advanced beyond their reach. Many of the theologians involved in the discussion already have some science background and are eager to pursue interdisciplinary questions that are at the same time personal questions. These theologians, though relatively few in number, adapt fairly easily to the terminology and categories that usually characterize the dialogue.

Many theologians would tend to view the VO/CTNS series as primarily scientists, some of them theologians in their own right, reflecting on theological issues. Some of the theologians whose names they recognize, when they turn to their contributions, may appear to be speaking another language. I find this all somewhat ironic, as theology, by its very nature, must be in dialogue with all other disciplines, while the individual (or as Rahner called them, 'regional sciences') do not have the same need. Yet it seems to be the bulk of rank and file theologians whom we struggle to bring onboard.

If future projects are to make more headway in the conversion of theologians to the dialogue (assuming this is a worthwhile goal) I would suggest that the following points may be helpful: (1) More established theologians not already active in the dialogue should be invited to participate. This should not just be new names, but some new voices/perspectives should be sought out. (2) They should be encouraged to bring their own terminology and categories to the question at hand. That is, they should be encouraged to address the questions raised by the dialogue as theologians. (3) There should be a greater recognition that theology is not a unified field with a single approach. It would be helpful to note and reflect upon nuances between approaches taken by Catholic and Protestant theologians, 'liberal' and 'evangelical,' Reformed and Lutheran, pre- and post-Rahnerian, etc. In many instances theological differences play a significant role in how the natural world is viewed and these different theological strands need to come into dialogue with one another as well as with the natural sciences.

Of course these three things all already occur to some extent in the VO/CTNS series. My suggestion is that more of a focus in these directions would not only make the discussions more appealing and accessible to theologians, but the conversations would benefit from a deeper and more theologically focused and nuanced approach at key points. Now, if only this can be achieved without frightening off the many fine scientific contributors already fully engaged in the discussions!

IV. RESOURCES FOR FURTHER RESEARCH

CONTRIBUTORS

Philip Clayton, Ingraham Professor of Theology, Claremont School of Theology and Professor of Philosophy and Religion, Claremont Graduate University, Claremont, California, USA.

George V. Coyne, S. J, Former Director, (retired) Vatican Observatory, Vatican City State, Italy.

George F. R. Ellis, Distinguished Professor of Complex Systems, Department of Mathematics and Applied Mathematics, University of Cape Town, Rondebosch, South Africa.

Niels Henrik Gregersen, Professor of Systematic Theology, University of Copenhagen, Copenhagen, Denmark.

Nancey Murphy, Professor of Christian Philosophy, Fuller Theological Seminary, Pasadena, California, USA.

Arthur Peacocke, Former Director, Ian Ramsey Centre, Oxford, England, Former Warden Emeritus of the Society of Ordained Scientists, Former Dean of Clare College, Cambridge, England.

Robert John Russell, Ian G. Barbour Professor of Theology and Science in Residence, Graduate Theological Union, and Founder and Director, The Center for Theology and the Natural Sciences, Berkeley, California, USA.

William R. Stoeger, S.J, Staff Astrophysicist and Adjunct Associate Professor of Astronomy, Vatican Observatory, Vatican Observatory Research Group, Steward Observatory, University of Arizona, Tucson, Arizona, USA.

Thomas F. Tracy, Phillips Professor of Religion and Chair, Department of Philosophy and Religion, Bates College, Lewiston, Maine, USA.

Keith Ward, Regius Professor of Divinity Emeritus, University of Oxford, Oxford, England.

Kirk Wegter-McNelly, Assistant Professor of Theology, School of Theology, Boston University, Boston, Massachusetts, USA.

Wesley J. Wildman, Associate Professor of Theology and Ethics, School of Theology, Boston University, Boston, Massachusetts, USA.

Mark Worthing, Dean of Theology and Senior Lecturer in Theology and Ethics, Tabor College, Adelaide, Australia.

Name Index

Subject Index

D

F

G

H

M

Q

www.ingramcontent.com/pod-product-compliance
Ingram Content Group UK Ltd.
Pitfield, Milton Keynes, MK11 3LW, UK
UKHW051852160425
457516UK00009B/376